Introducing and
Implementing
Autodesk® Revit®
Building

Introducing and Implementing Autodesk® Revit® Building

LAY CHRISTOPHER FOX
JAMES J. BALDING, AIA

THOMSON
DELMAR LEARNING

Autodesk®

Australia • Canada • Mexico • Singapore • Spain • United Kingdom • United States

Autodesk®

Introducing and Implementing Autodesk® Revit® Building
Lay Christopher Fox / James J. Balding, AIA

Autodesk Press Staff:

Vice President, Technology and Trades SBU:
Alar Elken

Editorial Director:
Sandy Clark

Senior Acquisitions Editor:
James DeVoe

Senior Development Editor:
John Fisher

Marketing Director:
Dave Garza

Channel Manager:
Dennis Williams

Marketing Coordinator:
Stacey Wiktorek

Production Director:
Mary Ellen Black

Production Manager:
Andrew Crouth

Production Editor:
Stacy Masucci

Technology Project Manager:
Kevin Smith

Technology Project Specialist:
Linda Verde

Editorial Assistant:
Tom Best

Library of Congress Cataloging-in-Publication Data:
Card Number:

ISBN: 1-4180-2056-7

NOTICE TO THE READER

CONTENTS

WELCOME

> "Get the right tool for the job."
>
> —*Mr. Natural*

The above words entered the popular lexicon in the free-wheeling 1960s, spoken by a metaphoric cartoon character who combined confident disregard of hide-bound conventional thinking—a common commodity in those days—with a wise, practical outlook, rare enough in any generation. His advice is timeless, not old school.

In today's design world the tools of the trade are computerized, and applications or releases appear every couple of years. How do we recognize the right tool amidst today's clamor of constant change? What marks a real advance in technology that will benefit designers and the firms that employ them? How do we take advantage of changes in tools while constantly involved in the rigors of meeting expectations and deadlines? Are there definitive answers to these questions?

This book has a double purpose:

- to introduce Autodesk Revit Building as the next generation in architectural design, a fully 3-dimensional building modeler of exceptional power, depth, and ease of use, that functions well at all phases of the design process, from concept sketches all the way through construction documentation

- to prepare individuals, firms and schools for the process of implementing this new tool into their current work flow and processes. Successful implementation will require planning, training, and a transition phase from current Computer Aided Design applications. We will use AutoCAD, the worldwide standard, as a reference.

CAD AND REVIT BUILDING: WHAT ARE THE DIFFERENCES?

CAD stands for Computer Aided Design (or Drafting). This acronym explains succinctly the approach that mid-priced design software (what individuals and small firms can afford) has embodied until the last few years: the computer has been used as an aid to an age-old design process that hasn't changed substantially even as the tools became electronic. CAD software still requires the user to design line-by-line for the most part, and lines representing a wall in a plan do not create or appear in elevation, section, or 3D views of that wall.

Revit Building is a robust, affordable, and efficient software application created specifically for architectural designers. It is not based on the drafting language of lines, arcs, and circles. Revit Building is a building modeler database, with drafting tools included for those inevitable cases where simple linework is the most efficient way to illustrate a design concept or detail. Revit Building differs fundamentally from CAD; it is not a drafting application that has been expanded to attempt limited 3D content and views. Revit Building models consist of objects that are completely definable in size, structure, appearance, data content, and relation to other model components. All possible views of the model update with each command, including sheets and schedules. Much of current design process routine—especially propagating changes throughout a set of construction document pages—disappears as a result. Designers can spend the bulk of their time designing, not transcribing.

An example of the effectiveness of the 'right tool for the job' comes from recent headlines. As of press time the *Freedom Tower* project for New York City, possibly the most complicated skyscraper proposal in the world today, is being developed in Revit Building, from the subway interchanges beneath the foundation to the electricity-generating wind turbine matrix at the top of the tower. On May 1, 2005, very late in the design process, the New York City Police Department announced that it was dissatisfied with security measures in the site and building design, and put the whole project on hold. Public outcry and consternation included "Where do we go from here?" editorials in the New York Times. On June 8, 2005, only five weeks later, the Police Department announced its satisfaction with design changes put in place by the architect, and the review board certainly did not base its decision on napkin sketches. If you can imagine yourself sitting in the hot seat on a project of that size, scope and public scrutiny, are you convinced that your present design software tools would allow you to respond to a major last-minute challenge as quickly and as well?

WHAT DOES PARAMETRIC DESIGN MEAN?

Good design of any kind is composed of harmonious relationships, which are based on rules, or parameters. Revit Building objects are placed in relation to all other objects in a model, and parameters govern the relationships. Designers can now establish relationships among building components that will hold through design revisions unless specifically removed. Suppose that an array of windows in a wall works best if the windows are equidistant from one another, the end windows are six feet from the corners, and the sills are two feet from the floor. Revit Building will recognize and apply those constraints, and adjust the positions of the windows if the wall length or floor height changes. The potential use of constraints flows throughout a model.

Revit Building carries parametric specifications much further than the previous simple example of windows in a wall. All types of components can carry information (job address, object size or placement specifications, supplier, cost, department, electrical requirements, etc.) according to supplied or user-defined parameters. This information can be collected in schedules or diagrams, and exported to other applications.

Revit Building's use of parameters underlies every aspect of the user interface. There is no drafting-oriented organization such as layers to control the appearance of objects; much of the learning curve for CAD applications therefore does not pertain to new Revit Building users.

WHO SHOULD READ THIS BOOK?

If you currently work in architectural design and your firm is considering or moving towards Revit Building, this book will explain the basics of the Revit Building tools that you, your co-workers, or your employees will be using every day. An important part of implementing Revit Building in an active design company's workflow is blending this powerful new tool with the often vast store of valuable information in existing drawing files. The first chapters of the text deal with importing AutoCAD information into Revit Building and exporting DWG files from Revit Building, so that the new technology can be transparent to clients, collaborators, and other departments or offices in your firm. We make mention throughout the text (see particularly Appendix A) of the effects Revit Building will have on workflow, process, personnel, and attitudes in organizations of nearly any size making this particular technology transfer.

If you are a student of architectural design, this book will show you the very latest software tool in the marketplace. You will be among the first to design buildings parametrically, entirely in three dimensions, without having had to learn the mechanics of representational lines, arcs, and circles beforehand. A few words of caution that your professors will no doubt echo: you will never become a good designer if you don't understand and practice good drafting techniques. Architects who can't successfully create clear sketches or readable detail pages do not win prizes, or even commissions.

FEATURES OF THIS TEXT

WHAT YOU WILL FIND IN THIS BOOK

The second edition of **Introducing and Implementing Revit Building** has been extensively revised and expanded from the 2004 first edition.

Each chapter in the book covers specific, logically-sequential skill development using Revit Building in a planned progression from basic modeling through more complicated refinements, data extraction, and illustration. All work is based on tutorial exercises, with conceptual explanations, tips, notes, and cautions where appropriate. Each chapter ends with review questions suitable for quizzes.

Chapter 1 provides an overview of Revit Building's emerging position in the market of architectural design software, and contains a tutorial exercise showing how to create, revise, and document a simple building model. Chapter 2 covers template files, their uses, and pre-loaded content. This chapter also shows how Revit Building can import information from AutoCAD during the creation of a custom titleblock. These chapters should be of particular interest to company principals, supervisors, or CAD managers.

Chapters 3 through 10 use a hypothetical multi-building, multi-phase project on a college campus as the framework for their exercises. Chapter 3 covers the creation of a site plan with building shell, using walls, floors and other building element objects combined with imported information, and also shows how to export information from a Revit Building project file to AutoCAD. Chapter 4 covers sketching, massing objects, transformation of massing into building components, and file linking. Chapter 5 develops modeling techniques in the context of multi-user design teams, and introduces project phasing. Many of Revit Building's most powerful modes and tools are covered in these exercises.

Chapters 6, 7, and 8 concentrate on various aspects of annotation, within the context of the hypothetical project workflow. Chapter 6 covers customization and management of the appearance of annotation, from tags to titleblocks. Chapter 7 works through Revit Building's extensive and powerful scheduling capabilities. Chapter 8 moves from schedules into area and room plans, with customizable color fills. These chapters show—in broad strokes, concentrating on design development—how to create, manage, and enhance the "information" side of Revit Building, including streamlined data export to external applications.

Chapter 9 covers graphic output: printing, perspective views, rendering, and animations. Chapter 10 wraps up with Revit Building's tools for organizing and managing design options, and a look at Revit Building family creation, itself a topic worthy of its own book.

WHAT YOU WON'T FIND IN THIS BOOK

This text is not a Revit Building help manual or command dictionary. Its scope covers all aspects of design in Revit Building—sketching, mass modeling, conceptual design, design development, construction documentation, data export, and illustration. Space limitations prevent us from covering every tool available in the software, or all the possible control options for the tools and methods we do illustrate. Where appropriate the text shows or mentions alternate ways of accessing tools or completing tasks.

STYLE CONVENTIONS

Text formatting and style conventions used in this book are as follows:

Text Element	Example
Step-by-Step Tutorials	1.1 Perform these steps
Menu Selections	Select File>Import/Link>RVT
Keys you press are in SMALL CAPS	Press CTRL or ESC
User input is in **bold**	Change the Eye Elevation value to **55′ 6″**.
Files and Paths are shown in *italic*	Content/Imperial Library/Doors

HOW TO USE THIS BOOK

The order of chapters has been planned to follow normal design firm flow and process. Chapters 1 and 2 are introductions of particular value for instructors, firm principals, and CAD managers, although the exercises presume no extensive design experience. If you are new to Revit Building, we recommend that you complete the exercises in order. If certain chapters do not pertain to the work you or your firm does, feel free to skip topics as you see fit.

FILES ON THE CD

Certain tutorial exercises use startup files. These files are on the CD enclosed with this book. Users can load files from the CD into convenient locations on their own systems, or as specified by their instructor(s). Revit Building project files are on the CD in unzipped *rvt* format; Family files are in *rfa* format. Additional supplement files are also in native format (*doc*, etc.).

WE WANT TO HEAR FROM YOU

We welcome your suggestions and comments regarding *Introducing and Implementing Autodesk® Revit® Building.* Please send your correspondence to:

The REVIT BUILDING Team
c/o Autodesk Press
Executive Woods
5 Maxwell Drive
P.O. Box 15015
Clifton Park, NY 12065
Web sites: www.autodeskpress.com
                www.archimagecad.com

REVIT BUILDING RELEASES AND SUBSCRIPTIONS

Autodesk has announced plans to make substantive new releases of Revit Building roughly once a year, with an interim (.x) release halfway through the cycle. The latest build of Revit Building will be available as a download from the Autodesk website, or you may request a free CD. Without a license, Revit Building works for a 30-day demonstration period. After that time you will need to purchase a subscription in order to save work or print.

The exercises in this book were tested with Release 8.0, which appeared in March of 2005. While the authors have tried to coordinate exercises and illustrations with the latest available information, there may be inconsistencies between this book and the behavior of the software you have installed. Exercise files on the CD are in 8.0 format. Check the CD or websites for additional information not available at press time.

ABOUT THE AUTHORS

Lay Christopher Fox is an independent architectural drafter/illustrator, author, and educator. He has co-authored two books on Architectural Desktop® and one on Revit® Building prior to this text. He has written columns on AutoCAD® and Revit Building for *AUGIWORLD Magazine,* the bi-monthly newsletter of Autodesk User Group International (AUGI®). Chris is president of the Rochester (NY) Area AutoCAD Users Group, an adjunct professor at Rochester Institute of Technology (RIT)—teaching AutoCAD to engineering students—and was an instructor of AutoCAD, Architectural Desktop, and Inventor® at the former Autodesk® Training Center on the RIT campus. He has been a speaker at Autodesk University (2003). He received his undergraduate degree in literature at Harvard, and has pursued technical training in Rochester, NY, and Adelaide, South Australia.

Jim Balding is a licensed architect with more than 17 years of experience integrating technology into the architectural field. He is currently employed with Wimberly Allison Tong & Goo (WATG) in Newport Beach, CA. Jim earned his Bachelor of Environmental Science degree from the University of Colorado, Boulder. He has been a member of the Autodesk® Revit Building® Client Advisory board since its inaugural meeting and is currently serving as the Revit Building Product Chair for AUGI®. He is also currently serving as the South Coast Revit Building Users Group (SCRUG) president. Jim has spoken at several technology conferences and is one of the top-rated speakers at Autodesk University. He has developed a successful Autodesk Revit Building implementation strategy and is currently bringing the seven offices of WATG up to speed in its use.

ACKNOWLEDGEMENTS

The authors would like to thank all of those who have contributed to this project and to our personal and professional development over these many years.

First, thanks to the editors at Delmar/Thomson Learning/Autodesk Press: Jim Devoe, John Fisher, Stacy Masucci, Tom Best, and Linda Verde. Particular thanks to John Fisher for his patience and understanding.

Special thanks to those who reviewed the text and exercises:

> Michael Gatzke—Des Moines Area Community College, Ankeny, Iowa
> John Knapp—Metropolitan Community College, Omaha, Nebraska
> Robert Mencarini—Autodesk and RGM Architecture, Maplewood, New Jersey

The authors would also like to acknowledge the help of others who took the time to review the book in its formative stages and assist with feedback. You went above and beyond the call of duty:

> Scott Davis, Greg Cashen, Justin Kelly, Carl Walls

The international Revit Building community of users, while small in numbers compared to users of CAD products, is lively and committed to mutual growth. Recognition should go to Chris Zoog for starting a thriving web discussion group where users new and experienced share comments, concerns, and content. This forum has since moved to the site of Autodesk Users Group International (www.augi.com), along with the Revit Building Users Group International.. Revitcity (www.revicity.com) is another source for shared content, forums and general information regarding Revit Building. Many discussion groups exist at Autodesk's discussion group site, news.autodesk.com.

Jim Balding not only wrote a significant part of the book text, but he worked as Technical Editor on the rest. Little did he know what he was getting into—many thanks from Chris.

Chris would like to make particular acknowledgement to John Clauson and David Harrington of AUGI, who encouraged me to keep writing newsletter articles for years, and Elise Moss (also of AUGI), who generously and graciously let me assist her on a number of book projects.

Jim would like to thank the many people that have had an influence on his career as an architect and leader in architectural technology. Many thanks to: Chuck Cutforth, Ed Conway, M. G. Barr, Gerald Cross, Curt Kamps, Charlie Wyse, Jim Grady, Perry Brown, and last but certainly not least, Larry "Figurehead" Rocha. Tom (Buck) and Collette, thank you very much for all of your love, support, and understanding while I was under your wings. I love you both very much.

Finally, we would both like to recognize the Revit Building Team—Leonid Raiz, Irwin Jungreis, Marty Rozmanith, Steve Burri, Dave Heaton, Rick Rundell, David Conant, the development and support staff, and the hard workers too numerous to name at the Factory. Your pride and inspiration shine through, and do you honor.

DEDICATIONS

This book is dedicated to my wife, Diane, and sons, Judson and Luke. Without their understanding and patience there would be no book at all. Thank you for saying, "That's ok, we can do it another time," when there was too much work and too little time to do everything. I love you.

—Jim

This book is dedicated to my wife, Sally Goers Fox, a bright-eyed, intrepid bird of passage, with all my love and strength forever. Coming in on a wing and a prayer...

—Chris

CHAPTER 1

The Very Basics

INTRODUCTION

This chapter addresses the concerns of architectural design firms and technical schools or university architectural departments who seek to meet the current and future needs of the architectural design marketplace through the adoption and implementation of Autodesk® Revit® Building.

Students and designers new to Building Information Modeling will find the summary and "quick tour" exercise in this chapter a helpful introduction to the basic modeling and documentation tools in Revit Building, in preparation for more advanced concepts in later exercises.

OBJECTIVES

- Understand Revit Building's position as an architectural design tool
- Prepare for successful implementation of Revit Building in your current work process
- Experience an overview of the software

BUSINESS SUMMARY

WHAT'S REVIT BUILDING?

Autodesk, Inc. is the world's premier producer of mid-price design software. Its general-purpose AutoCAD drafting package is the world standard. It has sold millions of copies worldwide, and there is an entire industry of third-party developers dedicated to creating task-specific or industry-specific design applications based on the AutoCAD engine. Autodesk has developed and successfully marketed "Desktop" vertical applications to push AutoCAD into mechanical, architectural, civil, and land-development specialties.

Architectural Desktop is now in release 2006, with approximately 250,000 licenses worldwide. In April of 2002, Autodesk acquired Revit Technologies, Inc. (formerly Charles River Software), whose sole product was Revit, a parametric building modeler package. "Revit" is taken from architecture's slang for "revise it," a nearly constant refrain in building design. Autodesk has established Revit as its non-AutoCAD based building design and documentation system and its strategic platform for software serving the building industry. The name change to Revit Building for the 8th release in mid-2005 indicates that the Revit platform is expanding as planned. Why the change from AutoCAD, and what does it mean for architectural design firms using AutoCAD and Architectural Desktop?

Autodesk's move away from the DWG format (AutoCAD's output file) for architectural models is consistent with a previous transition the company made in the mechanical design field. Autodesk developed Inventor, a parametric modeler of 3D solids that uses file formats incompatible with the DWG, and has used it to replace the previously successful and widely accepted Mechanical Desktop. With Revit, Autodesk purchased a product (and company) rather than attempt to replicate the effort that had been put into Revit by its developers. Here again Autodesk has been consistent. Architectural Desktop itself is an outgrowth of a purchase—Autodesk bought out Softdesk software in the mid 1990s. Softdesk had developed AutoArchitect, an add-on package to AutoCAD for architectural documentation. Autodesk reworked AutoArchitect and released it as Architectural Desktop; other packages Softdesk had developed turned into Autodesk's Building Systems.

Autodesk executives have recognized the limitations inherent in the DWG format, whether for coordinating large mechanical assemblies or the complexities of building design and construction documentation. The results of their aggressive search for the next phase of design software have been the development of Inventor and the purchase of Revit.

Firms using AutoCAD or Architectural Desktop, that may have a huge inventory of DWG files and an ongoing commitment to developing/purchasing/applying peripheral applications to enhance parts of their work process such as schedules or space planning, will undeniably face a transition period while they come to grips with this change in the tools of the trade. This book is designed to show what changes actually lie in store for firms and institutions that decide to embrace the new shape of architectural design, how to manage and make best use of the parametric capabilities of Revit Building, and what job skills, office practices, and client benefits are likely to develop in answer to the new conditions.

If you are reading this book because you have come to the conclusion that maintaining your company or school's current investment in AutoCAD is not as important as learning to make effective use of the next generation of tools, then you understand the costs of doing business and are willing to pay the price of keeping ahead of the pack rather than the price of falling behind.

SIMPLICITY MEANS SPEED

One of the biggest differences between Revit Building and any other CAD architectural design software currently in widespread use lies well under its surface, but shows up in the entire interface and organization of its information handling. Revit Building is a complete architectural design software application with drafting components, not drafting software that has been modified or expanded to make architectural design possible. This means a number of major differences from CAD packages, AutoCAD in particular.

- Every object created in Revit Building is classified according to its function (walls, doors, structural members, components such as furniture, etc.); the relations between objects are also based on architectural function. Walls "host" doors and windows and the necessary openings, for example. Geographic positioning, size, and other properties of objects can be explicitly controlled by constraints or made subject to parameters, so that relationships can be established which will survive changes. A door can always center on a given wall, and windows on either side can be equidistant from one another, no matter how long the wall or the size of the windows. This eliminates many steps checking and rechecking dimensions during design iteration (as shown in Figures 1–1 and 1–2).

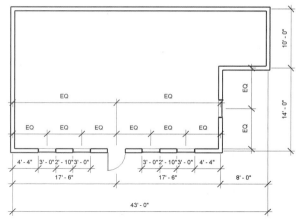

Figure 1–1 *Doors and windows spaced equally in walls of certain lengths*

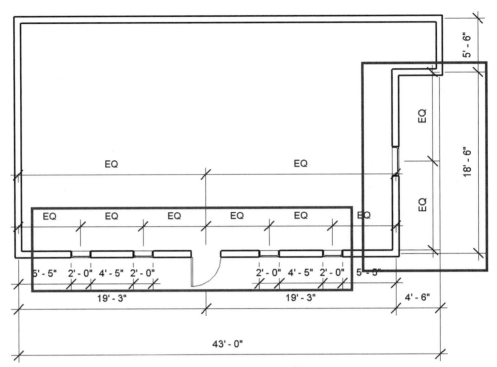

Figure 1–2 *Drag walls and change window size—the equal spacing still holds*

- The visibility of objects is controlled in views (plan, elevation, section, 3D) according to properties of the views and the objects. There are no layers or their equivalent used, which means vast simplification of the interface. Revit Building's appearance is largely black lines on a white background, but this can be changed to as many colors or patterns for as many different purposes as the user may desire. The default visual settings will suffice for the great majority of users, and the visible lineweights correspond to printed lineweights. The function and detail level setting of each view control the detail visible— no material symbols or section hatching in walls at coarse scale or in reflected ceiling plans, but hatch visible in floor plans at medium and fine scale (as shown in Figures 1–3 and 1–4). This eliminates a large amount of setup and correlation between the screen view and the plotted results—no plot configuration files or layer standards are necessary.

Parameter	Value
View Scale	1/8" = 1'-0"
Scale Value 1:	96
Display Model	Normal
Detail Level	Coarse
Visibility	Coarse
Model Graphics Style	Medium
Advanced Model Graphics	Fine

☐ Coarse
▣ Medium
▓ Fine

1/8" = 1'-0"   ☐ ⊗ 0⟍◪ ⌣

Figure 1–3 *The view detail level control in the View Properties dialog, and on the View Control Bar*

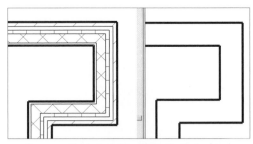

Figure 1–4 *The same walls—medium detail level in the left view, coarse detail right*

- The views in Revit Building are extremely powerful—a fundamental organizational principle. All come from the existing model, so changes in the model are automatically propagated throughout all possible views of it (see Figure 1–5). An elevation will always contain exactly the number of windows in the exact location that the plan shows, and a wall section will always show the correct components of the wall no matter how many times the wall type changes, with no manual updating procedure required of the user. This saves significant amounts of design time and eliminates chances for error over the course of a project.

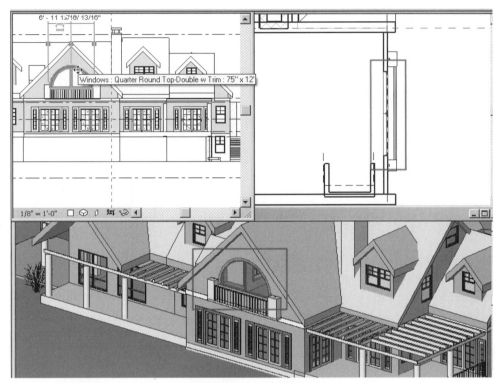

Figure 1–5 *Plans, elevation, and shaded view. A change to the model in one view (note the window currently selected in the elevation) shows immediately in the other views.*

- Revit Building includes "traditional" line-arc-circle drafting tools and editing commands, which are used, as one would expect, for sketching custom components or 2D detailing. There is no visible coordinate system in Revit Building, but a system of work planes on which one drafts or sketches. These work planes utilize reference planes that can be governed by parameters when creating components, so part of a column can always be twice as wide (or as long) as another part, or the column will always run from its base level to the one above (plus or minus an offset value if desired) even as floor heights change. Design intent can be specified to carry through the iterative process that is architectural design.

Revit Building also separates itself from other available architectural applications, including Architectural Desktop, in a number of ways:

- Plotting sheets and annotations are smart—Figure 1–6 shows how a plan tag will renumber itself according to the sheet and detail number of its reference as the sheet package develops, eliminating a source of much error and time spent checking annotations in the latter stages of documentation.

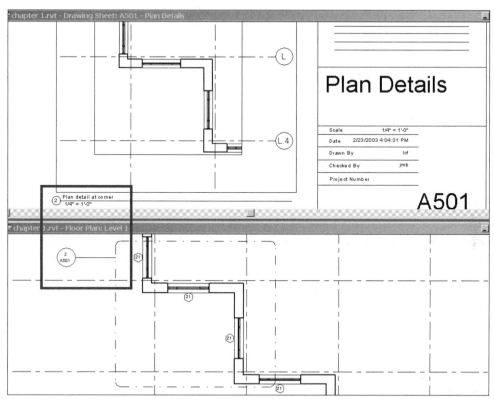

Figure 1–6 *A callout in the plan view knows its referenced sheet and detail number. Move the detail to another sheet, and the plan symbol updates immediately wherever it is visible.*

- Nearly any object with parametric properties can be scheduled, and nearly all objects are or can easily be made parametric. Schedules (which are views) are live and reflect changes in quantities or properties.

- Room and area tags are easy to create and apply, and feed directly into space planning calculations, schedules, and colored views.

- Revit Building includes a rendering engine that makes lighted, colored views, with material maps applied, possible at any time in the design process, for any part of the model. Still views and animated walk-throughs are quick to generate. Early visualizations of conceptual models are so easy some Revit Building users have resorted to subterfuge so as not to respond "too quickly" to client requests for renderings.

All this boils down to greater ease of use for more users than design systems based on drafting methods. There have been totally object-based architectural design packages available before, but they were not as easy to use, nor did their interfaces provide the ready visual understanding and ease of plotting that Revit Building does, and those applications are no longer viable in the market.

QUICK PRODUCTIVITY

Revit Building's entire system has been carefully crafted to provide a design environment that architects will readily understand and find easy to use—all software strives for the elusive goal of providing an intuitive experience for the user, and Revit Building, in its basics, comes very close to the ideal.

As a result, designers who may not be CAD-literate can expect to create models, edit them, change their appearance, and create specifications in considerably less time than it takes to learn a drafting package and use that drafting package to create a building design. Architectural Desktop requires facility with AutoCAD—Revit Building does not, and much productive time can ride on that simple difference.

Experienced CAD designers who are not thrown for a loop by not having layers to control, or by not having X-Y-Z coordinates to enter when defining the location of an object, can find themselves productive in a surprisingly short training cycle. Competent architectural drafters who are not simply CAD jockeys will find their drafting skills as much in demand as ever. There is a world of 3D content yet to create for this young program.

SIZE AND STRENGTH MATTER

Revit Building is a robust application, capable of generating complex and complete building models of considerable size in its current form. Its hardware requirements are substantial but fully in line with those of current CAD workstations. Revit Building is designed to hold a single building model in each file. It contains a mechanism (Worksets) that allows teams to parcel out and work on parts of a model and correlate their updates to a central file. File linkages (corresponding to external references in AutoCAD-based products) allow for assemblies of models or campus layouts combined in a single file location.

Since Revit Building contains a rendering module, architects can produce detailed presentation images without leaving the application. The controls of the AccuRender module within Revit Building are easy to learn, production and capture of useful images or animations is quick, and no updating of the model or file linking is necessary. For firms with a separate graphics department apart from building design teams, Revit Building models export readily into DWG format for import into VIZ. The latest release of VIZ now recognizes most Revit Building object properties.

For those worried that Revit Building might represent a "flavor of the month" that will soon disappear and leave early adopters worse off than before, Autodesk's position in the global market as a premier supplier of design and visualization applications provides a stable development and support environment for Revit Building.

Architecture is a notoriously conservative industry. As late as the mid-1990s nearly half of the architectural firms in the U.S. had next-to-no investment in CAD technology in their offices. Before the development of Architectural Desktop, when AutoCAD was Autodesk's flagship design product, each new release of the software occasioned hand-wringing and consternation as firms agonized over the decision whether to keep CAD stations and standards current or not. Now that the software decision before today's firms involves a new file format, different design techniques, changing office roles, and fluid client relationships, the basic question is actually simpler than before.

Autodesk stated from the moment it acquired Revit that this application is its building modeling solution of the future. The company obviously does not expect wholesale rejection of its initiative from the architecture industry. Nor does Autodesk expect quick or eager acceptance of this announced move away from the only CAD standard many architectural firms have ever known, and the myriad third-party applications developed by an entire industry devoted to extending AutoCAD and the DWG.

Autodesk has substantial resources to help firms making a commitment to Revit Building understand and manage the transition. Neither the product nor the company is going to evaporate any time soon. Autodesk, while aggressive when necessary in acquiring technology, is a company dedicated to developing and maintaining a growing stable of industry-leading design products. The company takes its leadership position seriously enough to know the dangers of out-pacing the market. Revit Building is Autodesk's proposition to those architectural firms that also take leadership seriously.

NO LONGER THE NEW KID ON THE BLOCK

Autodesk has come out with more than 16 releases of AutoCAD and 6+ releases of Architectural Desktop. Revit Building is currently in its eighth version after almost four years of public release. With no underlying CAD framework in prior release (as AutoCAD underlies Architectural Desktop) providing a user base, Revit Building has achieved a modest but quickly-growing adoption to date—currently about one-tenth as many licensed seats as Architectural Desktop. To reach the majority of the market will clearly take years. Whether your firm or institution is an aggressive front runner or a wait-and-see trend adopter, you should take some issues into account in your transition and implementation planning.

TIP Autodesk makes Revit Building available as a free download from the www.autodesk.com/revit website, and will send a CD on request. Revit Building will work in demo mode without a license, and for a limited time as a full function trial version. You will be unable to save work after the trial period expires until you purchase a subscription.

FILE COMPATIBILITY

Given the relative user base numbers, the issue of file compatibility will be daunting to some firms, though it need not be a reason to avoid or delay exploring Revit Building's capabilities. Revit Building is built to interface easily with AutoCAD products. "Interoperability" is the word Autodesk uses to emphasize the two-way connectivity. Revit Building can link to AutoCAD files as easily as AutoCAD does. Revit Building imports and links to 2D DWG files with full layer, linetype, and color information, and handles a growing variety of 3D solid information. Revit Building outputs 2D or 3D DWG files that are tidy and complete, with layers logically assigned.

TRAINING

Executives facing the decision to begin using Revit Building will need to budget for training as part of the implementation process. Revit Building ships with a series of tutorials and the company provides free live training sessions online, so for those individuals who can successfully self-teach using the provided lessons there will be no cost for materials or instruction. Autodesk will provide a training team for larger firms making an investment in Revit Building. Resellers and Autodesk Training Centers have begun providing scheduled classes and on-site instruction sessions by certified instructors.

PROJECT SIZE

Large building projects are difficult to manage no matter what software you use to design and document them, and with Revit Building's single-file building model, larger projects have proportional challenges. Your first projects in Revit Building should be relatively modest, perhaps in the 100,000 square foot range. With careful configuration of template files, work sharing, and model linking, firms are presently doing projects in Revit exceeding a million square feet. Each release of Revit Building has improved performance in project size, and with Revit Building 8 an experienced Revit Building team can confidently tackle just about any building design. As we mentioned in the Author's preface, the Freedom Tower project for New York City, possibly the most complicated skyscraper proposal in the world today, is being developed in Revit Building, from the subway interchanges beneath the foundation to the electricity generating wind turbine matrix at the top of the tower.

Revit Building's file is based on a single building model that carries all the relevant information about the building structure and contents, rather than a system of individual files created per floor, for instance, and brought back by reference into a shell or assembly file. Revit Building's system of Worksets allows for a growing team to work on a model as a design develops by "checking out" parts of the model with a central file holding and coordinating data. This system has the advantage of simplifying what can grow into a bewildering nest of references in a large model.

DATA OUTPUT

Revit Building is very efficient at presenting building or component information in schedules. The data tabulated in schedules can be exported to a spreadsheet or database such as MS Access via ODBC, the Microsoft-developed process for sharing field information between applications. This is a limited and cumbersome process compared to the Data Extract function in AutoCAD, and while it allows for updating an external file based on changes in the Revit Building model, the process at present cannot be reduced to a one- or two-click operation, nor automated.

Revit's Export Schedule capability does allow for quick conversion of any scheduled information into *.txt* format for import into other applications, and holds promise of developing into a live link between the Revit building model and external data-parsing applications.

PROGRAMMING INTERFACE

Revit Building does not contain an exposed programming interface, as does Auto-CAD. One of AutoCAD's strengths from the very beginning was the decision early on to embed a LISP interpreter in the application and make this interpreter available to users. This allowed developers of all stripes to augment basic "vanilla" AutoCAD with routines of sometimes startling ingenuity to aid in creating CAD content, to automate repetitive tasks, and to read values from the drafted geometry. In recent years with the rise of Microsoft Visual Basic as a programming language common to many Windows applications, AutoCAD has developed Visual Basic for AutoCAD as a way for the application to interface with and react to other applications. As a single-purpose building modeler rather than a general-purpose drafting tool, Revit Building does not have the same need for external hooks as AutoCAD, and it has been carefully designed to eliminate many of the repetitive tasks that AutoLISP is often used to speed up. All building model objects are complex, with properties that can placed into and read from schedules. Revit Building contains an API (advanced programming interface) that is available for members of the Autodesk Developer Network. There is, at present, no capability for the user to create scripts, macros or toolbar customization within the program.

WHAT ELSE IS COMING?

Autodesk has made it clear that Revit is the company's platform for architectural building information modeling. The original program has become Revit Building. A second application, Revit Structural, with bi-directional links to Architectural Desktop, Autodesk Building Systems and external (third-party) analysis applications, appeared as this book was going to press in the summer of 2005. Plans are underway for Revit MEP (Mechanical, Electrical and Plumbing), so that the Revit building model can provide the same types of specialized output as DWG-based files.

PREPARE FOR SUCCESSFUL DESIGN OFFICE IMPLEMENTATION

SHOCK TROOPS

Only the foolishly optimistic would expect to proclaim a new company standard without working to overcome inertia and objections. Introducing new concepts and tools to a work arena is a process, not an event, and takes preparation and education even more than vision and enthusiasm.

Most firms have members who are eager to learn, willing to push the limits of their knowledge, ready to try new things. All firms have members who are most productive when comfortable with routine, not very interested in expanding their skills if risk is potentially involved, and slow to adapt to new circumstances. Quick and ready learners are not necessarily better designers or more useful team members than steady-as-she-goes types (and may not be the company's AutoCAD or ADT experts), but can be valuable as "early adopters" who spearhead an effort to learn, utilize, and broadcast an important shift in work standards.

- Assemble a team whose members are willing and capable of working with new software and techniques
- Develop a plan for training
- Set a schedule for individual and group training
- Establish a forum for communication: team members to each other, to CAD manager, and to project manager
- Use an initial project as a focus for the training and development of team work practices

PROJECT BY PROJECT

Decide beforehand on the first project for the spearhead team to work on in Revit Building. Considerations for this project may involve building size, client requirements, and schedule.

- Understand your company's design process, including your current standards
- Understand the scope of the project
- Clearly define and communicate the project goals
- Establish project and team evaluation criteria for later review
- Expect and allow for differences in workflow from current company process

REMEMBER YOUR ROOTS

Revit Building is built to take maximum advantage of AutoCAD—"interoperability" can be key. Revit Building reads AutoCAD files either via link (equal to external reference) or import, and will retain layer, color, and linetype information. It will also export linework from its models to AutoCAD on a by-view basis. Its export DWG files use layer, color, and linetype information per default built-in standards, which are fully customizable to your current company standards.

- Team members can import existing site data in DWG, DGN (MicroStation), or DXF format. Use Revit Building's topographic tools to build the 3D site surface.

- Link or import existing plans for reference in creating new project content

- Scan in and trace over hand concept sketches

- Link or import DWG/DGN/DXF files received from consultants

- Export DWG/DGN/DXF files to consultants

- Link or import standard details and symbols from company archives during CD phase

EXPAND THE TEAM

Once a first project has been completed or nearly completed in Revit Building, the company can build on that experience to develop the use of Revit Building as useful enterprise software.

During post-project analysis from the first Revit Building project, let other members of the company know about the successes of the spearhead team, and be sure to evaluate it honestly and communicate problem areas. Was training adequate? Were expectations clearly expressed? Was support available in a timely way?

Target the next projects, divide the Revit Building team between the projects, and plan and schedule training for new team members. In this way, knowledge and experience with Revit Building can spread the most effective way—from user to user. People learn most efficiently when actually working with their new knowledge, and when there is a free flow of information—tips, tricks, questions, and answers pertaining to work in hand.

STUDENT'S PULL-DOWN MENU

The exercises in this book are designed to give you a thorough grounding in the most powerful and complete building modeler on the market today. The exercises and review questions do not presume that you have any experience in architectural design or drafting. They do presume that you can follow step-by-step instructions explicitly and carefully. Wherever necessary, explanations will be provided as to why certain steps should happen in a certain order, so that you understand how this modeling program works most efficiently.

You should be aware by now that modeling is not the same as drafting, even though many views (plans, elevations, sections) in Revit Building may look just like plans, elevations and sections that you may have created before now in other design software applications. Building components are complicated objects with multi-part definitions that control how they appear, behave and interact with all other parts of a project model.

The ease with which designs can be developed and presented using Revit Building has changed the former definitions of "basic" and "advanced" skills. You will be presented with tasks such as file importing and setting various element properties along with instructions on creating basic wall layouts, for instance. This mix not only reflects the skills and usages you will need to master in your work life, but is designed to make you comfortable with the interface and controls that make Revit Building the useful application that it is.

QUICK TOUR—DESIGN, VIEW, AND DOCUMENT A BUILDING

This book contains exercises and lessons that are substantially different from the tutorial lessons provided with the software. First up is a short exercise to give you a very quick look at Revit Building's most basic tools and integrated display capabilities.

REVIT BUILDING COMMANDS AND SKILLS

Walls

View Properties

Doors

Windows

Editing—Copy, Mirror, Element Properties

Dimensions—EQ toggle

Add Level

Filter Select

Floor

Roof

Section

Sheet—Drag and Drop Views

View Scale

Activate View

3D View

Shade View

EXERCISE 1. LAY OUT A SIMPLE BUILDING

1.1   Launch Revit Building from the desktop icon. (You can also start Revit Building by picking Start> All Programs>Autodesk Revit Building 8.) Revit Building will open an empty project. The View Window (drawing area) will open to a Level 1 plan view (see Figure 1–7). Four elevation symbols will be visible in the View Window. Menus on the Menu Bar at the top of the screen and in the tabs of the Design Bar control modeling and other tools. The Project Browser located between the View Window and Design Bar shows available views in customizable tree form.

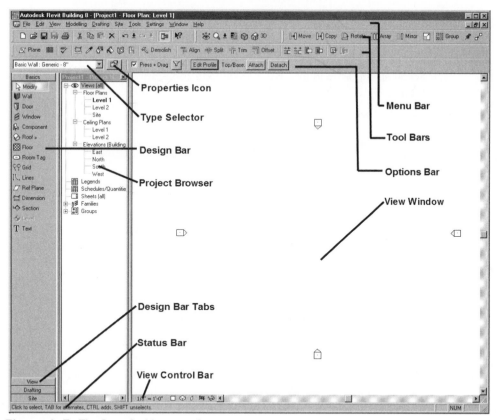

Figure 1–7 *The Revit Building interface*

1.2 From the Basics tab on the Design Bar, pick Wall. The cursor will change appearance to a pencil symbol and the Type Selector drop-down box on the Options Bar will display a generic wall type, as shown in Figure 1–8.

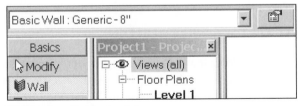

Figure 1–8 *The Wall tool on the Basics tab*

1.3 Pick the down arrow at the right of the Type Selector, as shown in Figure 1–9, to expand the list.

Figure 1–9 *Expand the list of default wall types*

1.4 Scroll up the list and pick Basic Wall: Exterior – Brick and CMU on MTL Stud as shown in Figure 1–10.

Figure 1–10 *Brick and CMU wall type*

1.5 Accept the default Height and Location Line values on the Options Bar. Check the Chain Option so that your picks will draw walls continually until you stop the command. Leave the Straight Line tool depressed. Leave the Offset value at **0′ 0″** (see Figure 1–11).

Figure 1–11 *Settings for the walls; select the Chain option*

1.6 Place the cursor in the View Window left of center and above the middle, but inside the elevation markers. If you pause the cursor a Tooltip will appear, reading Click to enter wall start point, as shown in Figure 1–12. This message will also appear in the Status Bar at the bottom of the screen.

Figure 1–12   *About to draw walls*

1.7 Left-click to start the wall. Pull the cursor to the right. This will create the wall with its exterior face to the top of the screen. A temporary dimension will appear showing the length of your wall, and an angle value. If you pull directly right as shown in Figure 1–13, a dashed green reference line will appear and a Tooltip will verify that the wall is horizontal. Pull the cursor until the dimension reads **40'-0"** and left-click to set the wall endpoint.

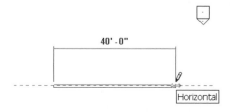

Figure 1–13   *The first wall, drawn left to right*

1.8 Pull the cursor down to start the second wall. If you need to zoom in closer to see your work, right-click to get a context menu with zoom options, and pick Zoom In Region. Then pick two points as appropriate (see Figure 1–14). You can also use the scroll wheel if your mouse is so equipped.

Cancel
Find Referring Views
Zoom In Region
Zoom Out (2x)
Zoom To Fit
Previous Scroll/Zoom
Next Scroll/Zoom
View Properties...

Figure 1–14 *Right click to access zoom controls*

1.9 As you pull the cursor down, the temporary dimension, angle readout, reference line, and Tooltip all appear, as shown in Figure 1–15. Make this wall 40′ long as well, and then left-click.

 NOTE The actual dimensions of the walls we are creating in this exercise do not matter. Nor do we really care if the walls are perpendicular to one another. Revit Building is designed for quick sketching and contains many tools to establish/revise dimensions and relationships later in the design process. Ironically, this can be a point of confusion for skilled drafters. These tools will be illustrated extensively in later lessons.

Figure 1–15 *The second wall is perpendicular to the first*

1.10 Pull the cursor to the left 40′ and left-click. Revit Building will pick up alignment points, snap to them, and display them in the Tooltip, as shown in Figure 1–16.

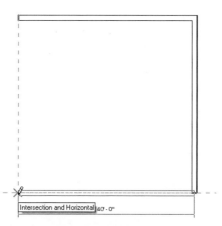

Figure 1–16 *The third wall ends even with the first one*

> 1.11 Pull the cursor up and Revit Building will snap to your start point. Left-click to finish this simple exterior (see Figure 1–17). To exit the Wall tool, hit ESC twice, or pick Modify from the Design Bar, as shown in Figure 1–18. This puts the cursor back into select mode.

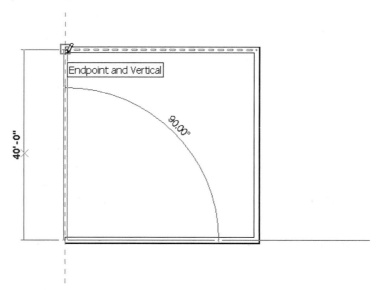

Figure 1–17 *Finish the outside walls*

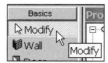

Figure 1–18 *The Modify command is at the top of all the Design Bar tabs. Use it to terminate many command routines.*

1.12 Put the cursor in the View Window and right-click. The bottom option is View Properties (see Figure 1–19). Select (left-click) View Properties to bring up Revit Building's main visibility control, as shown in Figure 1–20.

Figure 1–19 *Select View Properties from the right-click menu*

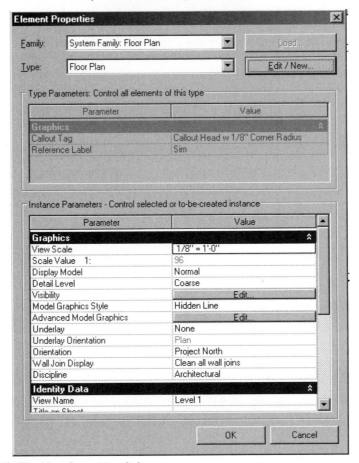

Figure 1–20 *The View Properties dialogue*

1.13 Click in the Value field for Detail Level and change it from Coarse to Medium, as displayed in Figure 1–21.

Detail Level	Coarse ▼
Visibility	Coarse
Model Graphics Style	Medium
Advanced Model Graphics	Fine

Figure 1–21 *Three levels of view detail*

1.14 Click the Edit button in the Visibility value field to bring up the Visibility/ Graphic Overrides dialogue. It has three tabs. Click on the Annotation Categories tab and clear Elevations as shown in figure 1–22. This will turn off the Elevations in this view to simplify it. Click OK twice to exit the Visibility and View Properties dialogues to reveal the model.

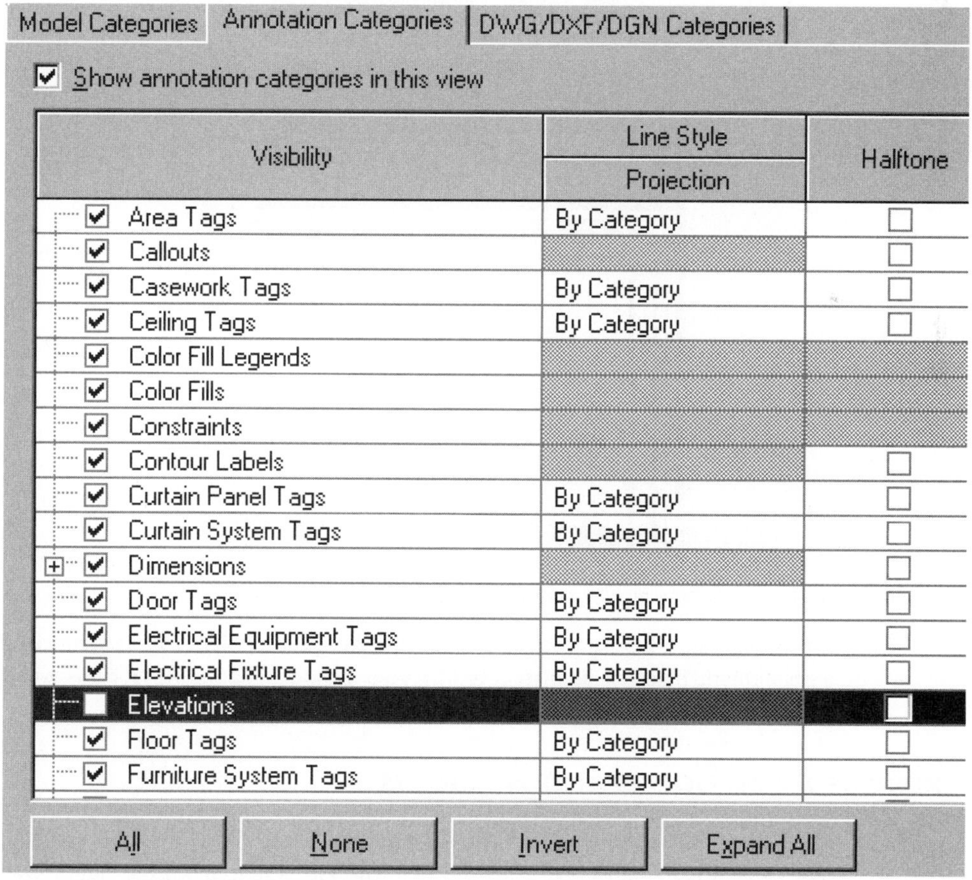

Model Categories	Annotation Categories	DWG/DXF/DGN Categories

☑ Show annotation categories in this view

Visibility	Line Style / Projection	Halftone
☑ Area Tags	By Category	☐
☑ Callouts		☐
☑ Casework Tags	By Category	☐
☑ Ceiling Tags	By Category	☐
☑ Color Fill Legends		
☑ Color Fills		
☑ Constraints		
☑ Contour Labels		☐
☑ Curtain Panel Tags	By Category	☐
☑ Curtain System Tags	By Category	☐
⊞ ☑ Dimensions		☐
☑ Door Tags	By Category	☐
☑ Electrical Equipment Tags	By Category	☐
☑ Electrical Fixture Tags	By Category	☐
☐ Elevations		☐
☑ Floor Tags	By Category	☐
☑ Furniture System Tags	By Category	☐

All	None	Invert	Expand All

Figure 1–22 *Turning off the elevation symbols in this plan view*

1.15 Right-click the context menu and pick Zoom to Fit, as shown in Figure 1–23. Figure 1–24 shows a portion of the adjusted view, a complex wall structure with hatching.

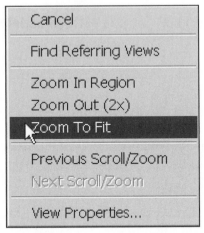

Figure 1–23 *The quickest way to zoom to the edges of the model*

Figure 1–24 *The walls now display complex structure with hatching*

1.16 From the Menu Bar, choose File>Save and save the project as *quicktour.rvt* in a convenient place or as instructed by your trainer. File saving options will be explained in detail in later exercises.

EXERCISE 2. ADD DOORS AND WINDOWS

2.1 Pick Door on the Basics tab of the Design Bar. Pick Single Flush: 36″ x 84″ from the available choices. Locate one in the left wall 4′ from the upper wall. Revit Building will snap to the walls and read the wall location as you move the cursor. Holding the cursor to the interior or exterior side will set the swing and hand (see Figure 1–25).

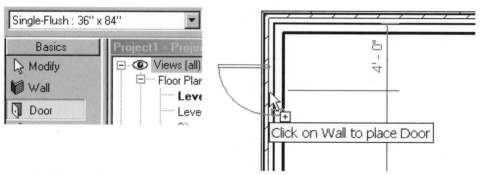

Figure 1–25 *Locate an exit door near the upper wall*

2.2 Set a second door 4' from the bottom wall. If you need to adjust the swing of a door, control arrow sets appear when it is selected. Click the appropriate control arrow to toggle the swing or hand (see Figure 1–26). You can do this while in the door placement routine.

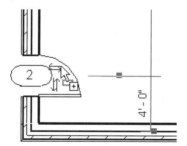

Figure 1–26 *Locate a second door*

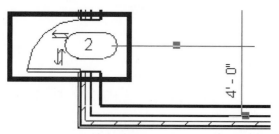

Figure 1–27 *Adjust the door swing if necessary*

2.3 The default list of doors does not include a double door. While the door tool is active, click the Load button (shown in Figure 1–28) on the Options Bar to bring up a content file browser dialogue as shown in Figure 1–29. Select the *Doors* folder in the *Imperial Library* folder, and then pick Open.

Figure 1–28 *Look for content from the library*

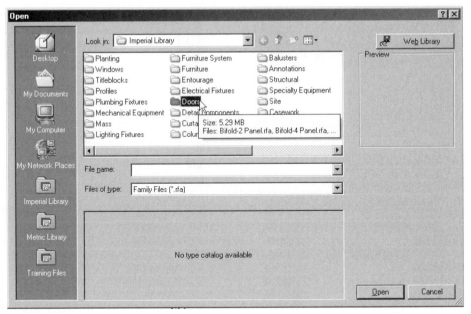

Figure 1–29 *The content library browser dialogue. You can create your own folders, or browse the Web.*

2.4 In the *Imperial Library/Doors* folder, select *Double-Glass 1.rfa* as shown in Figure 1–30, and then pick Open.

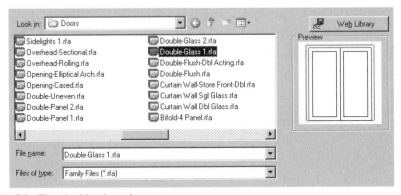

Figure 1–30 *The double glass door type*

2.5 Locate an instance of the double glass door midway between the other two doors (Revit Building will snap to their center points) as shown in Figure 1–31, then hit ESC or pick Modify to terminate the Door tool.

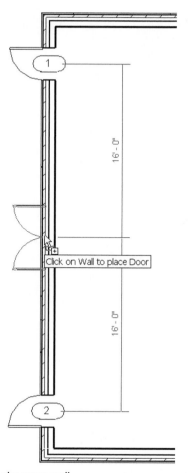

Figure 1–31 *Three doors in the west wall*

2.6   Pick the Window tool button in the Basics tab on the Design Bar.

2.7   Select Fixed: 24″ x 48″ from the list of available windows.

2.8   Locate an instance of this window in the west wall, 8′ from the centers of the lower and middle doors (see Figure 1–32). If you hold the cursor to the exterior side of the wall, the dimensions will appear to the right and the glazing will be on the left (exterior) side.

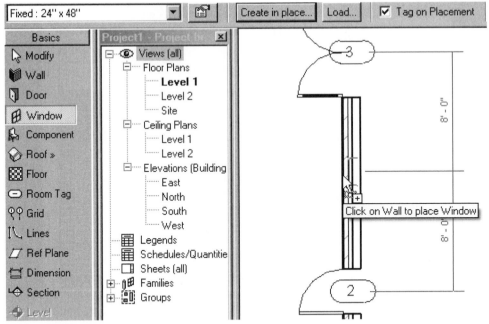

Figure 1–32 *The first window, equidistant from the door opening centers*

2.9 Add a second window midway between the center and upper doors.

2.10 Change the window selection to Fixed: 36" x 72" and place 5 instances in the lower wall. Do not be concerned about their location during placement. Use the control arrows if necessary to keep the glazing on the exterior side of the wall. Hit ESC or pick Modify to terminate the window placement.

EXERCISE 3. EDIT THE MODEL

3.1 Pick Dimension on the Basics tab of the Design Bar, shown in Figure 1–33.

Figure 1–33 *The Dimension command, used for locating components precisely*

3.2 The dimensioning symbol will appear next to the cursor (see Figure 1–34). Hold the cursor over the left-hand wall. It will snap to the wall centerline, which will highlight, and the Tooltip will read out the wall properties. Pick the wall, and then hold the cursor over each window in turn from left to right—a center mark for each window will highlight to show when the window is selectable—and choose the window. Pick the right wall last to complete the dimension string.

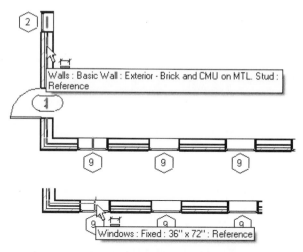

Figure 1–34 *Starting the dimension string—pick the wall, then windows left to right*

3.3 Pull the cursor above the wall, inside the building, and left click to place the dimension string. After you pick, the grips for the dimensions will appear, and the letters EQ with a slash through them (see Figure 1–35). Pick the EQ symbol—it is a toggle to lock the objects dimensioned in an equidistant relationship (see Figure 1–36).

Architectural drafting practice locates dimension strings on the exterior of building walls for purposes of clarity in busy plan pages. In our simple model we want to keep the picks easy for right now. It would be a simple matter to relocate the string later, in the construction document phase of a real project.

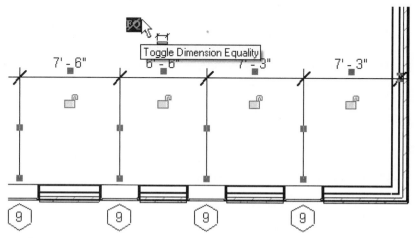

Figure 1–35 *Setting the EQ toggle*

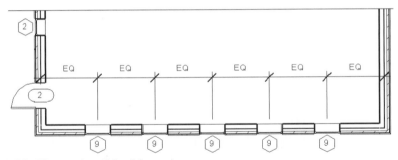

Figure 1–36 *The results of the EQ toggle*

3.4 Pick Modify from the Basics Tab of the Design Bar to terminate the Dimension tool.

In the Select mode, where the cursor appears as an arrow, Revit Building allows you to select more than one entity by holding down the CTRL key while picking. This builds a selection set upon which you can perform an action. Selected items will appear in red. To remove an item from a selection set, hold down the SHIFT key and choose it.

Revit Building follows the Autodesk standard for selection windows—picking a point in the view window **not** over an entity and holding down the cursor left button starts a selection window with different properties depending on its orientation. Dragging **left to right** starts an enclosing window that selects only entities completely enclosed within its border, which is a solid line on the screen. Dragging from **right to left** starts a crossing window that selects any entity within or touching its border, which is a dashed line.

3.5 Place the cursor below and to the left of the left-hand window. Left click and drag to the right to make a window around the windows in the lower wall, but do not include the entire wall. Release the cursor button to make the selection, as shown in Figure 1–37.

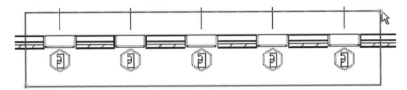

Figure 1–37 *Windows highlighted in the enclosing window*

3.6 When the windows are highlighted, pick the Mirror command on the Toolbar, shown in Figure 1–38.

Figure 1–38 *The mirror command*

    3.7 Pick the center of the center door in the left wall to define the mirror line (see Figure 1–39). A reference line will appear to assist your pick. Windows will appear in the upper wall.

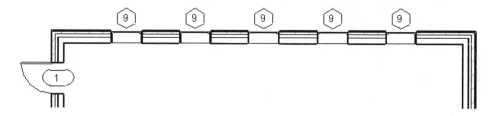

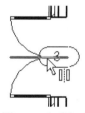

Figure 1–39 *Use the door as the mirror midpoint to create windows in the upper wall*

    3.8 Save the file.

EXERCISE 4. ADD ANOTHER LEVEL, UPPER WINDOWS AND A FLOOR

    4.1 In the Project Browser, find Elevations (Building Elevations). Double click on West to open up that view.

    4.2 Right-click and Zoom in Region around the left door. Note the appearance of the vertically compound wall structure at close range.

    4.3 Right-click and pick Previous Scroll/Zoom (or Zoom to Fit) to restore the entire elevation on the screen.

    4.4 Pick the Level tool on the Basics tab of the Design Bar, as shown in Figure 1–40.

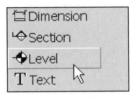

Figure 1–40 *The Level tool*

4.5 Place the cursor over the Level 2 marker line at the left side of the building face. Revit Building will snap to it and read out an offset dimension as you move the cursor up, as shown in Figure 1–41, so long as you keep the cursor over the model.

4.6 Pick to start a new level **8′** above Level 2. You can type **8** to set the vertical offset at **8′** if you have any problem getting the temporary dimension to read the desired distance. Pull the cursor to the right of the building. Left-click and the new Level 3 will appear. Note that the Project Browser now shows a floor plan and ceiling plan for Level 3.

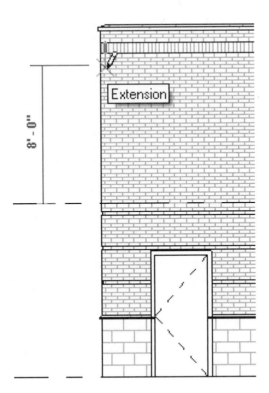

Figure 1–41 *The new level started*

4.7 Hit ESC or pick the Modify button to terminate the command. Zoom to Fit in the view.

4.8 Select the dashed line for Level 2. Select its left end as shown in Figure 1–42 and drag it closer to the left side of the building model. The end of Level 1 will move with it.

Figure 1–42 *Select the left end of Level 2*

4.9 Repeat this adjustment for the right ends of Levels 1 and 2, and for the left end of Level 3, as shown in Figure 1–43. Snap lines will appear as components align. This adjustment will make later work with this view easier.

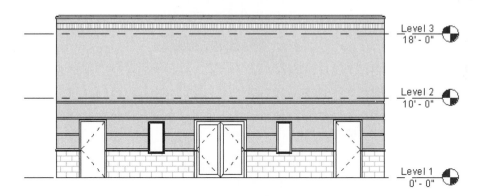

Figure 1–43 *The West Elevation with Level lines brought close to the model*

4.10 Use a window pick box as before (pick left to right) to select the windows and middle door, as shown in Figure 1–44.

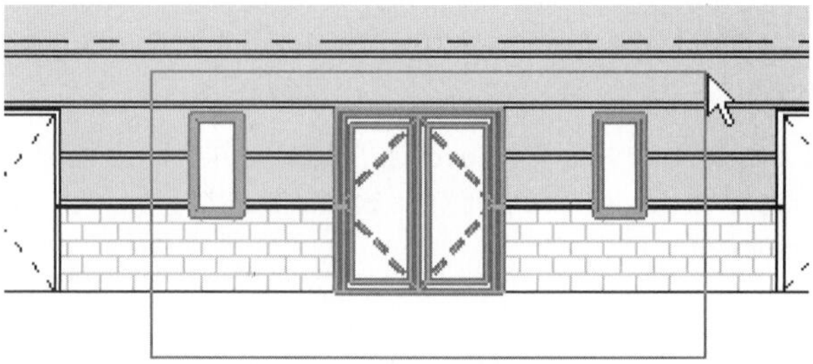

Figure 1–44 *Select windows and a door*

4.11 Choose the Filter Selection icon from the Options Bar. In the dialogue that opens, clear Doors and pick OK. (See Figures 1-45 and 1-46.)

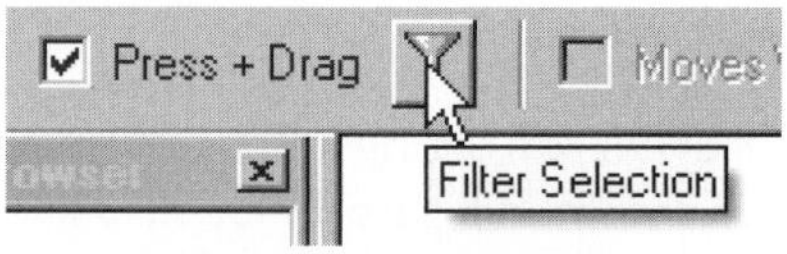

Figure 1–45 *The Filter Selection tool can be much quicker than picking individual objects*

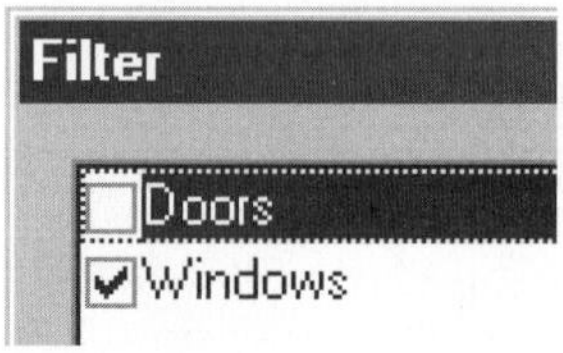

Figure 1–46 *Remove doors from the selection*

4.12 The two windows will remain highlighted. Choose Copy from the Toolbar, as shown in Figure 1–47.

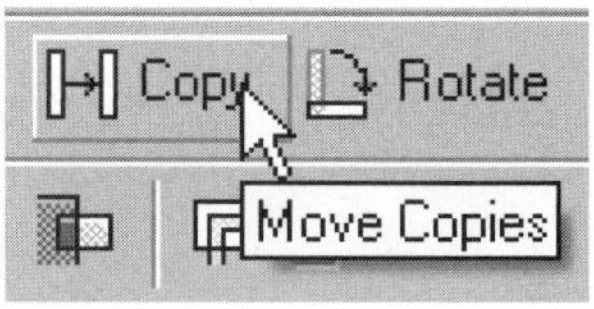

Figure 1–47 *Copy the two windows*

4.13 Revit Building will need a start point and end point to define the relative distance and direction for the copy. Select the left end of Level 1, as shown in Figure 1–48.

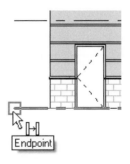

Figure 1–48 *Start the copy by referencing Level 1*

    4.14 Pull the cursor straight up until it snaps to Level 2, and left click, as shown in Figure 1–49.

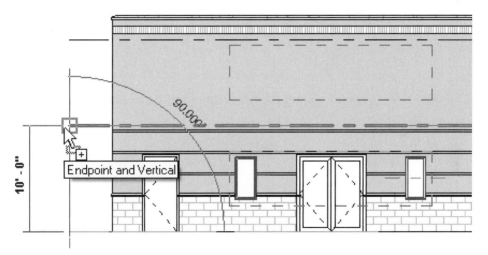

Figure 1–49 *Select Level 2 to makes copies of the two windows one level up from the originals*

    4.15 New windows appear above the originals (see Figure 1–50).

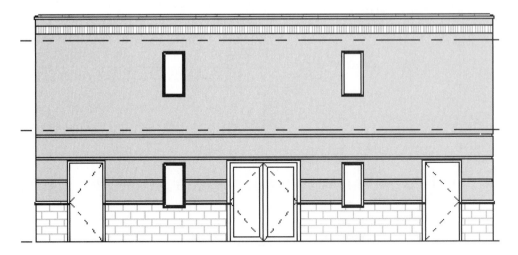

Figure 1–50 *Newsecond story windows aligned with the first story windows*

4.16 Double click Floor Plan Level 2 in the Project Browser to open that view.

4.17 Pick the Floor tool on the Basics tab of the Design Bar, shown in Figure 1–51.

Figure 1–51 *The Floor tool*

4.18 The walls in the View Window will gray out. The Design Bar will go into Sketch mode, with the Pick Walls tool activated. The Status Bar reads Pick walls to create lines. Leave the Offset value at **0′ 0″** and the Extend into wall (to core) option checked on the Options Bar (see Figure 1–52).

Figure 1–52 *Floor Options*

4.19 Hold the cursor over the interior surface of the left wall—the wall will highlight and the wall properties will appear in a Tooltip and in the status bar, as shown in Figure 1–53.

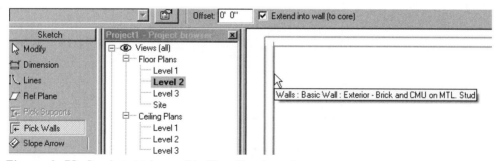

Figure 1–53 *Ready to pick a wall in Floor Sketch mode*

4.20   Pick the wall and a magenta sketch line will appear on its interior surface. Use the control arrows that appear to adjust the sketch line position to the interior side if necessary.

4.21   Pick the remaining three walls to complete a sketch of the floor edges, as shown in Figure 1–54.

Figure 1–54 *The floor sketch is complete*

4.22   Pick Finish Sketch (see Figure 1–55). Pick Yes in the question box that appears, so that the floor structure will meet the wall structure and finishes will be trimmed back appropriately in section views. When sketch mode completes the Basics tab will reappear in the Design Bar. The View Screen will not change appearance.

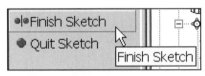

Figure 1–55 *Finish Sketch*

EXERCISE 5. ADD A ROOF AND VIEW IT IN SECTION

5.1 Double-click on Level 3 in the Project Browser. Pick the Roof>>Roof by Footprint tool in the Design Bar, as shown in Figure 1–56.

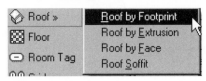

Figure 1–56 *Roof by Footprint*

5.2 Revit Building shifts into Sketch mode, as with the Floor tool. This will be a flat roof below the parapet cap on the exterior walls. On the Options Bar, clear Defines Slope and check Extend into wall (to core), as shown in Figure 1–57.

Figure 1–57 *No slope for this roof, which will extend into the wall past the finish layer*

5.3 Pick the interior surface of the walls, as with the Floor command. The walls will highlight in turn as the cursor passes near them, and reference lines will appear. Adjust the position of the sketch lines to interior faces using the control arrows, as before.

5.4 Pick Finish Roof. Pick No to the Attach Walls to Roof question box, and Yes to the join/cut/overlap question box. The exterior walls will extend past the roof to form a parapet and will not attach to its underside, but will trim their finish layers at the floor structure. The view will not change appearance when the roof is created.

5.5 Double-click on Floor Plans Level 1 view in the Project Browser to open that view in the View Window.

5.6 Pick the Section tool on the Basics tab of the Design Bar, shown in Figure 1–58.

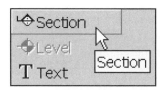

Figure 1–58 *The Section tool*

5.7 Pick a start point for the Section line to the left of the upper window in the left wall (see Figure 1–59).

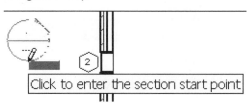

Figure 1–59 *Starting the Section*

5.8 Pull the cursor to the right horizontally, and pick a spot to the right of the wall to locate the Section (see Figure 1–60). The section line can be selected and moved if necessary so that it passes through the window.

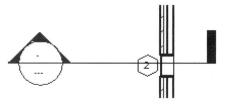

Figure 1–60 *Finishing the Section*

5.9 The Project Browser will show a + sign to the left of a new Sections (Building Section) branch. Click on the + sign and Section 1 will appear in the Project tree, as shown in Figure 1–61.

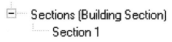

Figure 1–61 *The new Section in the Browser*

5.10 Double-click on Section 1 in the Project Browser to open that view.

5.11 Right-click and pick View Properties (always the bottom option in the right-click context menu). Change the Detail Level to Medium.

5.12 Pick on the Level 2 floor. Use the Type Selector drop-down list to change its type from Generic to Steel Bar Joist 14″ – VCT on Concrete.

5.13 Pick on the Roof. Use the Type Selector drop-down list to change its type from Generic to Steel Truss – Insulation on Metal Deck – EPDM. Note the new appearance of these components.

EXERCISE 6. DOCUMENT THE MODEL

6.1 Right-click on Sheets in the Project Browser, and pick New Sheet, as shown in Figure 1–62.

Figure 1–62 *Creating a new sheet for plotting*

6.2 Pick OK to accept the default E1 30 x 42 Horizontal title block for the new sheet.

6.3 The new sheet appears in the View Window. Pick Level 1 in the Project Browser (do not double-click), hold down the left mouse button, and drag it into the sheet (see Figure 1–63). Release the button and the outline of the view at scale with a title below it will appear. Move the cursor and click to place a viewport into the Level 1 floor plan on the sheet (see Figure 1–64).

Figure 1–63 *Starting to drag a view onto the sheet*

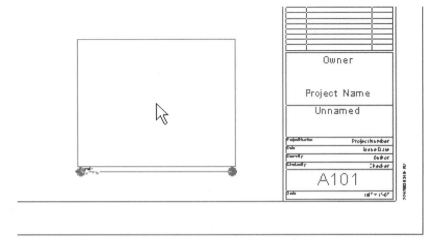

Figure 1–64 *The view outline appears under the cursor*

6.4 The viewport is small on the page. Press ESC to terminate the placement. Select Floor Plan Level 1 in the Project Browser again, right-click and select Properties.

6.5 In the Properties dialogue for the Floor Plan as shown in Figure 1–65, change the View Scale to 1/4″ = 1′-0″ and choose OK.

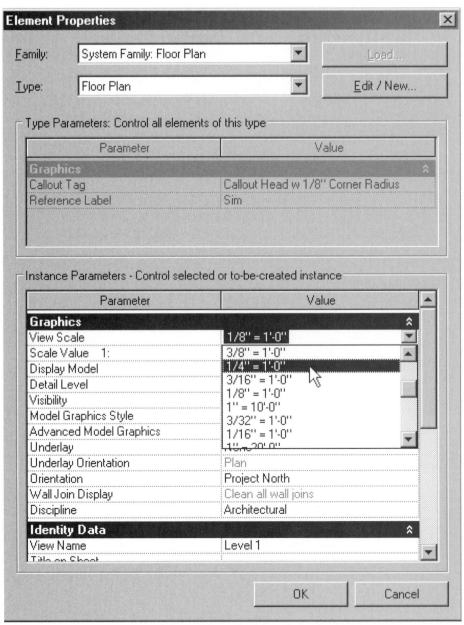

Figure 1–65 *Changing the view scale without opening the view*

6.6 Repeat this scale change for the West Elevation and Section 1 views in the Project Browser.

6.7 Select Floor Plan Level 1 in the Project Browser and drag it onto the sheet, as before. Note the new size indicated in the lower right corner of the sheet.

6.8 Select and drag views West Elevation and Section 1 from the Project Browser to the sheet and arrange them as shown in Figure 1–66. Views will snap to alignment lines to aid in placement. Align the Elevation view title with the title of the Plan view, and align Level 2 of the Section with Level 2 of the Elevation view.

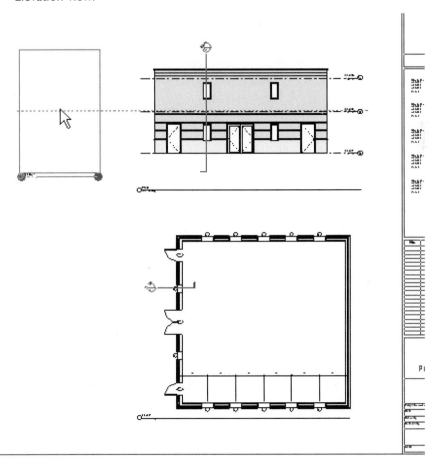

Figure 1–66 *Views will snap to other view titles and to level level lines for viewport alignment*

6.9 Zoom in over the plan viewport so you can see the left wall components clearly. Place the cursor over the viewport so that its border becomes highlighted. Right-click and pick Activate View, as shown in Figure 1–67. This will allow you to change the model from the sheet view.

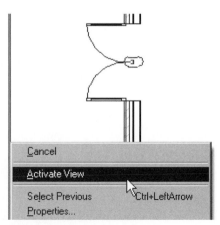

Figure 1–67 *Making the viewport active to work in it*

6.10 Pick one of the windows in the left wall, hold down the CTRL key, and pick the other so both are selected.

 TIP If you have difficulty picking the windows you can zoom in closer. You can also use the TAB key on your keyboard to cycle through the available choices until the window you want to modify becomes highlighted and you can select it.

6.11 When both windows are selected, use the Type Selector drop-down list to change their type to Fixed: 16″ x 72″ as shown in Figure 1–68.

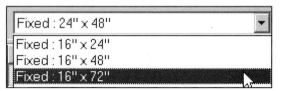

Figure 1–68 *Changing the window type*

6.12 Pick the Properties icon to bring up the Element Properties dialogue shown in Figure 1–69. Change the Sill Height for these two windows to 1′ (you can simply type 1—Revit Building's default Imperial unit is the foot). Pick OK.

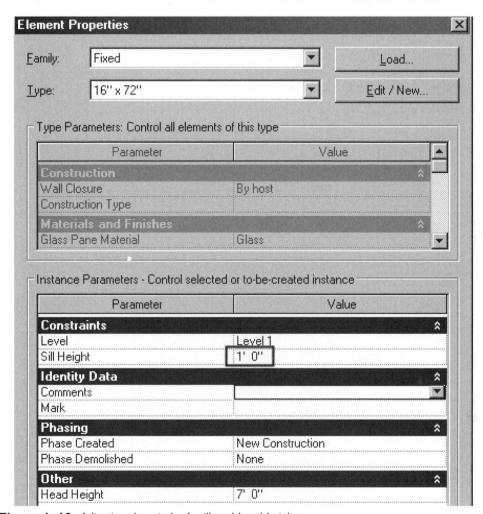

Figure 1–69 *Adjusting the window's sill and head height*

6.13   Right-click and pick Deactivate View.

6.14   Right-click and pick Zoom to Fit.

6.15   Examine the Elevation and Section views to verify that the windows have changed.

6.16   Click the Default 3D View button on the Menu Bar, as shown in Figure 1–70. This will open a Southeast isometric view of the model.

Figure 1–70 *The button to create a 3D view*

6.17 Pick View>Shading with Edges from the Menu Bar, as shown in Figure 1–71.

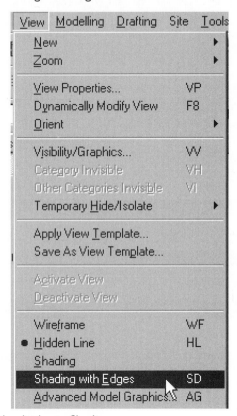

Figure 1–71 *Change the display to Shading*

6.18 Pick View>Orient>Southwest from the Menu Bar. Pick View>Orient>Save. Choose OK to accept the default name.

6.19 Double-click Sheets>A101 – Unnamed from the Project Browser to return to the plotting sheet. Note that 3D Views now has a + sign to the left of it in the Browser. Click on the + sign to reveal the saved view 3D Ortho 1 and the default 3D view, named simply {3D}.

6.20 Pick and drag view 3D Ortho 1 onto sheet A101. Release the left mouse button to see the outline of the new viewport. It will come in small in proportion to the other views. Locate it in a convenient place to the left of the View Window.

6.21 Pick the new viewport. Pick the Properties icon on the Options Bar and change its scale to 1/4″ = 1′-0″. Click OK. The viewport will change size, and the title under it will now need adjusting. Drag the right end of the title to correspond with the new size of the view.

6.22 Move the view to the lower left corner of the sheet. Its title will snap into alignment with the title of the Plan view. (See Figure 1–72.)

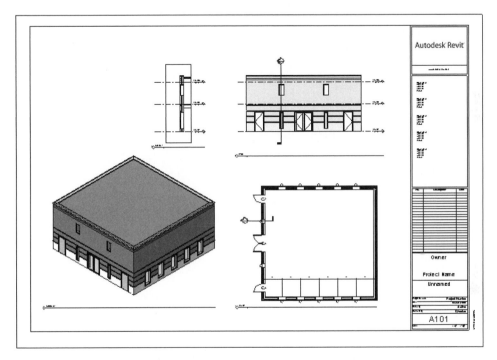

Figure 1–72 *The plotting sheet with plan, elevation, section, and shaded isometric views*

6.23 Save the file.

SUMMARY

Congratulations! You have created, edited, and documented a building shell in Revit Building, using its intelligent components and their innate properties entirely. This sheet is ready to print and you have not drawn any 2D lines, but worked entirely with objects, views, and their associated properties.

Admittedly, this model is quite basic, but you have gotten a taste of Revit Building's simplicity, ease of use, and power. The remaining chapters will show how Revit Building's strengths in modeling, representation, documentation, data extraction and change propagation function, and prepare you to implement this new architectural design tool in a work environment.

REVIEW QUESTIONS – CHAPTER 1

MULTIPLE CHOICE

1. To make objects visible or not, use controls found in

   a) the Design Bar

   b) the properties of each view

   c) the Project Browser

   d) the Options Bar

2. Hatch lines inside walls in a plan

   a) are always visible if you zoom in close

   b) are only visible in the Level 1 Floor Plan view

   c) are visible when the View Detail level is set to medium or fine

   d) have to be drawn in by hand

3. Zoom controls are found

   a) on the upper Toolbar

   b) in the View menu

   c) in the right-click context menu

   d) all of the above

4. Revit Building will allow you to edit the swing and hinge controls of a door

   a) during placement

   b) if it is selected by picking after placement

    c) by using the View properties

    d) a and b, but not c

5. For construction documentation purposes Revit Building provides

    a) Sheets with Views placed on them at scale

    b) Annotations that keep track of their Sheet placement

    c) both a and b

    d) Revit Building can't produce documents

TRUE/FALSE

6. Sections and Elevations require update procedures to reflect model changes.

7. When creating floors and roofs, Revit Building uses Sketch mode to allow the user to pick walls or draw lines.

8. Views placed on Sheets do not require update procedures to reflect model changes.

9. A floor object can be offset from its base Level.

10. Entities cannot be selected in 3D views.

 Answers will be found on the CD.

CHAPTER 2

Under the Hood:
Standards and Settings

INTRODUCTION

This section is addressed to managers, supervisors, and team leaders in design firms. College professors or those leading courses of instruction will also find the concepts covered and exercises useful.

Students will benefit from examining content and structure in template files, importing a CAD file, creating a custom titleblock, and exporting a file to CAD.

OBJECTIVES

- Understand the mechanics and value of template files in Revit Building
- Experience Revit Building's input/output interface with AutoCAD and other applications
- Discuss Revit Building's file management systems: import/export, linking, worksets

CAD MANAGER/INSTRUCTOR SUMMARY

If your responsibilities include deploying the design software your organization uses, keeping designers equipped with software tools, organizing files, and maintaining standards, you have headaches enough without having to adopt (or worse, half-adopt) a new standard. As your company or institution considers using Revit Building, what does Revit Building have in store for you?

The bad news is that Revit Building is different enough from AutoCAD and Architectural Desktop that even the people who listen to you and make an effort to follow your directives, instructions, specifications, and standards will be confused at times by this new software. The good news is that Revit Building is much simpler than AutoCAD and ADT in significant ways—*not* having to use layers to govern the appearance of objects on screen and paper, for example, does away with an entire, often complicated, set of standards that nobody ever follows completely anyway.

Revit Building does contain many of the same mechanisms that you currently use—text and dimensions have styles; reference files, content files, and central files all use coherent path setups; drawing sheets work the same way as Paper Space layouts, only more efficiently. Once you understand the strengths and operational underpinnings of this new modeler you will be able to serve the needs of your designers, project leaders, clients, and students.

TEMPLATE FILES HOLD STANDARDS AND SETTINGS

Revit Building uses template files to hold an array of settings, exactly as does AutoCAD. It works with two types of template files: rft files for family templates and rte files for project (building model) files. Families are a major organizing system in Revit Building. They do the work both of Styles in AutoCAD or Architectural Desktop and Multi-View blocks in ADT. Families will be treated in later exercises.

When first opening, Revit Building uses a *default.rte* template file that corresponds to the *acad.dwt* template file for AutoCAD—a basic setup with minimal predefined content. Just as AutoCAD and Architectural Desktop allow the user to specify and modify other template files with enhancements for specific purposes, Revit Building provides template files with significantly developed views, content, and sheet setups for specific types of building models. These templates are starting points to create company-standard, project-specific, or client-specific templates. As with AutoCAD products, completed Revit Building project files can be converted to templates.

As a CAD Manager or instructor you may or may not be familiar with Revit Building at this point. Whether or not you have created and documented more building models than the designers or students you are responsible for, in this next exercise you'll look under the hood and examine some of the settings and setups that will make your job easier.

STUDENT OVERVIEW

In Chapter 1 you worked through a series of exercises in which you created a simple building model using objects that Revit Building supplied for you. These predefined objects and the views in which you manipulated them, while complete enough to use from start to finish on a project, are only a portion of the working setups available to you.

In the exercises in this chapter you will examine template files in some detail. You will work with lines, text, labels and parameters while creating a title block using imported content. You will adjust various settings that affect the display of imported, drawn and exported items.

TEMPLATE INSPECTION

REVIT BUILDING COMMANDS AND SKILLS

Templates

Predefined Views

Predefined Sheets

View Phase Filters

View Graphic Overrides

Predefined Schedules

Titleblock

Import File

Import Settings—Lineweights

Revit Building Lineweights

Parameters

Labels

Export File

Export Settings—Layers

EXERCISE 1. EXAMINE TEMPLATES

THE DEFAULT TEMPLATE—THE VERY BASICS

1.1 Launch Revit Building. Revit Building will open to a Floor Plan View of Level 1 (the words **Level 1** will be highlighted under the Floor Plans section of the Project Browser tree under Views (all), as shown in Figure 2–1). Study the contents of the Project Browser to the left of the View Window. You can open predefined Floor Plans for Levels 1 and 2, Ceiling Plans for Levels 1 and 2, and four Elevations (graphically indicated on Level 1 Floor Plan by elevation markers). There are no legends, schedules or sheet views defined.

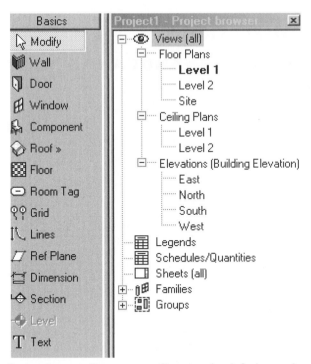

Figure 2–1 *The Project Browser for an empty file using the default template*

1.2 Click the + symbol next to Families in the Project Browser. Scroll down the list of Families and expand (click the + symbols next to) Structural Beam Systems, Structural Columns, Structural Foundations, Structural Framing, and Walls. Expand Structural Beam System under Structural Beam Systems, W-Wide Flange-Column under Structural Columns, Continuous Footing under Structural Foundations, W-Wide Flange under Structural Framing, and Basic Wall under Walls (see Figure 2–2).

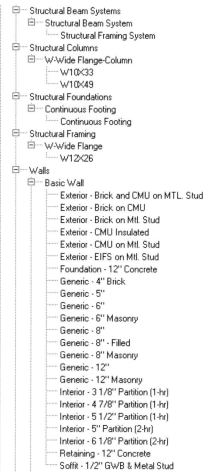

Figure 2–2 *Structural Content and Walls in the default template*

1.3   Place the cursor over the point of the West Elevation symbol in the View Window.

1.4   Right-click and pick Go to Elevation View to open the West Elevation view, as shown in Figure 2–3). This has the same effect as double-clicking the West View listing in the Project Browser.

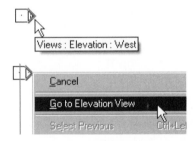

Figure 2–3 *Locate, right-click, and pick to open the West Elevation*

> 1.5 The West Elevation shows the two Levels in this file, at 0′-0″ and 10′-0″, shown in Figure 2–4.

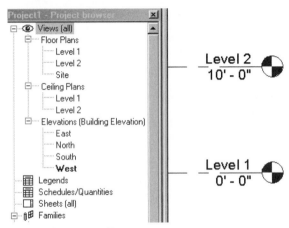

Figure 2–4 *The levels in this empty file*

EXAMINE THE RESIDENTIAL (IMPERIAL) TEMPLATE—PREDEFINED VIEWS AND SHEETS

> 1.6 From the File menu, pick File>New>Project, as shown in Figure 2–5, to open the New Project dialogue.

Figure 2–5 *Opening a new project*

> 1.7 In the New Project dialogue, pick the Browse button to choose a template file for the new project, as seen in Figure 2–6. Note that the path to your Templates folder will be different than what is shown.eps

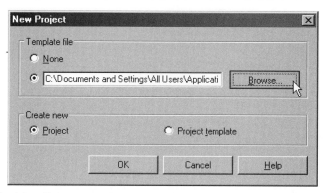

Figure 2–6 *The New Project dialogue*

> 1.8 In the Choose Template dialogue, pick *Residential-Default.rte,* as shown in Figure 2–7. Pick Open to return to the New Project dialogue. Pick OK to create a new project using this residential template.

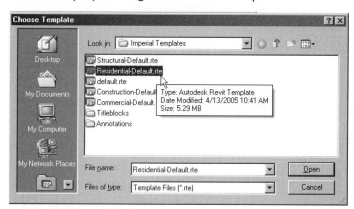

Figure 2–7 *Choose the Residential-Default template*

> 1.9 When the new residential project file opens, study the Project Browser (shown in Figure 2–8). There are many more Floor Plans than in the default file, including Footing, Framing, and Electrical Plans. Levels 1 and 2 have been renamed to First Floor and Second Floor, and a Basement and Roof Level added. There are Schedules predefined as well: Door, Lighting Fixture, Room Finish, and Window.

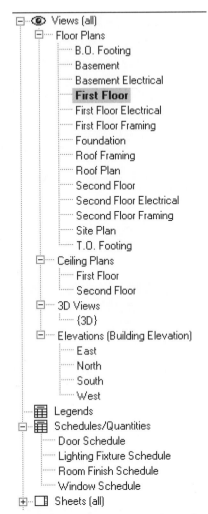

Figure 2–8 *Floor Plans and Schedules in the Residential template*

    1.10 Click the + next to Sheets to expand the list of Sheets. This template includes Plans, Elevations, Detail Sheets, Sections, and a Schedule page.

    1.11 Click the + next to A0 – Basement Plan and A1 – First Floor Plan, as shown in Figure 2–9. These expand to show that the Basement Floor Plan and First Floor Plan have already been placed on these two sheets. Similar views have been placed on other Floor Plans, Reflected Ceiling Plans, Roof Plans, and Electrical Plans.

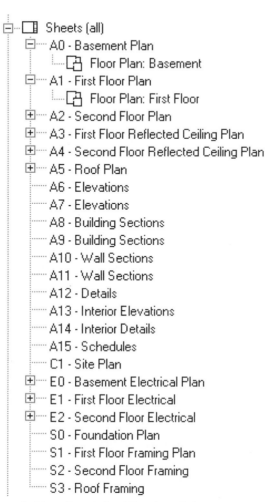

Figure 2–9 *The expanded list of predefined Residential sheets*

1.12 Double-click on A1 – First Floor Plan in the Project Browser to open that view. It has been defined to hold a default Revit Building D size 22 x 34 titleblock, with the sheet name and number entered.

1.13 Click the + symbol next to Families in the Project Browser. Scroll down the list of Families and expand Structural Columns, Structural Framing, and Walls (see Figure 2–10). Expand Pipe-Column under Structural Columns, and Basic Wall under Walls. Study the different content available in this template compared to the default file.

Figure 2–10 *Columns, Structural Framing, and Wall types in the residential template*

1.14 From the Window menu, pick Window>Project 1 - Floor Plan: Level 1 to open up this view. Note that the Project Browser now shows the views, sheets, and families appropriate to this file.

1.15 From the Window menu, pick Window>Project 2 – Floor Plan: First Floor to return to the residential file.

1.16 Double-click on East Elevation in the Project Browser to open that view (see Figure 2–11). With the cursor in the View Window, right-click and pick Zoom in Region. Pick two points around the Elevation symbols at the right of the screen to see the predefined Levels in this empty file. There is a defined Foundation Level 1'-3" below the First Floor level, a Second Floor at 9'-0", and a Roof level at 18'-0".

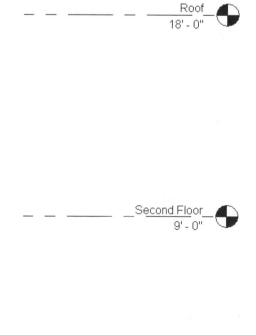

Figure 2–11 *Default Residential levels: Foundation, First Floor, Second Floor, and Roof*

EXAMINE THE CONSTRUCTION (IMPERIAL) TEMPLATE – VIEWS WITH PHASES, SCHEDULES

1.17 From the File menu, pick File>New>Project as before.

1.18 In the New Project dialogue, pick Browse.

1.19 In the Choose Template dialogue, pick *Construction-Default.rte*. Pick Open.

1.20 In the New Project dialogue, click OK.

1.21 When the new construction project opens, note that the Elevation symbols have been turned off in the Level 1 View window. Study the Project Browser: note the Floor Plans located below Level 1 and the list of 3D Views.

1.22 Pick 3D View 02 – Demo in the Project Browser, right-click, and pick Properties (always located at the bottom of the right-click menu), as shown in Figure 2–12.

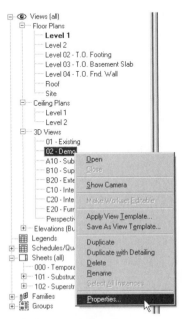

Figure 2–12 *Opening the View Properties dialogue*

1.23 In the Element Properties dialogue that appears, as in Figure 2–13, scroll down and note the Phase Filter and Phase that have been applied to this view. This file has been set up to use Phases—building elements will be differentiated according to the time they come into existence or disappear. Phases can be named and filtered, in this case by function: **demo, substructure, superstructure, exterior enclosure, interior construction, interior finishes** and **furnishings**. Each Phase can appear differently (or not at all) in each view set up for the project file. Pick Cancel to leave the Element Properties dialogue for the current view

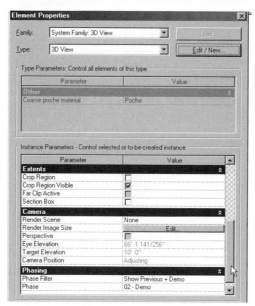

Figure 2–13 *The Phase settings for this 3D view*

1.24 From the Settings menu, pick Settings>Phases, as shown in Figure 2–14.

Figure 2–14 *Opening Phase Settings*

1.25 In the Phasing dialogue that opens, study the first tab, Project Phases, as shown in Figure 2–15. The other two tabs are Phase Filters and Graphic Overrides. This file has nine phases defined in it: **Existing**, **Demo**, five **New Construction** phases, **Furnishings**, and **Project Completion**.

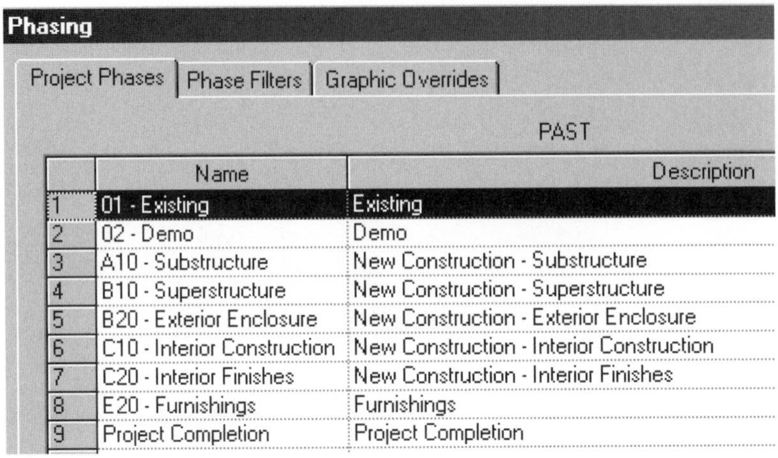

Phasing

Project Phases | Phase Filters | Graphic Overrides

PAST

	Name	Description
1	01 - Existing	Existing
2	02 - Demo	Demo
3	A10 - Substructure	New Construction - Substructure
4	B10 - Superstructure	New Construction - Superstructure
5	B20 - Exterior Enclosure	New Construction - Exterior Enclosure
6	C10 - Interior Construction	New Construction - Interior Construction
7	C20 - Interior Finishes	New Construction - Interior Finishes
8	E20 - Furnishings	Furnishings
9	Project Completion	Project Completion

Figure 2–15 *Project Phases in this template*

1.26 Pick the Phase Filters tab shown in Figure 2–16. Phase Filters sets up view conditions in which phases are displayed according to their categories, displayed with overrides, or not displayed. In this way each element can be viewed in different ways in different views. Revit Building supplies the Temporary phase for elements that are created and demolished during the project.

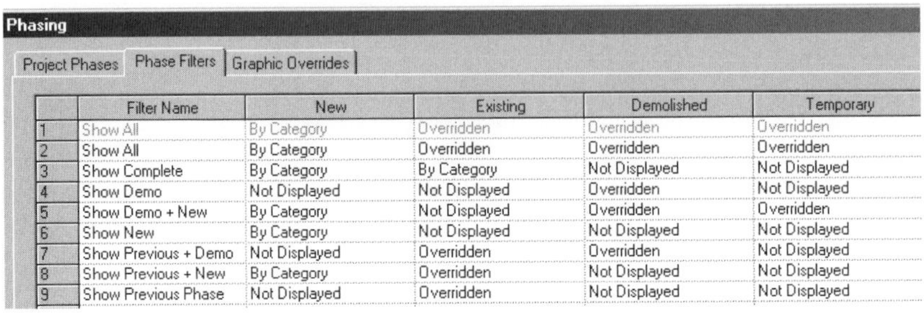

Phasing

Project Phases | Phase Filters | Graphic Overrides

	Filter Name	New	Existing	Demolished	Temporary
1	Show All	By Category	Overridden	Overridden	Overridden
2	Show All	By Category	Overridden	Overridden	Overridden
3	Show Complete	By Category	By Category	Not Displayed	Not Displayed
4	Show Demo	Not Displayed	Not Displayed	Overridden	Not Displayed
5	Show Demo + New	By Category	Not Displayed	Overridden	Overridden
6	Show New	By Category	Not Displayed	Not Displayed	Not Displayed
7	Show Previous + Demo	Not Displayed	Overridden	Overridden	Not Displayed
8	Show Previous + New	By Category	Overridden	Not Displayed	Not Displayed
9	Show Previous Phase	Not Displayed	Overridden	Not Displayed	Not Displayed

Figure 2–16 *Phase Filters*

1.27 Pick the Graphic Overrides tab shown in Figure 2–17. Here the visual properties of Phases are applied—line weight in plan and section, color, linetype, and material for 3D views.

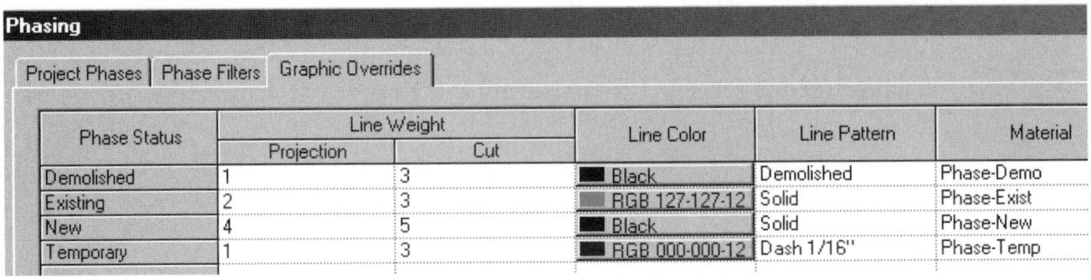

Phasing

Project Phases | Phase Filters | Graphic Overrides

Phase Status	Line Weight		Line Color	Line Pattern	Material
	Projection	Cut			
Demolished	1	3	Black	Demolished	Phase-Demo
Existing	2	3	RGB 127-127-12	Solid	Phase-Exist
New	4	5	Black	Solid	Phase-New
Temporary	1	3	RGB 000-000-12	Dash 1/16"	Phase-Temp

Figure 2–17 *Graphic Overrides for Phasing*

1.28   Pick Cancel to exit the Phasing dialogue.

1.29   Pick the + symbol next to Schedules/Quantities to expand the list in the Project Browser, as seen in Figure 2–18. This template includes examples of nearly all the available standard schedules.

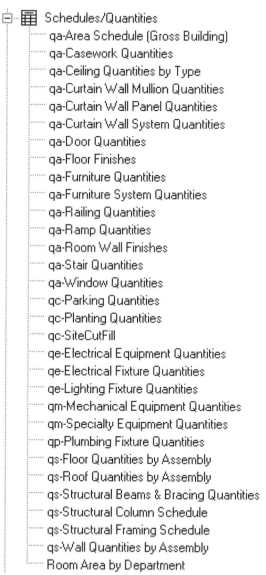

Figure 2–18 *Predefined schedules in this template*

1.30   Expand Elevations in the Project Browser. Double-click Elevations>East to open that view. Zoom in around the Elevation symbols in the right of the

View Window. Study the predefined Levels and their heights in this template as shown in Figure 2–19.

1.31 Click the + symbol next to Families in the Project Browser to expand the list. Click the + symbol next to Structural Columns and Walls to expand those two Family Categories.

1.32 Click the + symbol next to Concrete—Rectangular—Column, W—Wide Flange—Column, and Basic Wall to expand the list of preloaded content in those Families. Study the list of available components in this template.

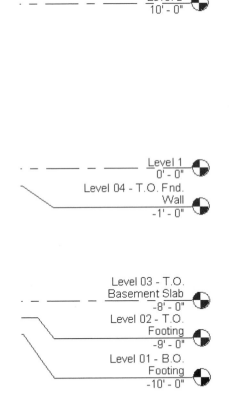

Figure 2–19 *Predefined Levels in the Construction template*

EXAMINE THE COMMERCIAL (IMPERIAL) TEMPLATE

1.33 From the File menu, pick File>New>Project to open the New Project dialogue.

1.34 Pick the Browse button to choose another template.

1.35 In the Choose Template dialogue, pick *Commercial—Default.rte* in the list of templates in the *Imperial Templates* folder and click the Open button. Click OK in the New Project dialogue.

1.36 In the Project Browser, study the list of default Floor Plans. Double-click Elevations>East. Zoom in around the Elevation symbols in that view (see Figure 2–20). Note the location of the predefined levels in this template.

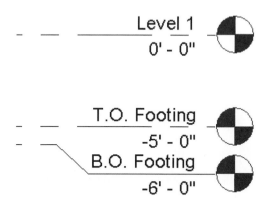

Figure 2–20 *Levels in the commercial template*

1.37 In the Project Browser, note the list of predefined sheets. Click the + symbol next to Families to expand the list of categories.

1.38 Click the + symbol next to Structural Columns, Structural Foundations, Structural Framing, and Walls to expand those categories.

1.39 Click the + symbol next to Structural Columns>W—Wide Flange—Column, Structural Foundations>Continuous Footing and Pile Cap—Rectangular, Structural Framing>W—Wide Flange, and Walls>Basic Walls to expand those families (shown in Figure 2–21). Study the list of available components in this template.

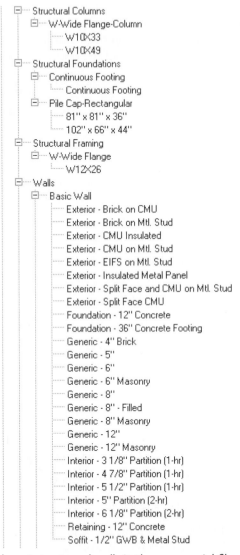

Figure 2–21 *Structural components and walls in the commercial file*

1.40 Double-click Level 1 to return to that view. You are still in the commercial project file. Do not close any open Revit Building files.

EXERCISE 2. COMPARE SPECIFIC CONTENT IN DIFFERENT TEMPLATES

2.1 From the Window menu, pick Window>Close Hidden Windows, as shown in Figure 2–22. This will close all the views that you have open except the current one in the current file and the last open view in other open files. Keeping open windows to a minimum will free up display system resources.

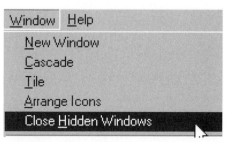

Figure 2–22 *How to close hidden view windows*

2.2 In the Project Browser, click the + symbol next to Families>Profiles to view the list of predefined and named profile shapes available for sweeps or edges in this template. If necessary, widen the Browser pane by holding the cursor over its right border until the cursor changes to a two-headed arrow, then click and pull the pane border to the right so you can see the expanded tree view (see Figure 2–23).

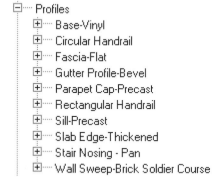

Figure 2–23 *Profiles in the commercial template*

2.3 From the Window menu, pick the available view from *Project 1* (the default template). When that file is loaded, click the + symbol next to Families>Profiles to view the list of profile shapes in the default template, as shown in Figure 2–24.

Figure 2–24 *Profiles in the default template*

> 2.4 From the Window menu, pick the available view from *Project 2* (the residential template). When that file loads, click the + symbol next to Families>Profiles to view the list of profile shapes in the residential template.
>
> 2.5 From the Window menu, pick the available view from *Project 3* (the construction template). When that file loads, click the + symbol next to Families>Profiles to view the list of profile shapes in the construction template.

EXAMINE VIEW PROPERTIES IN A TEMPLATE

> 2.6 From the Window menu, pick the available view from *Project 2* (the residential template). When that file loads, double-click Floor Plan>First Floor in the Project Browser to make that view active if it is not already the open view.
>
> 2.7 On the Basics Tab of the Design Bar, pick Wall to start the Wall tool.
>
> 2.8 In the Type Selector on the left of the Options Bar, click the down arrow in the drop-down box and select Basic Wall: Exterior – Wood Shingle on Wood Stud, as shown in Figure 2–25. Do not change the Height or Location Line properties.

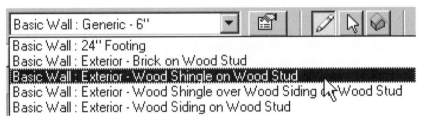

Figure 2–25 *Exterior wall properties*

2.9   In the sketch lines area of the Options Bar, pick the Rectangle button (see Figure 2–26) so that by picking two points in the View Window you can draw four connected walls with their exterior sides correctly situated.

2.10   Pick a point to the upper left of the View Window and pull the cursor down and to the right to start defining the new walls. Use the temporary dimensions that appear to make the new walls approximately 30' wide by 25' long (see Figure 2–26). The actual wall lengths do not matter.

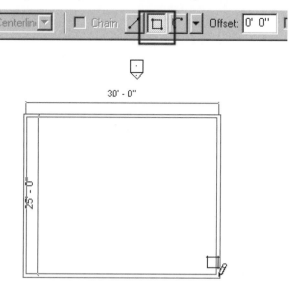

Figure 2–26 *Drawing four connected walls*

2.11   With the Wall tool still active, change the wall type to Basic Wall: Interior – 4 1/2" Partition. Change the sketch line type to Line and pick the Chain checkbox to sketch new walls continuously until you exit the command.

2.12   Draw a wall from the midpoint of the left wall (approximately) horizontal to the middle of the interior space, then up vertically to the middle (approximately) of the upper wall (see Figures 2–27 and 2–28). The actual wall dimensions or locations do not matter.

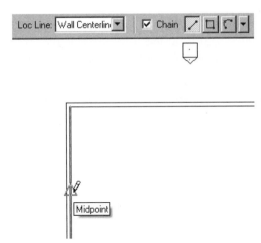

Figure 2–27 *Starting interior walls*

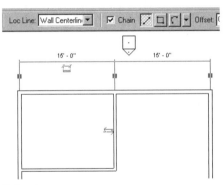

Figure 2–28 *Interior walls in place*

2.13 Pick Component from the Design Bar to terminate the Wall tool and start the Component placement tool. Use the drop-down list to select Outlet Duplex: Single, as shown in Figure 2–29.

Figure 2–29 *Pick Duplex Outlet in the Component command*

2.14 Place an outlet at the approximate midpoint of the right side interior wall, facing left into the smaller room you have created with the new walls (see Figure 2–30). The actual placement does not matter.

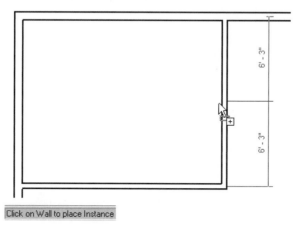

Figure 2–30 *Locating an electrical outlet*

2.15 A Revit Building warning, shown in Figure 2–31, will appear to let you know that the settings for this Floor Plan view will not let you see the new electrical fixture. Pick OK to close this warning.

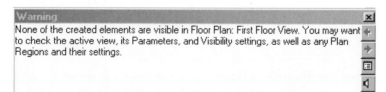

Figure 2–31 *Revit Building warns you about this view*

2.16 Double-click Floor Plans>First Floor Electrical to open that view. You will see the new electrical outlet correctly placed. Note the light appearance of the walls.

2.17 With the cursor in the View Window but not directly over any part of the model, right-click and pick View Properties (always the bottom choice in this menu).

2.18 In the View Properties dialogue that opens, pick the Edit button in the Value field to the right of the Visibility Parameter field.

2.19 In the Visibility/Graphics Overrides dialogue that opens, study the Model Categories tab (see Figure 2–32). Note that for all component categories except Curtain Systems, Electrical Components and Electrical Fixtures, the

Halftone graphic override has been checked. This accounts for the dim appearance of the walls in this view.

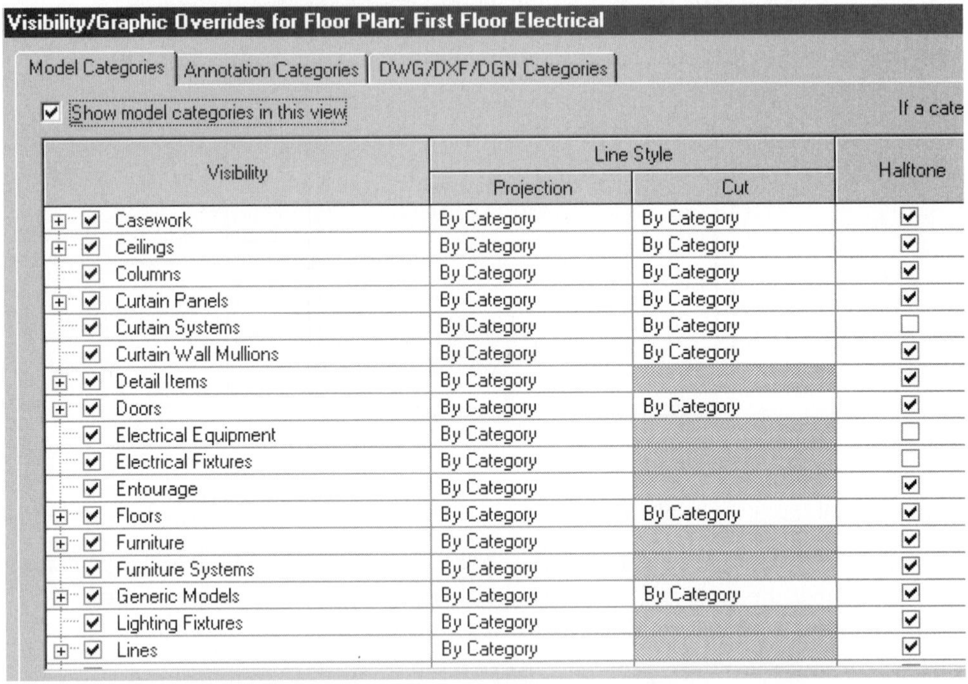

Figure 2–32 *Halftone Settings in the Electrical Plan*

2.20 Pick Cancel twice to exit the View Graphics and View Properties dialogues.

2.21 Pick Floor Plans>First Floor Plan in the Project Browser to highlight it. Right-click and pick Properties, as shown in Figure 2–33, to open the View Properties for the First Floor Plan view without having to open the view itself.

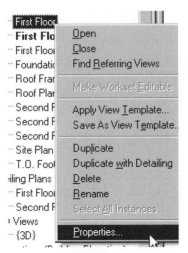

Figure 2–33 *Properties of the First Floor Plan*

2.22 In the View Properties dialogue that opens, pick the Edit button in the Value field to the right of the Visibility Parameter field.

2.23 In the Visibility/Graphics Overrides dialogue that opens, study the Model Categories tab, shown in Figure 2–34. Note that no component categories have the Halftone graphic override checked. Also note that the Visibility of the Electrical Fixtures category has been unchecked. In this template the Floor Plan view will not show outlets and electrical switches; a separate Electrical Plan has been created for that purpose. This is similar to the standard Ceiling Plans in all templates.

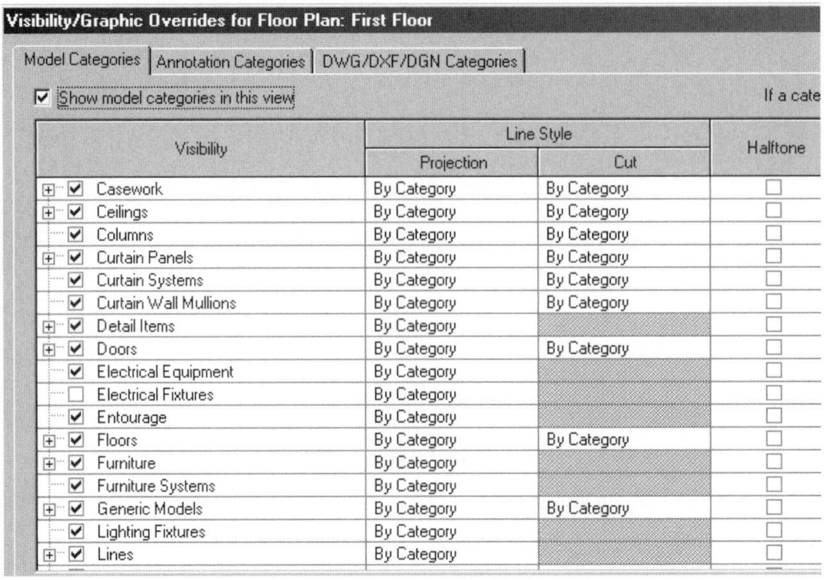

Figure 2–34 *Halftone and visibility controls for the First Floor Plan*

2.24 Pick Cancel twice to exit the View Graphics and View Properties dialogues.

2.25 From the File menu, pick File>Exit (always the bottom selection on this menu) to exit Revit Building. Do not save any of the open files.

Explore Templates On Your Own—Use What's Already Made For You

Revit Building contains other templates—imperial and metric—besides the ones you have just studied. All the templates are worth examining for useful view set-ups, sheets, families, and schedules that have been pre-loaded for users and managers to adopt and adapt. At present there is no mechanism in Revit Building—such as the Design Center file/content browser in AutoCAD 2000 and later releases—to allow drag-and-drop transfer of settings or content from one file to another. Revit Building does support standard copy/paste from one file to another, and has a Transfer Project Standards command to move settings, family types, line weights, and rendering materials maps from one open file to another. You'll look at this tool in future exercises.

SETTINGS AND PARAMETERS

Interface With CAD to Capture Existing Design Data

One of the cardinal rules for digital design productivity is to draw as little as possible, and use editing commands to populate your designs. That is, draw an object once and copy or array it many times, or draw half of a symmetrical design and mirror it to complete the pattern. This approach is not only quicker than repetitive drawing, but less error-prone.

Nearly every design firm on the planet has a library or archive of drawn work that designers mine repeatedly for reuse in new projects. Revit Building's ability to import from and export to AutoCAD and MicroStation means that firms setting out to use the power of new technology can make effective, leveraged use of their existing files and applications. If you have already drawn designs, components, symbols, or details in CAD that pertain to your new Revit Building project, there is at least one effective way to use them in your Revit Building project model or documentation.

Revit Building does not use layers or levels, the fundamental organizing mechanism of AutoCAD and MicroStation. CAD files use the concept of layers as corresponding to transparent overlays on a manual drafting table. All lines, text and other objects exist on named layers. Users create and manage layer names and colors, and many standards exist. Lines representing walls will, in a well-organized file, be on a layer named A-Wall (using one widely recognized standard). In a badly organized file, wall lines could be on a layer designated for text or roofing, allowing for confusion and error.

Revit Building creates objects and uses object category as its organizing principle. Walls are walls, and their component parts can never be designated as text or roofing, to keep to the previous example. Revit Building will readily read in layers/levels in linked or imported files, so that the user can benefit from the organization in the originating file. Revit Building exports DWG, DXF, or DGN files with properly constructed layers/levels, so that wall objects are translated into lines on wall layers and text objects become text on text layers..

Autodesk Revit Building can read and write ACIS® solids. This provides a way to export models from Architectural Desktop and Building Systems software and import or link 3D information into Revit Building. You can cut sections and perform visual interference detections in the imported file with this method.

The import process creates a single object in the Revit Building file, with control over the appearance of its contained elements—exactly like the Insert Block command in AutoCAD. You can explode imported objects with two levels of depth—all the way to simple lines, curves, text, and filled regions. Revit Building makes particular use of 3D line/polyline information from imported DWG, DXF, or DGN files to create editable Toposurface site objects. Revit Building does not recognize attribute definitions, but does recognize attribute values of inserted blocks in an imported DWG file and will explode those values to text. Revit Building uses Windows fonts, and contains an editable mapping file for routing AutoCAD shx font information (an older format for lettering contained in AutoCAD from the days before Windows) to Windows fonts. Revit Building will read proxy graphics for AEC Objects from Architectural Desktop, such as walls and floors.

These proxy graphics will come in as 3D and explode to 2D Model lines in Plan views.

Revit Building will export 2D information from 2D views or 3D information from 3D views using mapping files to assign layers or levels to Revit Building objects by category. Imported 3D information comes into AutoCAD as 3D polyline meshes that explode to 3D faces.

The next exercises will examine Revit Building's standard titleblocks and then use an imported AutoCAD file to create a highly formatted titleblock without extensive drawing or tracing.

EXERCISE 3. IMPORT AND REUSE AN AUTOCAD TITLEBLOCK

EXAMINE A STANDARD TITLEBLOCK—EDIT PARAMETERS

3.1 Open Revit Building to a new default project. It will have no sheets defined. Select Sheets in the Project Browser. Right-click and pick New Sheet, as shown in Figure 2–35.

Figure 2–35 *Adding a sheet to a new file*

3.2 In the Select a Titleblock dialogue that opens, pick the Load button.

3.3 In the Open file browser dialogue, make sure the Look In: value is the *Imperial Library* folder and select *Titleblocks*. Pick Open.

3.4 Pick each of the default titleblocks in turn to examine its appearance in the preview window. Select Cancel to return to the previous dialogue.

EXAMINE THE DEFAULT SELECTIONS

3.5 Pick OK to create a sheet with the E1 30 x 42 Horizontal default titleblock. Revit Building will open the new sheet view.

3.6 Click on the new sheet titleblock to select it. On the Options Bar, pick the Properties icon. Fill in the Parameter Value fields as shown in Figure 2–36. Use today's date for the Sheet Issue Date. Pick OK.

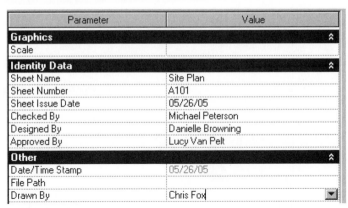

Figure 2–36 *Titleblock parameter fields*

3.7 With the cursor in the view window Zoom in around the lower-right corner of the title block. Note that some, but not all of the information fields (labels) have updated. Select the titleblock, and the linework will turn red to show it has been selected. Labels will turn blue.

3.8 From the Menu Bar, select Settings>Project Information. This will open the Type Properties Project Information dialogue. Fill in the project information as shown in Figure 2–37. You will need to pick the Edit button in the Project Address field to open the entry box for that information. When you have entered the address information, pick OK. Pick OK again to close the Project Information dialogue.

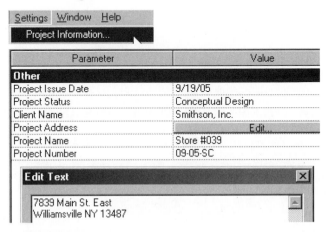

Figure 2–37 *Project Information Settings*

3.9 Study the labels in the titleblock again. You can edit the information in any label by selecting it. Pick the A101 Sheet Number label and hold the cursor over it. The Tooltip will show Edit Parameter. Select the label again and the data entry field will open. Change the Sheet Number from **A101** to **A1**, as shown in Figure 2–38; left-click outside the entry field (or hit Enter) and the label will update.

Figure 2–38 *Editing a titleblock label*

3.10 Click the drop-down arrow next to the Undo arrow on the Revit Building Toolbar, under the Drafting menu button. Scroll down the list of actions to Sheet. Your undo list may differ slightly from the one shown in Figure 2–39. Left-click inside the Undo drop-down list to accept the Undo command. Revit Building will delete the titleblock and sheet view.

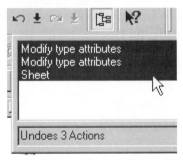

Figure 2–39 *Undoing a series of actions*

CREATE A NEW TITLEBLOCK THAT WILL USE AN OLD FILE

3.11 From the File menu, pick File>New>Titleblock, as shown in Figure 2–40.

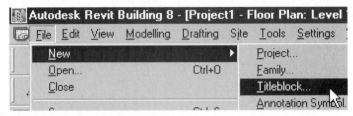

Figure 2–40 *Starting a new titleblock*

3.12 In the New Titleblock dialogue, pick the *E1 – 42 x 30.rft* template, as shown in Figure 2–41. Click OK.

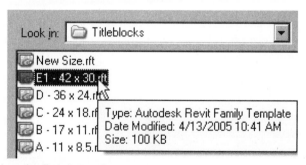

Figure 2–41 *Select the E size template*

3.13 Revit Building will enter the Family Editor. Lines defining a 42″ x 30″ outline will appear. These lines cannot be deleted. If you pick any of the lines, a dimension will appear. If you pick the value of a dimension an edit field opens, allowing you to change the size of the outline. Do not make any changes at this time.

3.14 From the Menu Bar select File>Save As to save the titleblock family file in a convenient place (or as instructed by your teacher) as *sample titleblock.rfa. Rfa* is the extension Revit Building uses for family files. Save titleblocks your company will use in projects in the *Content\Imperial* (or *Metric*) *Library\Titleblocks* folder on your network.

So far in this exercise you have examined a default titleblock supplied by Revit Building, and started to create a new one. Rather than draw this new titleblock from scratch, you will import an AutoCAD file. This file contains a titleblock consisting of lines, arcs (from filleted corners), text, and block attributes. The titleblock, which was originally created in Model Space, has been inserted into a Paper Space layout—the normal use for titleblocks in AutoCAD. Block attribute information is filled in by the user according to prompts built into the attribute definitions.

The imported file is empty of content except for the titleblock, to keep things simple. The titleblock linework is created in three colors, so it will plot heavy, medium, and light line weights. Revit Building has a mechanism to capture this intent and display AutoCAD lines in their desired line weights.

SETTINGS—DEFINE LINE WEIGHTS

3.15 From the Settings menu, select Settings>Line Weights. Since you are in the Family Editor for a Titleblock there is only one Tab, for Annotation Line Weights.

3.16 Change the line width for Line Weight 3 to **0.009"**. Change the line width for Line Weight 5 to **0.020"**. See the upper half of Figure 2–43. Click OK. If an information box appears click OK.

3.17 From the File menu, pick File>Import/Export Settings>Import Line Weights DWG/DXF, as shown in Figure 2–42.

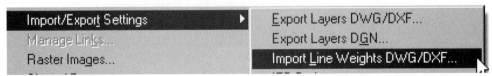

Figure 2–42 *Open the Import Line Weights dialogue*

3.18 Revit Building will open the Import Line Weights dialogue, which reads the contents of a line weights mapping file. The default file is *importlineweights-dwg-default.txt,* inside *Program Files/Autodesk Revit Building 8.0/Data.* This file gives all AutoCAD colors—listed by number—a default Line Weight 1. Change the line weight for color 1 (red—medium lines in the file you will import) to **3**; change the line weight of color 4 (cyan) to **2**; change the line weight of color 5 (blue in AutoCAD) to **5**. (See the lower half of Figure 2–43.)

This step applies Revit Building's line weights, which you adjusted to our own values in step 3.16, to incoming AutoCAD lines by layer, so long as the layers do not have their own lineweights applied.

Figure 2–43 *Editing the line weights in a titleblock file, and applying them to imported colors*

3.19 Select the Save As button to open the *Data* folder, shown in Figure 2–44. Save the line weight changes in a file named ***importlineweights-dwg-sampletb.txt***. Click Save. Click OK.

 NOTE This file then becomes the basis for imported line weights. You can use the Load button in the Import Line Weights dialogue to load other files.

Figure 2–44 *Save the line weight settings file*

3.20 From the File menu, select File>Import/Link>DWG>DXF>DGN, as shown in Figure 2–45.

Figure 2–45 *Import a file*

IMPORT THE FILE

3.21 Navigate to the folder where you placed AutoCAD content files from the CD included with this book. Select *sample titleblock inserted.dwg* as the file to import (shown in Figure 2–46). Pick the radio button to select Black and white under Layer/Level Colors for the display. Under Import or Link, make the Layers selection read All, and leave the Link box unchecked. Under Positioning, select Automatically place and Center-to-center. Pick Open.

82

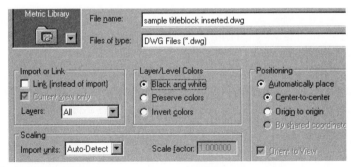

Figure 2–46 *Import settings*

3.22 If a warning appears, pick OK. When the imported file loads, Zoom In to examine the lower-right corner of the titleblock closely, as you can see in Figure 2–47. The line weights you specified are visible.

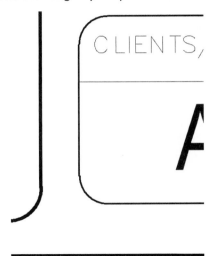

Figure 2–47 *Imported lines with weight*

The imported file is a single object, and therefore not editable. If you explode it to make the lines and text editable the linework will lose width information. Rather than do that, you will use the imported lines to make new linework that you can edit, then remove the imported linework as you no longer need it.

ADJUST THE VISIBILITY OF IMPORTED ENTITIES

3.23 Zoom to Fit to see the whole titleblock. Click on the import to select it. On the Options Bar, operation buttons appear. Select Delete Layers.

3.24 In the Select Layers/Levels to Delete dialogue that opens up, check layer 0 and layer Defpoints to eliminate them (shown in Figure 2–48), then pick OK.

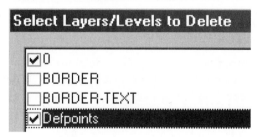

Figure 2–48 *Removing unneeded layers*

3.25 Type VG at the keyboard. This is a shortcut to open the Visibility/Graphics dialogue.

3.26 Since you are working in the Family Editor for a Drawing sheet, there are no Model categories available and the Model categories tab is blank in the dialogue that opens.

3.27 In the Visibility/Graphics Overrides dialogue for the Sheet view, select the DWG/DXF/DGN Categories tab. Click on the + symbol next to Imports in Families to expand that category if necessary.

3.28 Clear layer BORDER-TEXT to turn it off in this view. Click on OK to close the dialogue. The text will no longer be visible.

USING IMPORTED LINES TO MAKE NEW LINES WITHOUT DRAWING

3.29 Select the Lines command on the Family tab of the Design Bar to put Revit Building in sketch mode. Select the Pick (arrow) icon on the Options Bar.

3.30 In the Type Selector drop-down list, make Wide Lines the current line type. See Figure 2–49.

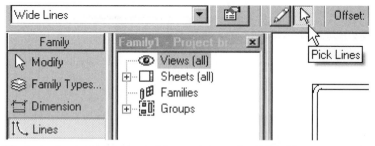

Figure 2–49 *Preparing to pick imported linework to make Wide Lines*

3.31 Select, in turn, the four sides of the large inner border of the title block that defines the drawing area of the page. Do not pick the outermost border lines. Each time you select a line in the imported file, dimensions and a lock symbol will appear, to enable you to edit the line you have just created. Carefully select the fillet arcs at the corners to finish the outline.

3.32 In the Type Selector drop-down list, make Medium Lines the current line type.

3.33 Select, in turn, the sides of the five title block text sections at the right side of the page. Carefully select the fillet arcs at the corners to finish the outlines. Zoom and Pan as necessary. See Figures 2–50 and 2–51.

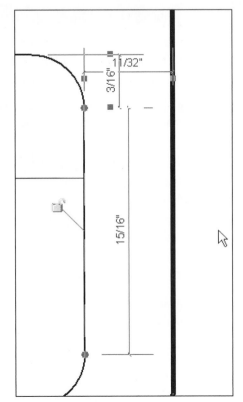

Figure 2–50 *A new line is created by picking a line in the imported file*

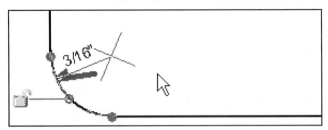

Figure 2–51 *Creating the filleted corners*

3.34 In the Type Selector drop-down list, make Thin Lines the current line type.

3.35 Select, in turn, the dividing lines inside the title block text sections.

3.36 When you have completed selecting imported linework to create new lines, check yourself. Zoom to Fit in the View Window. Type **VG** to open the Visibility/Graphics Overrides dialogue. Select the DWG/DXF/DGN Categories tab. Select the + symbol next to Imports in Families to expand that category. Clear layer BORDER to turn it off in this view. Click OK to close the dialogue.

3.37 Examine the linework for completeness. Note that the four outside lines making up the titleblock border (supplied by Revit Building) do not have fillet corners. Pick the Modify tool at the top of the Family tab on the Design Bar.

3.38 Hold down the CTRL key, select each of the four outside lines in turn and use the Type Selector drop-down list to change them from Title Block lines to Wide Lines.

3.39 Select the Lines tool on the Design Bar. Select the Fillet arc tool on the Options Bar, as shown in Figure 2–52. Use the Type Selector to make the current linetype Wide Lines, which will be applied to the fillets.

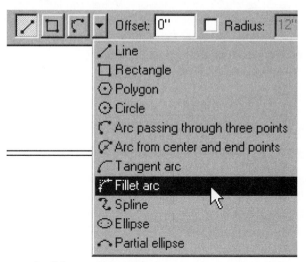

Figure 2–52 *Starting the Fillet arc line tool*

3.40 Zoom in around the upper-left corner of the title block. Pick the two outside lines in turn, close to their intersection point. A radius arc will appear and adjust its radius as you draw the cursor in and out from the intersection. When the radius value reads 1/2″ left-click to create the fillet.

TIP If you have trouble making the radius value read correctly by dragging the cursor, place the fillet where convenient, click the radius dimension value that displays to make it editable, and type in .5 or 1/2. See Figure 2–53.

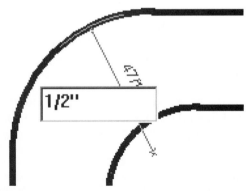

Figure 2–53 *Creating the first fillet*

3.41 Repeat this process for the other three corners of the titleblock. Pick Modify to terminate the fillet tool.

3.42 Zoom to Fit to check your linework. When the lines look complete, type **VG** to open the Visibility/Graphics Overrides dialogue. Select the DWG/ DXF/DGN Categories tab. Click the + symbol next to Imports in Families to expand that category. Check layer BORDER-TEXT to turn it on in this view. Click OK to close the dialogue.

3.43 Pick the titleblock text to select the import object. From the Options Bar, select Delete Layers. In the Select Layers/Levels to Delete dialogue that opens up, check layer BORDER to eliminate, then pick OK

3.44 Select Partial Explode to make the text editable.

3.45 Zoom in Region around the lower-right corner of the titleblock. Pick the A101 text to select it, then pick the Properties icon on the Options Bar. Revit Building has given this piece of text a name and assigned a font and size based on (but not necessarily the same as) the AutoCAD font. Pick the Edit/ New button to make the value field cells editable. Pick Cancel twice to exit the dialogue without making any edits. Save the family file to protect your work.

 TIP If your firm uses older AutoCAD shx fonts and you want to have them map automatically to Windows fonts during imports, open the file *shxfontmap.txt* in *Program Files\Autodesk Revit Building 8.0\Data* and enter your own mapping preferences.

LABELS AND PARAMETERS – MORE WORK THAN TEXT, BUT SMARTER

3.46   You want to replace text in this title block with Revit Building labels to give editable fields when this title block is loaded into files. Select the Label tool from the Family tab on the Design Bar. Click in an open area of the drawing to start to define a label.

3.47   The Select Parameter dialogue will open up, as shown in Figure 2–54. Study the list of predefined parameters for title block labels. This particular title block design uses two approval fields, so you will have to add parameters to the default list. Pick Add.

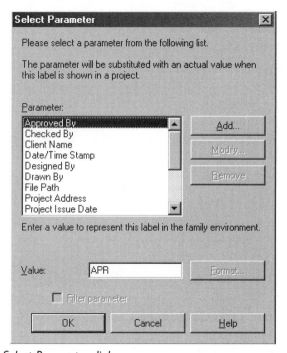

Figure 2–54 *The Select Parameter dialogue*

3.48   In the Parameter Properties dialogue that opens, click Select. If you get a warning that a shared parameter file has not been specified, click Yes to choose a file.

3.49   In the Edit Shared Parameters dialogue, pick Create to make a shared parameter file to hold the new parameters you are about to define. This file will be a *txt* file that you may very well edit later—save it in a convenient network location or as instructed by your teacher. Give it the name ***Shared Parameters.txt*** and click the Save button.

3.50 In the Shared Parameter dialogue, pick the New button under Groups to define a parameter group inside the new file. In the New Parameter Group dialogue shown in Figure 2–55, type in the name **Titleblock labels**. Click on OK.

Figure 2–55 *Naming the new parameter group*

3.51 In the Edit Shared Parameters dialogue, the name Titleblock Labels appears in the Parameter Group drop-down list. In the Parameters section select New.

3.52 In the Parameter Properties dialogue shown in Figure 2–56, give the first new parameter the name **Approved By 2**. There are numerous types available. Create this parameter as a Text field. Click OK.

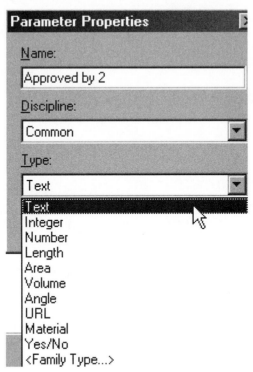

Figure 2–56 *Possible new parameter properties*

3.53 Add the following new text parameters: **Approved By 2 Date; Approved By Date**. Click OK each time. Click OK to edit the Edit Shared Parameters dialogue.

3.54 In the Shared Parameters dialogue, select Approved By 2 from the Parameters list. Click OK.

3.55 In the Parameter Properties dialogue, Approved By 2 will appear in the Name field. It will be grayed out to indicate that it is not editable in this dialogue. Click OK. The Select Parameter dialogue will open, with Approved By 2 now in the list of parameters available for our new label. You are now ready to add labels to the family. Pick Cancel to exit the dialogue. Hit ESC to terminate the label tool for now.

3.56 Delete the text entities for the sheet number and file name in the lower-right corner of the title block.

DEFINING AND WORKING WITH LABELS

3.57 Click on the Label tool in the Design Bar and place a new label inside the lower section of the box you just emptied of text, as shown in Figure 2–57. Select Sheet Number from the list, as you see in Figure 2–58. Leave the Value field at A101. Click OK.

Figure 2–57 *Preliminary location for the first label*

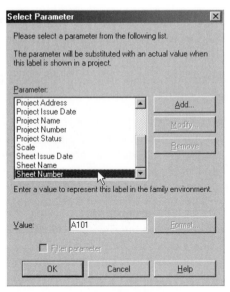

Figure 2–58 *Selecting the parameter for the first label*

3.58 In the Type Selector drop-down, the new label will be identified as Label: Tag 1. Click the Properties button.

3.59 In the Element Properties dialogue, pick Edit/New. In the Type Properties dialogue, select Rename.

3.60 In the Rename dialogue shown in Figure 2–59, give the label type the name **1/2″ Tag**. Click OK.

Figure 2–59 *Renaming the default label type*

3.61 In the Type Properties dialogue, change the Text Size to 1/2″. Make the Background value Transparent. Click OK.

3.62 You will now create two more label types without leaving the dialogue. In the Element Properties dialogue, pick Edit/New. In the Type Properties dialogue, select Duplicate. Give the new label type the name **1/4″ Tag**. Click OK.

3.63 In the Type Properties dialogue, change the size to 1/4″. Select Apply to set the field value.

3.64 Select Duplicate. Give the new label type the name **3/16″ Tag**. Click OK.

3.65 Change the size for the new label type to 3/16″. Click on Apply. You now have three label sizes defined.

3.66 Make the 1/4″ Tag the current Type and pick OK twice to return to the label placement task.

3.67 When you move the cursor into the upper section of the line box, Revit Building will find alignment planes from the previous label. Align the new label with the first one and left-click.

3.68 In the Select Parameter box, click Add, then Select, then File Path. Click OK three times to place the label.

3.69 Select Modify on the Family Design Bar. Pick the Sheet Number label. Select the Properties icon from the Options Bar.

3.70 Change the Horiz. Alignment value to Center. Change the Vert. Alignment value to Middle, as shown in Figure 2–60. Pick OK. Move the label to a suitable place in the linework. Repeat for the File Path label (see Figure 2–61). Note how Revit Building still finds alignment planes.

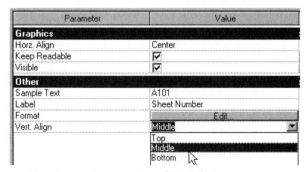

Figure 2–60 *Setting the alignment properties of the label*

Figure 2–61 *Aligning the File Path label above the Sheet Number label*

3.71 Erase the imported text objects reading **ENGINEERING, XX/XX/XX** (there are two of these), and **ARCH** in the box immediately above the area in which you have been working. You will replace this text with labels.

3.72 Select the Label tool. In the Type Selector, make 3/16″ Tag the current type. Click in a convenient place away from linework and text for a preliminary location for this label.

3.73 In the Select Parameter dialogue, select Approved By and click OK. The label will appear with control points and a rotation control. Put the cursor over the rotation control and rotate the label until it is vertical. Revit Building will snap to a vertical alignment.

3.74 Pick another spot to start placing a second label. In the Select Parameter dialogue, pick the Add button, then Select in the Parameter Properties dialogue. In the Shared Parameters dialogue select Approved By Date, then click OK twice to add this parameter to the list. The Select Parameter dialogue will still be active. In the Value field type **APR Date** for the placeholder text for this new label as shown in Figure 2–62, then click on OK. Rotate the new label to a vertical position as with step 3.73.

Figure 2–62 *Adding the Approved By Date parameter*

3.75 Select Modify and move the new labels to their correct positions, as shown in Figure 2–63.

3.76 Select Label from the Design Bar and add two new labels with these Parameters: Approved By 2 (Value **APR2**) and Approved By 2 Date (Value **APR2 Date**). Rotate and align these labels as with the first two.

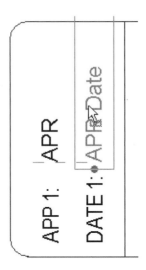

Figure 2–63 *Locating the Approval labels*

3.77 Delete the text object that reads **SITEPLAN**. Add a 1/2″ Tag label with the Parameter: Sheet Name to the right of the company name and address. Rotate it to a vertical position. Edit its Horizontal alignment to Center and Vertical alignment to Middle as before, locate it appropriately and use the field control dots to fill the allowed space, as shown in Figure 2–64.

Figure 2–64 *The lower half of the titleblock information panel*

3.78 Add 3/16″ Tag labels for Designed By, Drafted By, Scale, and Project Issue Date in the appropriate areas on the title block. Erase the original text.

3.79 Add a 1/2″ Tag label for the Client Name. To the right of it add a 1/4″ Tag label for the Project Name and a 3/16″ Tag label for the Project Address.

Change the Horizontal alignment of all those labels to Center and the Vertical alignment to Middle. Place the labels carefully after erasing the original text.

3.80 Adjust the remaining pieces of text as necessary. See Figure 2–65.

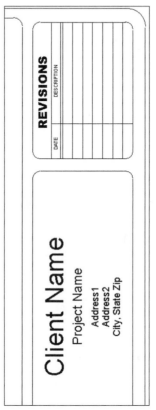

Figure 2–65 *The upper half of the title block information panel*

3.81 From the File menu, select Purge Unused. In the Purge Unused dialogue, you can expand the Other Styles items to see what you will be eliminating from the file and clear items you may want to keep. Click OK.

3.82 Save the Family. Select Load into Projects, as shown in Figure 2–66.

Figure 2–66 *Load your new titleblock family into the current project*

3.83 Revit Building returns to the open project file. In the project browser, pick Sheets, right-click, and select New Sheet. Select your new title block and click OK. The new sheet view will open with your new titleblock.

3.84 The labels with shared parameters that you created earlier in the exercise will be represented by question marks when the titleblock is selected, but they are valid labels, oriented properly. If you pick File>Shared Parameters from the Menu bar you will see that the new parameters have come into the project.

3.85 Close this file without saving it.

Exports from Revit Building into CAD and other applications

Revit Building writes out DWG, DXF, and DGN files to complete the interchange cycle with today's major CAD applications. You can export an individual view or multiple views. Exported 2D views produce 2D linework and 3D views produce a 3D model. Revit Building's 3D view settings (visibility, hidden line mode) are ignored—a hidden line view opens in 3D Wireframe mode in AutoCAD. The exported model will hide and shade in AutoCAD.

Revit Building maps object categories to preconfigured layer names. As with layer import map files, you can create your own export mapping files for projects or clients. You will examine this in the next exercise.

Revit Building exports schedule field information to delimited files for import into spreadsheet applications (such as Microsoft Excel and Lotus 123). Revit Building can export model components to ODBC databases (such as Microsoft Access) using specific table relationships (between type and instance, for example). ODBC export can then be updated throughout the project to the same database file to give progress reports. Revit Building ships with a separate PDF writer. Once this is installed, you can export views to PDF format as a print process. Revit Building exports DWF, Autodesk's proprietary format for Web-based viewing and sharing of design file information across many platforms. Revit Building will export various image types directly from any view or selection of views. These images will import into other applications, such as Microsoft Word. Revit Building has a walkthough creation routine that will export an avi animation file, which will play on any media player. You will treat these output options in future exercises.

These export capabilities make Revit Building powerful and useful throughout a building design project. If you need to send DWG or DGN information to an outside consultant using their layer standards, Revit Building makes this operation straightforward. If you need number-crunching information, Revit Building will supply this from the very beginning of conceptual design work, and gives you a consistent way to update the figures. If you need images, you can generate them at any time in more than one format. Once you have set up your various export routines the process is simple enough so that non-experts can readily generate supplementary files from a Revit Building model. (At present there is no one-button scripting capability to make admin-type output generation bulletproof.)

This will save designer time for designing.

EXERCISE 4. EXPORT LAYER SETTINGS

4.1 Launch Revit Building. From the File menu, pick File>Close to take the open empty project file out of memory. Select File>Open. Adjust the size of the File Open dialogue by dragging the edges (Microsoft standard behavior). Select the Training Files icon in the left pane of the dialogue. Open the Common folder, then select and open file *c_rvt8_Condominium.rvt* from the *Program Files\Autodesk Revit Building 8.0\Training\Common* folder (see Figure 2–67). The file will open to the default 3D view.

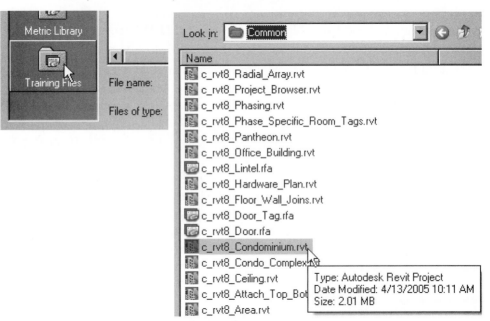

Figure 2–67 *Open a file from the Training/Common folder*

4.2 From the File menu, pick File>Import/Export Settings/Export Layers DWG/ DXF.

4.3 The Export Layers dialogue will open, displaying the current export layer map file (see Figure 2–68). *Exportlayers-dwg-AIA.txt* is the default. All of Revit Building's object categories are shown with layer and color information. This dialogue will allow you to create exact export specifications for the most demanding client, consultant, or government office. Once you have examined the categories and mappings, select Cancel to exit the dialogue.

Export Layers: F:\Program Files\Autodesk Revit Building 8\Data\exportlayers-dwg-AIA.txt				
Category	Projection		Cut	
	Layer name	Color ID	Layer name	Color ID
Area Polylines	A-AREA-BDRY	1		
Area Tags	A-AREA-IDEN	2		
Callouts	A-ANNO-SYMB	6		
Casework	A-FLOR-CASE	3	A-FLOR-CASE	3
Casework Tags	I-ELEV-IDEN	5		
Ceiling Tags	A-CLNG-IDEN	6		
Ceilings	A-CLNG-SUSP	5	A-CLNG-SUSP	5
Default host layer line style	*{A-CLNG-SUSP}*	5	*{A-CLNG-SUSP}*	5
Finish 1 [4]	*{A-CLNG-SUSP}*	5	*{A-CLNG-SUSP}*	5
Finish 2 [5]	*{A-CLNG-SUSP}*	5	*{A-CLNG-SUSP}*	5
Membrane Layer	*{A-CLNG-SUSP}*	5	*{A-CLNG-SUSP}*	5
Structure [1]	*{A-CLNG-SUSP}*	5	*{A-CLNG-SUSP}*	5
Substrate [2]	*{A-CLNG-SUSP}*	5	*{A-CLNG-SUSP}*	5
Surface Pattern	A-CLNG-GRID	1	A-CLNG-GRID	1
Thermal/Air Layer [3]	*{A-CLNG-SUSP}*	5	*{A-CLNG-SUSP}*	5
Color Fill Legends	G-ANNO	5		
Color Fills	A-AREA-IDEN	2		
Columns	A-COLS	4	A-COLS	4
Constraints	A-ANNO-NOTE	2		

Figure 2–68 *The Export Layers control panel*

EXAMINE REVIT BUILDING VIEWS AND CREATE EXPORTS

4.4 In the Project Browser, double-click on Floor Plans>First Floor to open that view. Look it over and make yourself familiar with it quickly.

4.5 Open the West elevation and look it over. Note the dashed green reference planes that are visible in this view. They will not plot, but will export.

4.6 In the Project Browser, click the + symbol next to Sections to expand the list in that view category. Double-click Section 1 to open that view, and examine it briefly.

4.7 From the File menu, pick File>Export>DWG, DWF, DGN, SAT. In the Export dialogue that opens up, use the Save In drop-down browser panel to locate a folder for the export files you are about to create.

4.8 In the File Name drop-down list, edit the file name prefix to read **condo**. Select Automatic and Long in the File Naming section. Revit Building will create files with this prefix and the name of the view as the rest of the file name in the folder you specified in the upper part of the dialogue. If you do not have AutoCAD 2000 or later available on your system, use the Save As Type list to select the application you will use to check the export files. Select Export each view or sheet as a single file.

4.9 In the Export range section of the Export dialogue, select Selected views/ sheets and pick the Select button, as shown in Figure 2–69.

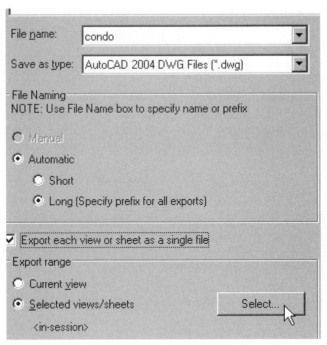

Figure 2–69 *Preparing the export file names and location*

4.10 In the Views list panel, click on 3D View: {3D}, Elevation: West, Floor Plan: First Floor, and Section: Section 1, as shown in Figure 2–70. Click OK to return to the Export dialogue. Choose No in the question box that opens.

View/Sheet Set

Name: <in-session>

☑3D View: {3D}
☐Elevation: East
☐Elevation: North
☐Elevation: South
☑Elevation: West
☑Floor Plan: First Floor

☑Section: Section 1
☐Section: Section Garage

Figure 2–70 *Pick the views to export*

4.11 Pick the Save button to generate the exports. Revit Building will work for a minute or two, depending on your system resources.

4.12 When Revit Building has finished the exports, pick File>Exit to leave Revit Building.

4.13 In your File Manager or local network Explorer, find the folder you specified for the DWG exports from Revit Building. Note that Revit Building has also created a *pcp* (plot configuration parameters file) for each export file (see Figure 2–71). These are editable text files, so that you can specify plotted lineweights based on the colors used in the Export layers file. Note also that the 3D file is considerably larger then the others.

Name	Size	Type
condo - Section - Section 1.pcp	22 KB	PCP File
condo - Section - Section 1.dwg	96 KB	AutoCAD Drawing
condo - Floor Plan - First Floor.pcp	22 KB	PCP File
condo - Floor Plan - First Floor.dwg	65 KB	AutoCAD Drawing
condo - Elevation - West.pcp	22 KB	PCP File
condo - Elevation - West.dwg	175 KB	AutoCAD Drawing
condo - 3D View - {3D}.pcp	22 KB	PCP File
condo - 3D View - {3D}.dwg	1,616 KB	AutoCAD Drawing

Figure 2–71 *Revit Building creates DWG and PCP files*

EXERCISE 5: EXAMINE THE EXPORTED FILES IN AUTOCAD

5.1 Launch AutoCAD if it is installed on your system. If you do not have access to AutoCAD, skip Exercise 5. Open the four new files Revit Building created in turn to examine their contents and layer structure.

5.2 Zoom Extents in Model Space in the *Condo-3D View- {3D}* drawing. Change the view from Shaded to Hidden Line to 2D Wireframe and view the results each time. In the 2D Wireframe view, pick a railing section on the interior stairwell and note its layer in the Object Properties toolbar. Right-click, pick Properties from the context menu, and note that the railing section is a block. Pick a few more entities and note what they consist of: blocks and polyface meshes.

5.3 Open the Layer Properties Manager dialogue box for this file and note the layers Revit Building has created (see Figure 2–72).

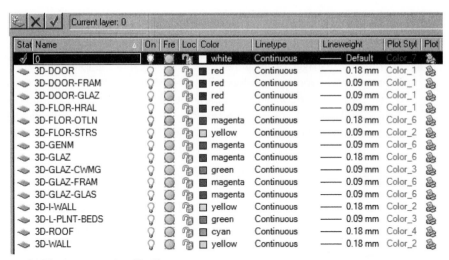

Stat	Name	△	On	Fre	Loc	Color	Linetype	Lineweight	Plot Styl	Plot
✓	0					white	Continuous	──── Default	Color_7	
	3D-DOOR		♀	○		red	Continuous	──── 0.18 mm	Color_1	
	3D-DOOR-FRAM		♀	○		red	Continuous	──── 0.09 mm	Color_1	
	3D-DOOR-GLAZ		♀	○		red	Continuous	──── 0.09 mm	Color_1	
	3D-FLOR-HRAL		♀	○		red	Continuous	──── 0.09 mm	Color_1	
	3D-FLOR-OTLN		♀	○		magenta	Continuous	──── 0.18 mm	Color_6	
	3D-FLOR-STRS		♀	○		yellow	Continuous	──── 0.09 mm	Color_2	
	3D-GENM		♀	○		magenta	Continuous	──── 0.09 mm	Color_6	
	3D-GLAZ		♀	○		magenta	Continuous	──── 0.18 mm	Color_6	
	3D-GLAZ-CWMG		♀	○		green	Continuous	──── 0.09 mm	Color_3	
	3D-GLAZ-FRAM		♀	○		magenta	Continuous	──── 0.09 mm	Color_6	
	3D-GLAZ-GLAS		♀	○		magenta	Continuous	──── 0.09 mm	Color_6	
	3D-I-WALL		♀	○		yellow	Continuous	──── 0.18 mm	Color_2	
	3D-L-PLNT-BEDS		♀	○		green	Continuous	──── 0.09 mm	Color_3	
	3D-ROOF		♀	○		cyan	Continuous	──── 0.18 mm	Color_4	
	3D-WALL		♀	○		yellow	Continuous	──── 0.18 mm	Color_2	

Current layer: 0

Figure 2–72 *Layers in the 3D file*

5.4 Close the Layer Properties Manager. Close this file without saving and examine the *Condo-Floor Plan-First Floor* file.

5.5 Use a crossing window to pick a section of wall including a window and window tag. Type **LI**, and then hit ENTER to list the contents of your selection set: it consists of lines, blocks, and Mtext.

5.6 Open the Layer Properties Manager for this file. Note the layers that Revit Building has created (see Figure 2–73). One layer has a DASH linetype. All layers except layer 0 have lineweights applied. The drawing has been set up to use color-based plot styles.

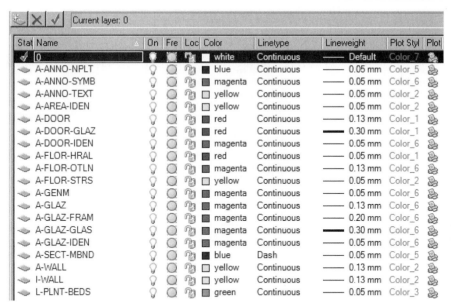

Figure 2–73 *Layers in the Floor Plan*

5.7 Close the Layer Properties Manager. Close the *Condo-Floor Plan-First Floor* file without saving and open the *Condo-Elevation-West* file.

5.8 Pick one of the dashed blue lines created from the reference planes in the Revit Building file. Note its layer in the Object Properties toolbar drop-down list—it has been created on an AIA-standard NPLT layer, to indicate that it should not be plotted.

5.9 Pick an area of brick facing and list it—a hatch on layer A-WALL-PATT, an AIA standard layer name.

5.10 Open the Layer Properties Manager and note the names and properties of layers in this file. One layer has an ALIGNING_LINE linetype. Pick the name of the linetype to open the Select Linetype dialogue box shown in Figure 2–74, and note the linetypes Revit Building has created in this file. Pick Cancel to close this dialogue. Pick Cancel to close the Layer Properties Manager.

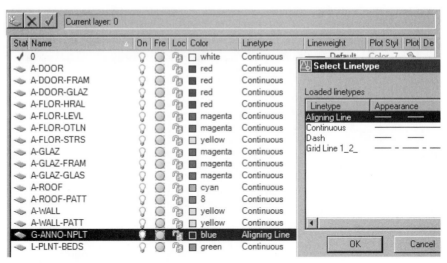

Figure 2–74 *Layers in the Elevation*

5.11 Close the *Condo-Elevation-West* file without saving and examine the *Condo-Section-Section 1* file.

5.12 In the *Section1* file, pick the text label for the Parapet elevation line. Right-click and pick Mtext Edit. Change the font, justification, or content of the text inside the editor to verify that it is editable Mtext. Close the Mtext editor.

5.13 Pick a section of stair or stair railing. Note that it is a block. Explode it and check the linework—everything has been drawn with color Bylayer. Each section of stair or railing and each window symbol, whether in elevation or section, has been created as a block. The blocks are named according to type, but each instance has a slightly different name as an identifier. The blocks will explode to linework as you just saw, and they will respond properly to the Refedit command.

5.14 Open the Layer Properties Manager for this file. Note the layer names and properties.

5.15 Close the file without saving and exit AutoCAD.

CONCLUSIONS TO DRAW: STANDARD SETTINGS SAVE STEPS

Proper understanding of template files and program settings benefit all users of complex applications such as Revit Building. Whether you are a beginning student or a seasoned user you can save yourself work by utilizing what has already been created for you in even the simplest file, and the more you know about the underlying framework the more quickly you can be productive.

The exercises in this chapter touched on import/export settings to illustrate the fact that very few building designs are created without having a relationship with previous work. In today's hi-tech environment, the ability to bring in data efficiently from sources outside of the current project file, and to send coherent, smart information from the project file out to other applications reliably, is considered part of a basic skill set.

Revit Building Works with CAD

CAD managers or others responsible for maintaining the flow of CAD information through a multi-platform office should take heart from this demonstration. Revit Building readily accepts AutoCAD content, and just as readily creates output to AutoCAD that captures in a legitimate and completely adjustable form the content of the Revit Building model. Firms with experienced AutoCAD drafters or detailers, or a valuable inventory of standard detail files, can proceed with implementation of Revit Building, knowing that Revit Building's modeling capabilities will not bypass the strengths of their workforce or archives.

Advanced Teamwork Tools: Links and Worksets

One of the early criticisms of Revit Building was of its single building model structure. Most users of AutoCAD are familiar with the xref system, whereby an unlimited number of files can be linked to compose a complex building model or a campus, with very little strain on system resources. External references also allow team members to work independently on parts of an assembled whole.

You have seen that Revit Building will import AutoCAD and other files. An import is not a live connection between files—in that way it corresponds to a block insertion in AutoCAD. Revit Building uses a system of live links that correspond exactly to external references in AutoCAD. Revit Building will create and maintain links with DWG, DXF, DGN, and RVT (Revit Building) files. Links with other Revit Building files are intended to be links between entire building models, as in a campus or site plan. Revit Building warns specifically that linking a vertical assembly of floors together to make a single building is **not** supported. This is the opposite of the way firms use AutoCAD and Architectural Desktop to create assembled building models, so be aware that designers moving from AutoCAD-based design to Revit Building will have to learn some new methods.

Revit Building has developed the Workset system for design teams working on a single building model. Worksets—divisions of the model using all sorts of criteria—provide a mechanism for individuals to check out parts of the model from a central storage location and return these parts to update the central file. This is also an opposing concept to the usual xref setup, and one that CAD managers will have to understand to make effective use of Revit Building and prevent loss of time or work on a project.

AutoCAD xrefs usually consist of many separate files, often on completely separated company systems and servers. The individual part files are collated in one or more files to create an assembly model or plotting setup, and often the designers or detailers that work on the parts and pieces never see the entire assembly. Revit Building's Worksets work the other way around. There is always a central file; worksets are subsets of the central file. Users editing worksets update them periodically in the central file without gaps or lapses in order for information to flow correctly to other members of the team. This system is not a return to the days of mainframes; Revit Building has mechanisms to allow work to take place remotely.

You shall look specifically at links and worksets in future exercises.

PROJECT MANAGER BULLETIN: REVIT BUILDING IS A CHANGE ENGINE

So far in this chapter you have examined topics that concern CAD managers, who have to maintain a filing system and set of company standards that exists quite apart from what may or may not happen with Revit Building in the organization. CAD managers, expert though they are in CAD, may not get the chance to become familiar with Revit Building during the period when their company is implementing its use. These managers must make sure that those who do use Revit Building can communicate effectively with the network, other departments or offices, and those in the company who will continue to use CAD.

There is an individual in most firms implementing Revit Building who needs to be aware of its effect on work habits and methods—the Project Manager. A Project Manager's primary responsibility is to assemble and shepherd the design team through its paces. No matter what your previous training and experience with Revit Building, there are bound to be a few surprises when people new to the program go live on a project. As with any foray into the unexpected, preparation and planning are key. You can't emphasize enough the value of organized training. Reasonable expectations, given the size, complexity, and deadlines of the trial project, are also critical for team success. Team makeup will play an important role in reaching project goals and developing the efficiency benefits that Revit Building makes possible.

Training

Managers are often responsible for scheduling training or enhancement instruction to keep designers aware of progress with CAD software. People who become relatively efficient with a particular set of tools or methods often become hide-bound and not interested in learning new practices and techniques, which results in loss of efficiency over time. Any manager who has tried to convince drafters that never mastered AutoCAD's Paper Space of the superior efficiency of multiple Layouts, will understand the difficulty of keeping workers up-to-date in this era of constant change.

The interface and tools make Revit Building considerably easier to pick up and start working with than nearly any CAD drafting program. It is quite possible for individuals to teach themselves enough in a short time to create the basics of designs and visualizations that are far more advanced than what they could have created in CAD in the same amount of time, even after training. However, Revit Building is a powerful, sophisticated apparatus that requires more than a rudimentary amount of knowledge to work effectively and efficiently. More and more firms are realizing that training is essential. The amount of training for experienced designers to become efficient in Revit Building is measured in hours or days, not weeks, but a training budget should be considered part of any implementation strategy.

Along with the necessary training, managers can help quite a bit by offering designers the encouragement that these brave souls deserve for tackling the job of learning an entire new software system in the stressful context of a live project.

Firms implementing Revit Building are reporting with regularity their conclusions that on the job training is not as effective as 16 to 32 hours of dedicated training time with Revit Building's online lessons and tutoring from users with previous experience.

Manage Expectations

Successful implementers also report that keeping the first project or two to an appropriate size for new users to handle goes a significant way to ensuring success. Revit Building is a complex, comprehensive design package. Nobody ever became proficient with AutoCAD, MicroStation, or Architectural Desktop overnight. Despite Revit Building's relative ease in creating complex models quickly, project and company management must allow the inevitable learning curve to run its course. Smart building models take a little more time for new users to create than line drawings, but provide downstream benefits.

Resistance and Skepticism

The willingness of designers and drafters to take on new challenges will also play a part in their ability to deal with the tensions of producing a project with new tools. Early adopters who relish the chance to learn and expand their skills will provide a different atmosphere than skeptics or those who work best entirely within their comfort zones. Change always meets some resistance, and new methods have to prove effective under fire, so to speak. It's the project manager's challenge to keep designers focused on solving problems, not on the perceived difficulties of working with unfamiliar tools.

Versatility is Valuable

Revit Building does not remove the need for designers to understand drafting. It does eliminate the primacy of drafting skills in today's building model design process, and will hasten the disappearance of the CAD jockey whose sole function is as a scribe for designers. A Revit Building model and documentation package requires far fewer repetitive tasks—designers spend more time designing, and the software takes care of providing views, updating annotations, extracting numbers, and the like. Designers with versatility rather than specialists in CAD skills will play larger roles in Revit Building projects.

Changing roles

What is a manager to expect from Revit Building? While each project will be different, and results will vary from firm to firm, certain trends are already clear.

Revit Building's ability to produce renderings within the application, and the relative ease of exporting models to other rendering programs, mean that design illustrations and visualizations will come into play earlier than before. This will mean earlier client response to conceptual design options, hardly ever a bad thing. Revit Building includes a suite of site development tools that allow building designers to specify site design more accurately and completely than before, without having to export plans to other applications.

Revit Building relies heavily on families, which function as combinations of Auto-CAD Styles and 3D Blocks. Families manage functional aspects such as titleblocks and building components such as walls or furniture. Revit Building provides a number of family templates for the creation of custom content. Skill at creating families for company use will prove as valuable as was skill at composing Auto-CAD library material. Creating and managing family content is a subject large enough to merit its own book.

Revit Building provides tools for data handling, including schedules for nearly every type of component, with customizable parameters for holding information that can be exported from schedules or shared between projects. Once schedules are set up in a project, updating and handling the information they contain becomes more of an administration task than a job requiring CAD or design skills to initiate. Revit Building's space-planning tools also lend themselves to use by administrative personnel.

The designed effect of Revit Building's modeling, documentation, and data handling structure is to move nearly all project tasks down the food chain, to eliminate repetitive, low value, but specialist work, and to return control of design projects to designers. This will eventually mean changes in company workforce structure. For managers implementing Revit Building in a first design project or two, you can expect to be pleasantly surprised at the ease with which designers will produce complex building models and documentation. The longer lasting value of those complex models will only become apparent over time.

SUMMARY

This chapter has taken a look at some of the more technical aspects of Revit Building—templates and titleblocks are important parts of any firm's design standards. Proper understanding and setup of templates will result in significant benefits. You examined Revit Building's titleblock structure both to illustrate labels, data fields, and parameters, and to show how Revit Building will accept and work with vector information from other CAD formats. Lastly, you saw how Revit Building will export model information to CAD.

REVIEW QUESTIONS – CHAPTER 2

MULTIPLE CHOICE

1. Template files can contain

   a) customized Levels

   b) customized views and sheets

   c) customized family content

   d) all of the above

2. Template files cannot contain

   a) custom menus

   b) custom Design Bar tabs

   c) custom lineweights

   d) a and b, but not c

3. Revit Building will export

   a) to specified layers

   b) the current view or selected views/sheets

   c) 2D and 3D information

   d) all of the above

4. Revit Building's name for an editable text field to place on a titleblock is

   a) dimension

   b) parameter

    c) label

    d) note

5. An imported CAD file comes in as

    a) individual line, dimension, and text entities

    b) a single object that can be exploded to individual entities

    c) a single object that can't be modified

    d) an image

TRUE/FALSE

6. Revit Building allows you to delete Layers from imported AutoCAD files

7. Once you save a template file, you can't edit it later

8. Revit Building does not recognize text in imported files that have been exploded

9. Text labels can be rotated

10. Project information automatically appears in certain titleblock label fields

 Answers will be found on the CD.

Design Modeling with Site and Building Elements

INTRODUCTION

This section begins with a series of related exercises using basic Revit Building techniques to create a project file. The next exercises show you how to bring existing AutoCAD files into Revit Building, and use those files as references to create a complex site object and complete building shell. Finally, you export specific parts of the Revit Building model out to AutoCAD.

OBJECTIVES

- Import CAD files as references for design project setup
- Create and subdivide a site object.
- Model a building shell and apply Phases
- Start detailing the shell using modeling and sketching techniques
- Export a view of the model into AutoCAD for development by others

REVIT BUILDING COMMANDS AND SKILLS

Settings

File import

Toposurface

Subregion

Levels

Align

Reference Planes

Walls

Roofs

Phases

Windows

Web content search

Type Editing

Profile Editing

Sketching: Line, Arc, Tape Measure

Editing commands: Move, Copy, Rotate, Align, Trim, Split

File export layer settings

File export

EXERCISE OVERVIEW

The exercises in the rest of this book will follow a hypothetical design project. The program of this project is as follows:

- Project setup tasks: Work with an existing lakeside site containing one multi-story building, an access road, and a parking lot. The building holds collegiate classrooms and offices. Model the existing building shell and site in Revit Building. The reference material consists of AutoCAD files—a site plan in 3D, and 2D exterior elevations of the existing building.

- Project Phase 1 tasks: Design a second multistory building near the first one. The new building is to hold offices, classrooms, labs, and a kitchen/cafeteria facility. Transform one wing of the existing building into an elevated passage to the new building. Move the parking lot underground; turn the space between the two buildings into a pedestrian courtyard; create a pedestrian plaza over the new parking garage. The reference materials consist of scanned sketches.

- Project Phase 2 tasks: Design a second building to contain offices, a dining facility, and a residence tower. Move the occupants of the offices in buildings 1 and 2 to building 3. Convert the office space in building 2 to classrooms. Convert building 1 into a residence hall. There is no reference material for this phase.

- Project Phase 3 tasks: Design a performing arts building to sit between buildings 2 and 3. This building is to have offices; rehearsal halls; an art gallery; and theater space with a stage, wing, overhead space, a projection booth, and raked seating. The building will feature a central atrium with skylights, cafe with seating, and a ceremonial staircase. There is no reference material for this phase.

The exercises will be designed to cover all the basic Revit Building skills necessary to complete a design and documentation project using multiple files. The exercises will develop skills and techniques as they would be used in the course of a typical design project; for this reason the exercises are best done in order. Since design projects of commercial or institutional size usually involve teams, the exercises are designed to be modular to a certain extent. In a class or training situation one group of students could be assigned the Phase 2 exercises and one group to Phase 3, for instance, and then the groups should exchange assignments.

Teamwork is an important part of every design project not produced by a single individual. Autodesk long ago developed wblock insertions and the external reference system for AutoCAD so that more than one person could work on a project and share information using a couple of different modes. Revit Building contains a similar import/link system for CAD files, and the single building model includes links to Revit Building files, so that designers can coordinate models in a multi-building or campus development. Revit Building contains Worksets and Phases, mechanisms for dividing and coordinating design tasks.

A project of the size and scope we have outlined takes many hours of work to complete, and a large portion of that work is inevitably repetitious, even with a design application as complete as Revit Building. Our exercises will touch on each task phase of the hypothetical project, but we will not carry these phases through to completion, once the appropriate concepts and techniques have been illustrated.

 INSTRUCTOR'S NOTE The scope of this hypothetical project will provide opportunities for auxiliary exercises for projects or extra credit. Students can explore alternate design ideas or flesh out areas of the model (classroom, office, or residence layouts).

For those individuals working these exercises on their own and not as part of a group, we admire your gumption and encourage you to work through the file linking and Workset exercises in later chapters so that you will understand how files and projects are subdivided and linked together. Even if you will work entirely by yourself for the foreseeable future, you will certainly find a use for a multi-file project, or Worksets.

To begin, in this chapter you will open a blank imperial file and import a series of AutoCAD files—a site plan and elevations for the existing building on the site. You'll use those files as background to make a Revit Building model of the existing building, and use that model to export an elevation of one face of the building that incorporates new design information out into AutoCAD.

EXERCISE 1. CREATE A PROJECT—ESTABLISH SETTINGS

1.1 Launch Revit Building. Revit Building will open to a Floor Plan view of Level 1 in an empty project file using the default Imperial template. Four Elevations appear in the View window. Place the cursor anywhere in the View window (but not over an Elevation or Elevation View control arrow), and right-click. This will open the right-click context menu (see Figure 3–1). View Properties is always the bottom command. Select View Properties to open the Element Properties Dialogue for Level 1 (see Figure 3–2). Study the properties of this view, noting the default View Scale (1:96, or 1/8″ = 1′-0″) and Phase information. You will have to scroll down to see the Phase field.

Figure 3–1 *You can reach View Properties with a right-click*

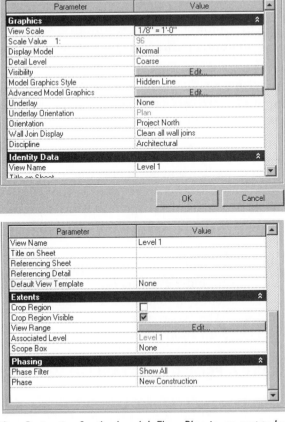

Figure 3–2 *The View Properties for the Level 1 Floor Plan in an empty Imperial file*

1.2 Double-click on Site in the Project Browser to open that view. This view also shows the four Elevations. Type **VG** at the keyboard to open the Visibility/ Graphics dialogue for the current view. **VV** opens the same dialogue.

TIP Revit Building contains a number of two-letter keystroke shortcuts for common commands and the user can create more. Shortcuts appear on drop-down menus (see Figure 3–3). We will introduce a number of them in the exercises in this and future chapters.

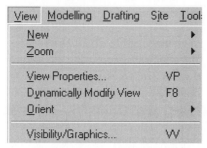

Figure 3–3 *Menu items and their associated keystroke shortcuts*

NOTE *Views* are crucial in Revit Building, and managing them is an important skill to develop early. The *View Properties* dialogue contains overall controls for the current view plus the *Visibility/Graphics* dialogue, which contains visibility settings for every object type in the building model within the current view.

1.3 Choose the Annotation Categories tab of the Visibility dialogue for the Site view and clear Elevations (see Figure 3–4). Click OK to exit this dialogue. The Elevations will disappear from the screen. You are removing them in this view to make future work less cluttered.

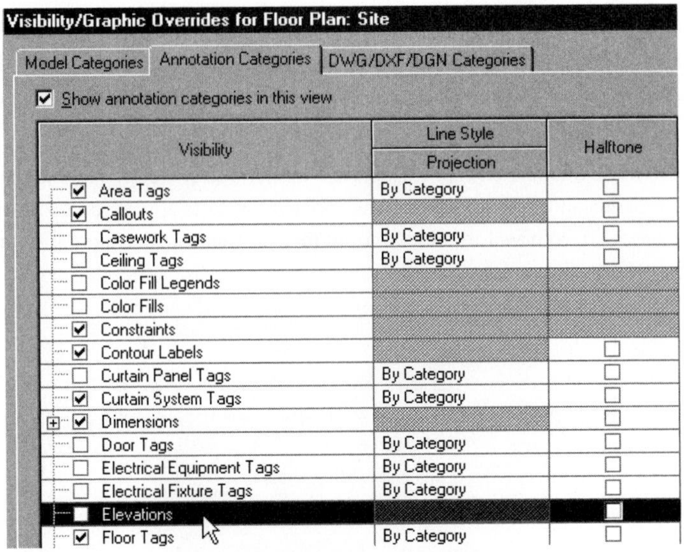

Figure 3–4 *Making the Elevations not visible in the Site View*

1.4 Type **VP** to open the View Properties dialogue for this view. Note the default View Scale (1:240, or 1″ = 20′ – 0″) and Phase settings. Select Cancel to exit this dialogue without making changes.

From the Menu Bar, pick Settings>Project Information (the top selection on the Settings menu) as shown in Figure 3–5. This will open a Type Properties dialogue for the Project Information System Family. Edit the value fields for Project Information Parameters as follows (or as specified by your instructor if in a class):

1. Project Issue Date   12/31/05 (or the ending date of your training)

2. Project Status   Conceptual Design

3. Client Name   Lakeshore College (or the name of your business or institution)

4. Project Address   4800 Madison Road, Diamond Lake TN 37052

5. Project Name   Dudley Drive Renovations

6. Project Number   506R1

Click OK to apply the values and close the dialogue. This information will appear on Titleblocks for Sheets.

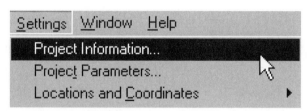

Figure 3–5 *Reaching Project Information*

1.5 From the Settings menu, pick Settings>Options (the bottom selection). In the Options dialogue that opens, on the General tab, set the Save Reminder interval value in the Notifications section to 30 minutes. Revit Building does *not* auto-save (see Figure 3–6).

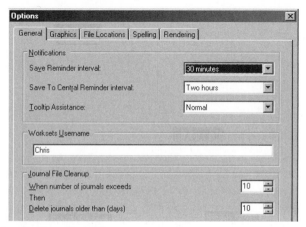

Figure 3–6 *Setting the Save Interval*

1.6 On the File Locations tab, set the Default path for user files as directed by your instructor if you're in a class (see Figure 3–7). Your path will be different from the one pictured. Click OK to exit the dialogue.

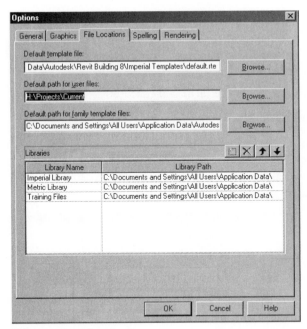

Figure 3–7 *The File Locations tab*

1.7 From the File menu, select Save As. In the Save As dialogue, pick the Options button. In the File Save Options dialogue that opens, set the number of backup files to save to 1, or as specified by your instructor (see Figure 3–8). The default setting is for 3 backups, identified by a 00x addition to the file name. Revit Building will save up to 50 copies of Workset enabled projects. This option is **not** a global Revit Building setting, and **must be adjusted for each new file**.

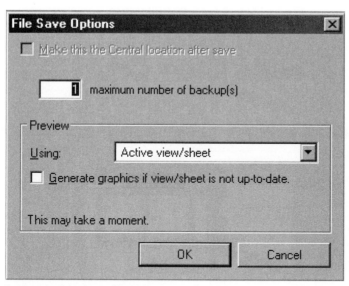

Figure 3–8 *Setting the number of backups*

1.8 Click OK to close the File Save Options dialogue. In the Save As dialogue, give this new project the name **Chapter 3 Phase 1**, and save it as type Project Files (*.rvt) in the location specified by the class instructor (see Figure 3–9).

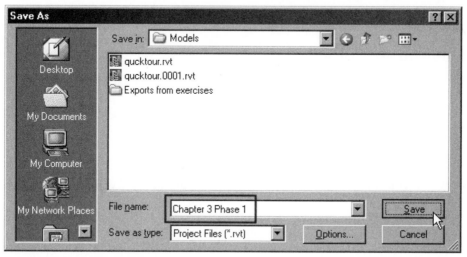

Figure 3–9 *Give the file a name and location*

EXERCISE 2. START A SITE PLAN USING IMPORTED CONTENT

2.1 Open the file *Chapter 3 Phase .rvt* from the previous exercise if you closed it after step 1.8, or continue with the open file. Make Site the current view.

From the File menu, pick File>Import/Link>DWG, DXF, DGN, SAT. In the Import/Link dialogue that opens, use the Look in: drop-down list to navigate to the folder that contains the file *site plan chapter 3.dwg* (from the CD). Pick that file so that the name highlights in the folder contents window and the name appears in the File name: field. A preview image will appear, as shown in Figure 3–10.

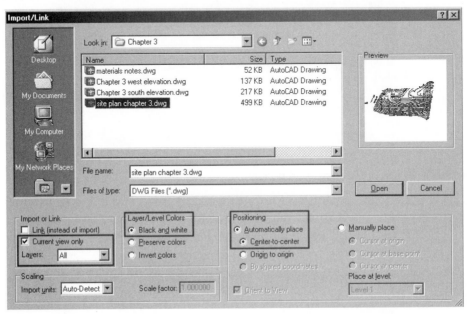

Figure 3–10 *Import settings*

2.2 Adjust the settings in the lower half of the dialogue: Check Current view only under Import or Link. This will place the import only in the current view, and it will not appear in elevations. Pick Black and white under Layer/Level Colors. Make the Import units under Scaling read Auto-Detect, and select Automatically Place>Center-to-Center under Positioning. (See Figure 3–10.) Click Open to start the import.

2.3 When the screen shows linework, type **ZF**, the keyboard shortcut for Zoom to Fit.

2.4 When the cursor is placed anywhere over the imported object, a tool tip will appear that identifies it as an Import Symbol, and a border will appear around the import.

Left-click to select the imported symbol. The cursor will change to a double-headed arrow, to signify that you can now drag the symbol to another position. The contour lines will change from black to red, Revit's color for selected items. The Type Properties drop-down list on the Options Bar will show the type category of the import (Import Symbol: site plan chapter 3.dwg), the Properties icon will become active, and other action buttons become visible on the Options Bar.

With the cursor over the selected symbol, right-click and study the right-click context menu. With an object selected, commands available on the Options Bar are also available by using right-click. In this case, Delete Selected Layers and Explode options appear in the context menu (see Figure 3–11). Move the cursor off the linework of the import symbol and left-click so it is no longer selected.

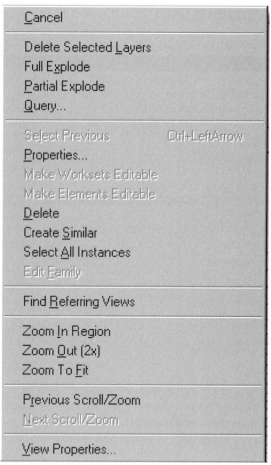

Figure 3–11 *Available commands for an Imported Instance*

2.5 Place the cursor in the Design Bar pane to the left of the screen. If Site is not in the stack of command tabs at the bottom of the Design Bar, right-click to bring up the list of available tabs and choose Site. This will add the tab to the Design Bar and make it the active one (see Figure 3–12).

Figure 3–12 *Making the Site tab available*

CREATE A TOPOSURFACE USING THE IMPORTED FILE

2.6 From the Site tab on the Design Bar, select Toposurface. The Design Bar shifts into Sketch mode. The top of the Design Bar now reads Toposurface, and the command list changes. Select Use Imported, then left-click over the Import Symbol to select it.

2.7 The Add Points from Selected Layers dialogue opens. All the layers in the AutoCAD drawing will be checked. Clear the choice box for layer topo lines, then select Invert. This will isolate the topo lines layer as the only layer to use for creating 3D points in a Toposurface mesh, as shown in Figure 3–13. Click OK to create the Toposurface.

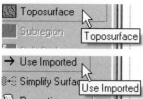

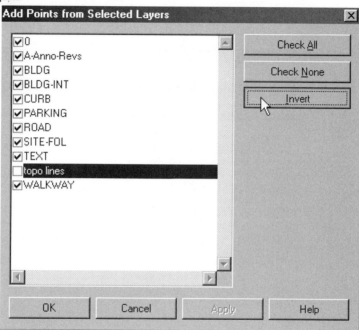

Figure 3–13 *Using points from one layer only*

2.8 Pick Properties in the Toposurface Sketch commands on the Design Bar. In the Element Properties dialogue, change Phase Created to Existing. Left-click the drop down arrow icon in the Materials value field to open the Materials dialogue. Scroll down to select Site-Grass. Click OK twice to exit the dialogues (see Figure 3–14).

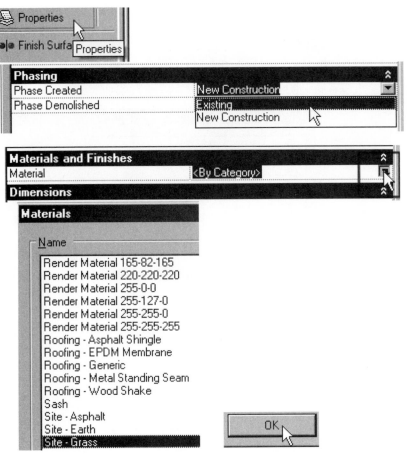

Figure 3–14 *Properties of the new Toposurface*

    2.9 Select Simplify Surface in the Design Bar. In the Simplify Surface dialogue, set the Accuracy value to **12″**, or simply **1** with no units of measurement, and Revit Building will change it to 1′ 0″ as shown in Figure 3–15. The default unit in project files from an Imperial template is the foot. Click OK. Select Finish Surface on the Design Bar.

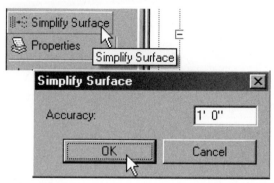

Figure 3–15 *Surface Accuracy*

2.10 Zoom in on the outline of the building.

2.11 Note the labels of the contour lines at the parking lots (see Figure 3–16). These are in feet. The contour lines in the imported file were created using AutoCAD polylines with their Z-elevations set to the corresponding value.

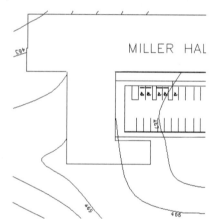

Figure 3–16 *Imported contour lines and their labels*

2.12 Type **VG** to bring up the Visibility dialogue. On the DWG/DXF/DGN Categories tab, click the + symbol next to *site plan chapter 3.dwg* to expand the list under it. Clear topo lines to turn off visibility of that layer, as shown in Figure 3–17. Click OK twice.

Visibility/Graphic Overrides for Floor Plan: Site

Model Categories | Annotation Categories | **DWG/DXF/DGN Categories**

☑ Show imported categories in this view

Visibility	Line
	Projection
☑ Imports in Families	By Category
⊟ ☑ site plan chapter 3.dwg	By Category
☑ 0	By Category
☑ A-Anno-Revs	By Category
☑ BLDG	By Category
☐ BLDG-INT	By Category
☑ CURB	By Category
☑ PARKING	By Category
☑ ROAD	By Category
☑ SITE-FOL	By Category
☑ TEXT	By Category
☑ WALKWAY	By Category
☐ topo lines	Override...

Figure 3–17 *Turning off the AutoCAD layer topo lines*

ADJUST PROJECT LEVELS TO FIT THE TOPOSURFACE

2.13 The contour lines in the import symbol will disappear, as that layer is no longer visible in this view. The new Toposurface is not visible. It was created from 3D information that places it out of the range of this view's default settings.

NOTE *All views in Revit Building except Drafting Views are 3D. Orthographic ("straight-on") views such as plans, elevations and sections do not show distance or depth, but in Revit Building they have depth settings that control what they display. Elevations and sections have graphic depth controls that also appear as properties. Plan view depth settings are found only in the View Properties dialogue.*

In order to make use of the newly created Toposurface, which is sitting at a height of 450′ to 470′, you will adjust the elevations of the Levels in the project file to bring them into the range of the Plan views.

2.14 Double-click on Elevations>East in the Project Browser to open that view. The Toposurface will occupy most of the View window, with the default Levels 1 and 2 visible at the bottom. Zoom in around the right side of the Level lines, so you can read and adjust their labels.

2.15 Pick the label text 0′ 0″ for Level 1. The level line will highlight red and a box will appear around the label. Pick the text 0′ 0″ again and the box will become an editing field. Change the Level's Elevation by typing in **466** (Revit Building will read this as 466′) and hitting ENTER.

2.16 The Level will disappear from the screen. Adjust the Elevation for Level 2 to **476**, as shown in Figure 3–18. It too will disappear from the screen when you hit ENTER.

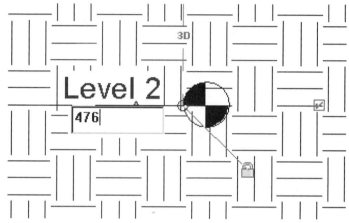

Figure 3–18 *Adjust the elevation of both levels*

2.17 Type **ZF**. Note that the Level lines are now above the contours of the Toposurface in this view.

2.18 Double-click on Site in the Project Browser. Contour lines are now visible in the Site view. Hold the cursor over a contour line and the tooltip will identify the object as Topography: Surface. The building outline in the import will not be visible, as in the left side of Figure 3–19.

2.19 Select a piece of text from the import so that the outline of the import appears. Select Foreground from the drop-down list on the Options Bar. The building outline will appear and you can select it. See the right side of Figure 3–19.

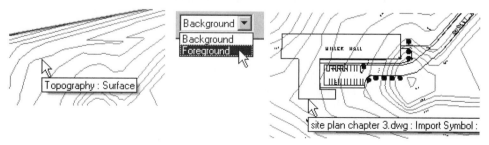

Figure 3–19 *Making the import visible in the toposurface*

2.20 Type **VG** to open the Visibility dialogue. On the DWG/DXF/DGN tab, expand the list under *site plan chapter 3.dwg* and clear all layer names except BLDG. This will simplify the linework for your next steps. Click OK. Save the file.

EXERCISE 3. SUBDIVIDE THE TOPOSURFACE

3.1 Open the file from the previous exercise or continue with the open project. On the Site tab in the Design Bar, pick Pad.

The Design Bar will switch to Sketch mode. The Pick Walls tool will be active by default. Select Lines. The Options Bar will show drawing commands. The cursor will appear as a pencil symbol; the Status bar and tooltip will display information pertinent to drawing. Make sure the Draw icon is active (depressed), and place a check in the Chain option on the Design Bar. See Figure 3–20.

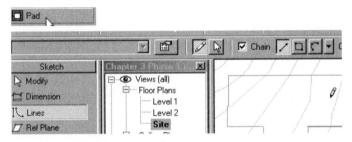

Figure 3–20 *Sketch a pad in the new surface*

3.2 Zoom in around the building outline to make that fill the screen. Trace the building outline starting in the upper left corner, going counterclockwise.

As you draw, Revit Building will display temporary dimensions to show you the length of the line you are creating, plus a direction vector that will snap to horizontal and vertical planes (see Figure 3–21). Revit Building will also snap to geometric points such as endpoints and midpoints as the cursor passes over them. You can type in distance values directly and Revit Building will apply them to lines or walls while you are sketching.

TIP You can always right-click to reach transparent Zoom commands: Zoom In Region, Zoom Out (2x), Zoom to Fit, and Previous Scroll/Zoom. This enables you to keep drawing while adjusting the View window.

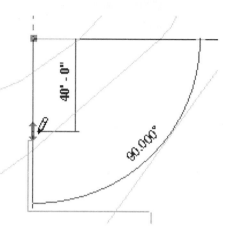

Figure 3–21 *Dimensions and alignment lines assist your sketch*

3.3 Sketch carefully around the building outline. You will encounter doubled lines and gaps in the imported plan. Use the alignment lines that appear to make your new sketch lines meet and intersect correctly.

3.4 Do not draw the last vertical segment. Pick Finish Sketch in the Design Bar. Since your sketched lines do not meet completely, Revit Building will display an Error – cannot be ignored message, as seen in Figure 3–22. The Expand>> button provides a tree view of the error or errors, which can be helpful for advanced users. Pressing the Show button directs you to the standard View commands and does not usually provide an informative picture of what you might have missed. Select the Continue button.

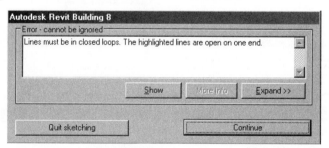

Figure 3–22 *Error message—your sketch will not work*

3.5 Select the Modify command on the Design Bar (always the top selection). As you move the cursor over the building outline, the lines you drew previously will highlight in magenta. This way you can test for missing lines (our deliberate fault in this sketch) or lines that overlap.

3.6 Choose the Lines command on the Sketch tab in the Design Bar. On the Options Bar, select the Pick icon. Hold your cursor over the line at the right side of the building outline that you did not trace over before. The line will highlight green to indicate that it is selectable.

3.7 Left-click to select the line. It will highlight with grips and a lock symbol (open) to indicate that you can lock the sketch line to the underlying symbol. See Figure 3–23.

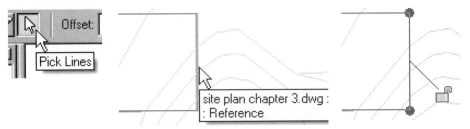

Figure 3–23 *Finish the Pad sketch by picking the last line*

3.8 On the Design Bar pick Pad Properties. In the Element Properties dialogue, change the Phase to Existing and the Height Offset from Level to −**10**, which will convert to −10'-0" (see Figure 3–24). This will set the pad below Level 1, effectively cutting a hole in the toposurface. Select OK.

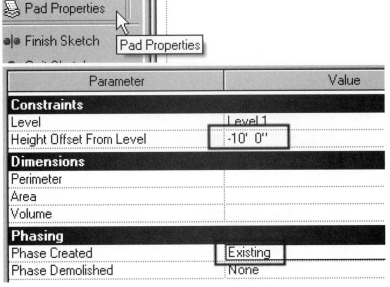

Figure 3–24 *Set Phase and Level Offset properties for the Pad*

3.9 Select Finish Sketch on the Design Bar. The Pad will appear.

3.10 Type **VG** to reach the Visibility dialogue. Pick the DWG/DXF/DGN Categories Tab. Expand the list under the imported dwg file. Clear layer BLDG and check layer ROAD to isolate that portion of the imported file.

CREATE A SUBREGION AT THE ROAD OUTLINE

3.11 Choose Subregion on the Site tab of the Design Bar. The Design Bar will switch to Sketch mode. The Toposurface will become gray.

3.12 On the Options Bar, select the Pick icon. Select the lines that make up the outline of the roadway and parking lot. You may have to Zoom and Pan to make your picks accurate.

If you pick the same line twice when picking existing linework to create sketch lines, Revit Building will issue a warning.

When you have picked all the existing linework of the roadway outline, select Draw and draw a line to close the sketch at the upper right end of the road. Pick Properties on the Design Bar to access the properties of the Toposurface you are sketching. In the Element Properties dialogue, make the Phase Created Existing and change the Material Parameter value to Site – Asphalt (see Figure 3–25). Click OK.

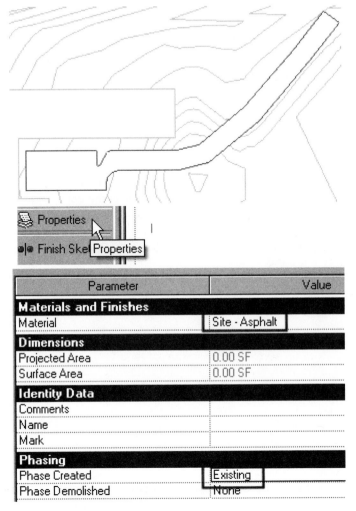

Figure 3–25 *Assigning a material to the new surface*

3.13 Pick Finish Sketch. If Revit Building displays an Error – cannot be ignored message as before, click Continue to return to your sketch. Select Modify from the Sketch tab on the Design Bar. Move your cursor over the roadway outline and Revit Building will highlight the new sketch lines in turn, so you can find any gaps.

You can use the Trim tool from the Toolbar to extend lines to meet one another to fill short gaps, rather than draw another line. See Figure 3–26.

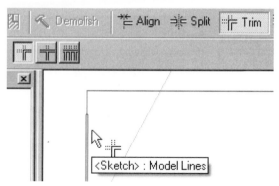

Figure 3–26 *The trim tool in action—note the appearance of the cursor*

3.14 When you have corrected any drawing errors, select Finish Sketch. The road surface subregion appears.

CROP THE SITE VIEW

3.15 Place the cursor over the Crop Region control in the View Control Bar at the bottom of the screen. Choose Show Crop Region, as shown in Figure 3–27.

Figure 3–27 *Activating the Site view Crop Region*

3.16 Type **ZF**. You will see the Crop Region around the site plan Toposurface. The cursor tooltip will identify it when you place the cursor over the Crop outline. Select the Crop Region to activate its control points (see Figure 3–28).

3.17 Use the Stretch controls at the midpoints of each side of the Crop Region to make it smaller, as shown in Figure 3–28. The area below and to the right of the existing building will be the location of the three phases of new work on this hypothetical project.

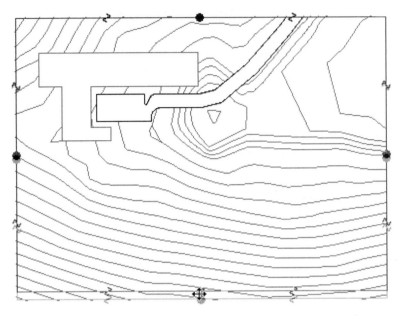

Figure 3–28 *Adjusting the Crop Region size*

3.18 Click outside the Crop Region once it has been condensed to your satisfaction to deselect it. Zoom to Fit to make the site plan fill the View Window.

3.19 Use the Hide Crop Region selection from the View Control Bar to simplify the Site view so you will not inadvertently pick the Crop Region during further work in this view (See Figure 3–27). Click OK.

3.20 Open the default 3D view. Open the View Properties dialogue to set the View Phase to Existing. Click OK. Use the Shading control from the View Control Bar to set the mode to Shading with Edges. Zoom in to see the pad (sitting below the contour surface) and asphalt road region.

3.21 From the View menu, select View>Orient>Save View. Accept the default view name and choose OK. (See Figure 3–29.) Save the file.

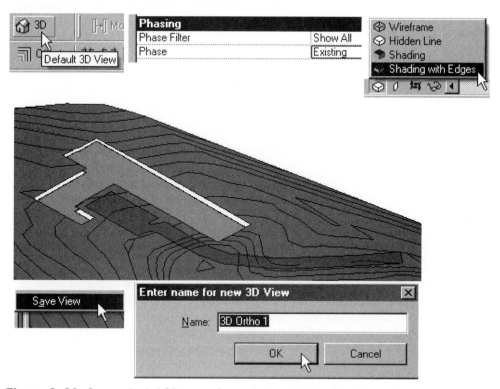

Figure 3–29 *Save a shaded 3D view of your changes to the site object*

EXERCISE 4. ADJUST AND ADD FLOOR LEVELS

4.1   Open the file from the previous exercise or continue in the open project. Double-click on Elevations>South in the Project Browser to make that view active. Zoom in to make the Level tags readable.

4.2   Select the Toposurface object so that it highlights red. Type **VH** at the keyboard. This is a shortcut for View Hide this Category, and is the same as clearing an object category in the View Graphics dialogue. The Toposurface will disappear.

4.3   From the File menu, pick File>Import/Link>DWG, DXF, DGN, SAT. In the Import/Link dialogue as before, and navigate to the folder that contains the file *Chapter 3 south elevation.dwg*. Select that file so the name appears in the File Name field of the dialogue. Adjust the Import or Link setting as before— check Current view only to bring in text. Select Black and white in the Layer/Level Colors section. In the Positioning setting portion of the dialogue, select Manually place and Cursor at origin (see Figure 3–30). Click Open.

Figure 3–30 *Place this import manually*

4.4 The imported file object will appear with the cursor at its lower-left corner. Place the import object to the right of the Level tags.

4.5 Select the Align tool from the Toolbar. This tool works with two picks: the first is the reference object, and the second object picked will move into alignment with the first. Pick the level line for Level 2. A dashed green reference line will appear. Pick the level line identified as 1ST FLOOR in the imported file. The import will align with Level 2, and a lock toggle (unlocked) will appear. Do not select the lock.

The Align tool is still active. Select the line identified as 2ND FLOOR in the import to make it a reference. Select the level line for Level 2. Level 2 will move up, its elevation field will read **479′ – 4″**, and the lock option appears (see Figure 3–31).

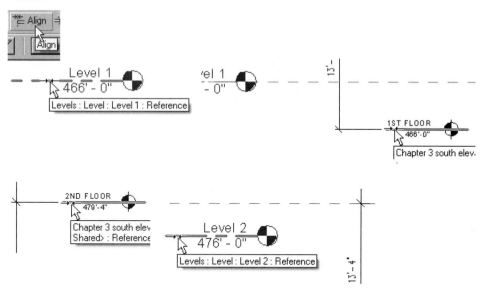

Figure 3–31 *Align the import with Level 1, then align Level 2 with the import*

4.6 Pick the Basics tab in the Design Bar to open its commands. Select Level to add more Levels to the Project. Move the cursor to the left of the View window above the Level 2 level line. Revit Building will read from that line and show a temporary dimension up from the level line. Type **13′4** at the keyboard to match the dimension in the import object. Revit Building will apply that dimension and show the level's Elevation bubble (see Figure 3–32). Move the cursor to the right and Revit Building will snap to an alignment point above the existing bubbles. Left-click to accept that alignment point and Revit Building will start another Level.

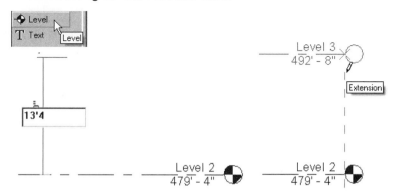

Figure 3–32 *Specify the Level offset from the one below*

4.7 Type the dimension offset between the Level 3 and the new level as shown in the import file (**14'8**) and either hit ENTER or left-click outside the new level's elevation value field. Align the tag as before (see Figure 3–33).

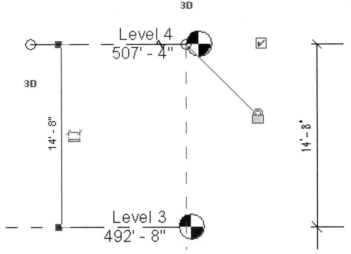

Figure 3–33 *Bubbles will snap to align*

4.8 Pick Modify from the top of the Basics tab in the Design Bar to exit the Level tool.

 TIP You can also hit **ESC** twice to exit the Level command completely. Many commands in Revit Building are multi-level routines, which allow the user to cancel a certain part of the procedure while still remaining in the command set. AutoCAD and Architectural Desktop users should be well familiar with the **ESC-ESC reflex.**

4.9 Select the name field for Level 1. The Level and associated fields will highlight. Select the name field again to open the value. Type **1ST FLOOR MILLER HALL** and click outside the field or hit ENTER. Select Yes in the question box that appears about renaming the plan views of this level. (See Figure 3–34.)

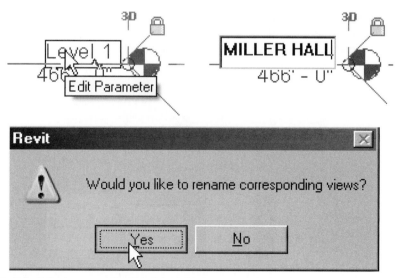

Figure 3–34 *Edit the name for Level 1 and rename its views*

    4.10 Edit the names for Levels 2, 3 and 4 to match, as shown in Figure 3–35. Accept the rename question each time.

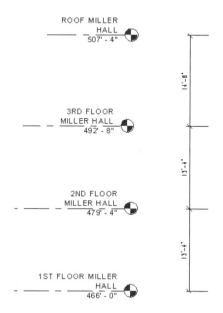

Figure 3–35 *Rename all the levels*

    4.11 Edit the length of the new elevation lines. Select the ROOF MILLER HALL level and use the blue control circle at its left end to drag the line to the left.

It will snap to the end of 3RD FLOOR MILLER HALL level. Continue to drag both level ends to the left (Zoom and Pan as necessary) and align them with the ends of the original levels. Repeat for the ROOF level. Note that the lines will align and you can move more than one (see Figure 3–36).

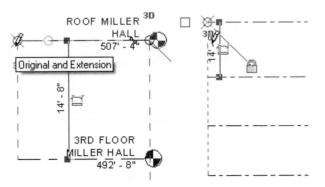

Figure 3–36 *Align the level line ends. They snap to each other.*

4.12 If Revit Building has not yet reminded you to save your file, save it now.

EXERCISE 5. CREATE WALLS

5.1 Open the file from the previous exercise or continue with the open file *Chapter 3 Phase 1.rvt*. Double-click on Site in the Project Browser to return to that view. Type **VG**. In the Visibility dialogue, select the DWG/DXF/DGN Categories tab. Expand the list under the site plan chapter 3.dwg category. Clear layer ROAD. Check layer BLDG.

5.2 Select the Model Categories tab. Clear Topography, as shown in Figure 3–37. Click OK

Visibility/Graphic Overrides for Floor Plan: Site

- ☑ site plan chapter 3.dwg
  - ☐ 0
  - ☐ A-Anno-Revs
  - ☑ BLDG
  - ☐ BLDG-INT
  - ☐ CURB
  - ☐ PARKING
  - ☐ ROAD
- ☑ Structural Framing
- ☐ Topography
- ☑ Walls

Figure 3–37 *Istolate a layer, turn off the topographny in the Visibility dialogue*

5.3 The imported building outline is now isolated in the Site view.

Open the View Properties dialogue. Set the View Phase to Existing (see Figure 3–38). Click OK. This will control the Phase of all elements created in this view.

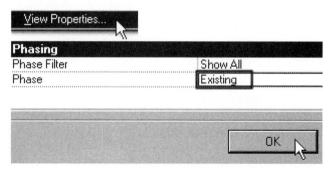

View Properties...

Phasing	
Phase Filter	Show All
Phase	Existing

OK

Figure 3–38 *Change the Site View Phase to Existing*

5.4 Pick Wall on the Basics tab of the Design Bar. The cursor will change to a pencil symbol, indicating sketch mode. The status bar at the bottom of the screen will read "Click to enter wall start point," which will also appear as the tooltip. Change the Height value for the walls you are about to create to ROOF MILLER HALL using the drop-down list. Check the Chain option. Leave other options to their defaults (see Figure 3–39).

Figure 3–39 *Set the height of the walls to a level*

5.5 Trace walls around the building outline clockwise. This will place the defined exterior face of the walls to the outside.

 NOTE There are sections of the building outline where lines do not terminate end to end. This is deliberate—all designers and drafters have at one time inherited incompletely translated files, unfinished models, or badly drawn sketches. Study the imported linework before you begin.

Revit Building's snap and alignment tools will allow you to draw a horizontal wall that terminates perpendicular to a desired line or wall even if the underlying elements you may be tracing over are not horizontal or do not define complete corners.

5.6 When the new walls make a complete outline, pick Modify on the Basics tab of the Design Bar. Type **VP** to reach the View Properties dialogue. In the View Properties dialogue, open the Visibility Dialogue. On the DWG/DWF/DGN Categories tab, clear the site plan chapter 3.dwg category, as shown in Figure 3–40. Click OK. This will leave the new walls isolated in this view.

Select the Edit button next to View Range in the View Properties dialogue. Set the View Depth to Unlimited. (See Figure 3–40.) This will allow you to view Foundation walls that you are about to create under the exterior walls you just placed. Click OK to exit the View Properties dialogue.

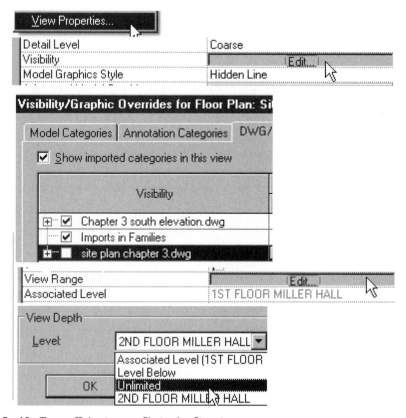

Figure 3–40 *Turn off the import file in the Site view*

    5.7 Select the Wall tool again. Change the type to Basic Wall: Foundation: 12″ Concrete. Set the Draw Option to Pick. Set the Depth to 10 (which will convert to 10′ – 0″), so that the foundation walls reach the pad you previously defined.

Carefully select the center lines of the exterior walls in turn to place foundation walls. A dashed line will appear when the cursor is in the correct position. See Figure 3–41. Place foundation walls all around the building footprint.

Figure 3–41 *Placing foundation walls*

TIP You can use the **TAB** key with the cursor placed over a wall to cycle through the selection choices (interior face, center line, exterior face). This will make your selection precise without having to zoom all the way in.

ADJUST THE ELEVATIONS AND LEVELS TO FIT THE WALLS

5.8 When the foundation walls are placed, double-click Floor Plans>1ST FLOOR MILLER HALL in the Project Browser to open that view. Type **ZF** to expand the view. Note the location of the Elevations. Since you created walls according to a location on an imported object, and not in the center of the project file, the Elevations are now inaccurate and should be relocated.

 NOTE If the Elevations are not visible in this view, type *VG* and use the Visibility dialogue to turn them on.

5.9 Use the cursor to pick two points around the North and South Elevations. The two Elevations will turn red to indicate they have been selected. Select the Move command from the toolbar. A dashed boundary will appear around the selected items, along with a temporary reference line, and the cursor will change to the Move symbol (see Figure 3–42).

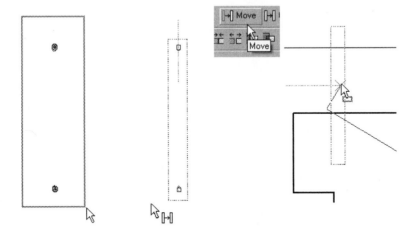

Figure 3–42 *Select and move the North and South Elevations*

5.10 Pick a start point above the North Elevation, then a second point above the middle of the north wall of the building outline to move the two Elevations. As you move the cursor to select a starting point, Revit Building will display a temporary dimension. As you move the items, Revit Building will show an angle vector and temporary dimension, which will allow you to enter a distance directly if you desire. Place the Elevations by eye, as shown in Figure 3–42.

5.11 Repeat the Move on the East and West Elevations, placing them outside the building outline roughly in the middle of the figure. Note that the South Elevation does not include the North–South wing and its extension (see Figure 3–43). You will keep that Elevation in its current place—the South Elevation view matches the import file by cutting a section through the north-south wing of the building—and create another.

Revit Building will readily duplicate Plan and Section Views by picking them in the Project Browser and using the right-click context menu, but not Elevations. You can create a copy of an Elevation in a Plan view and adjust the properties of the new Elevation separately.

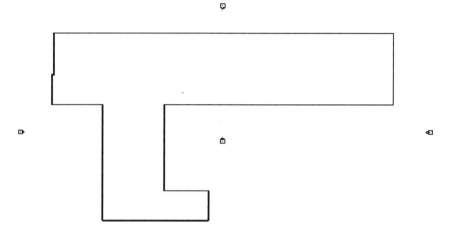

Figure 3–43 *The Elevations moved around the building outline*

5.12 Pick two points around the South Elevation to select it. Be sure to select both the Elevation and the view arrow. Pick the Copy command from the Toolbar. Create a copy of the South Elevation 120′ below the original, so that the copy includes the wing and extension, as shown in Figure 3–44. Revit Building will create a new Elevation named South1 in the Project Browser.

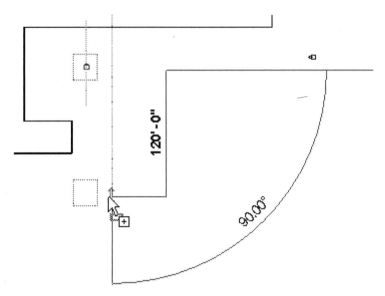

Figure 3–44 *Creating a copy of the South Elevation*

    5.13 Select the view arrow for each Elevation marker in turn. A blue line will appear indicating the extents of the view, which will be too small for this building. Use the control dots to extend the view extents line so that it includes the full width of the walls, as shown in Figure 3–45.

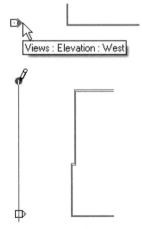

Figure 3–45 *Extending the Elevations*

    5.14 Double-click Elevations>South in the Project Browser to open that view. Type **ZF** to see the new walls and import symbol centered in this view.

    5.15 Select the level line for Level 1 and place the cursor over the left end. Use the TAB key to cycle through the operation options, until the Tooltip reads:Modify the level by dragging its model end. When it does, drag the level ends to the left of the building, as shown in Figure 3–46.

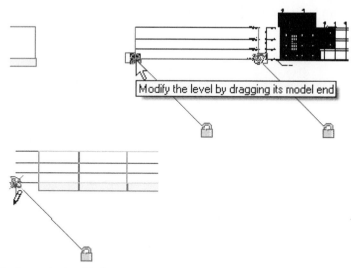

Figure 3–46 *Drag the level left ends.*

5.16 Repeat the process with the right ends: hold the cursor over the level end, cycle the options, then drag, as shown in Figure 3–47

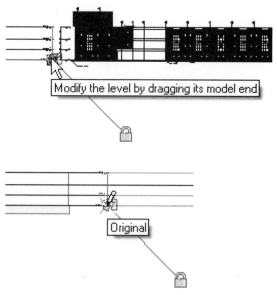

Figure 3–47 *Drag the level right ends*

5.17 Select the import symbol. It will highlight red and the cursor will change to a double-headed arrow, indicating that you can drag it. Drag the cursor (it will change appearance) to place the import symbol above the new walls, as shown in Figure 3–48. You will adjust the actual placement later.

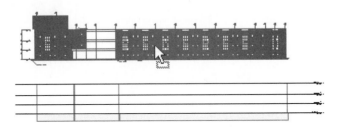

Figure 3–48 *Move the import above the model*

5.18 Select one of the level lines. From the Options Bar choose Propagate Extents. This will match level lines in parallel views to these, saving steps. In the dialogue that opens, select all the available views and choose OK. See Figure 3–49.

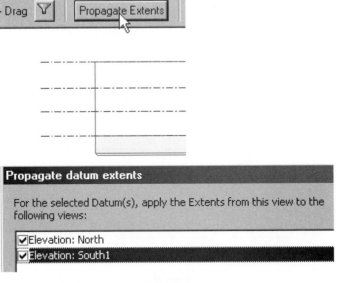

Figure 3–49 *Propagate level extents in other views*

5.19 Open Elevation View East. Select the Toposurface and type **VH**, as before, to turn the category off in the view. Zoom to Fit. Adjust the level extents as you just did in the South Elevation. Propagate the level extent into the West Elevation.

5.20 Open Elevation views West and North in turn, turn off the Toposurface and Zoom to Fit.

5.21 Return to the South Elevation view. From the Window menu, select Window>Close Hidden Windows to save system resources (see Figure 3–50).

Save the file.

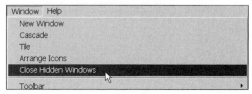

Figure 3–50 *Closing Windows on views you won't use for a while*

EXERCISE 6. MORE WORK WITH WALLS

ALIGNMENT USING REFERENCE PLANES

6.1 Open the file from the previous exercise or continue working in the open project file. Open Elevation view South.

6.2 Select the Align tool from the Toolbar. Select the left side of the building for the first reference, then carefully select the left edge of the wall in the imported elevation. Select the lock toggle (see Figure 3–51).

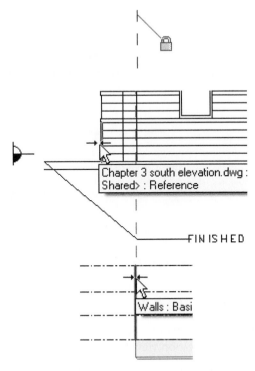

Chapter 3 south elevation.dwg :
Shared> : Reference

FINISHED

Walls : Basi

Figure 3–51 *Align the import to the model, then lock the placement. Read this from the bottom up.*

6.3 Type **ZF** to enlarge the model and import object in the View Window. Select Ref Plane from the Basics tab of the Design Bar. The cursor will change to a pencil symbol; the Status Bar and Tooltip will display information relative to the active command.

6.4 Zoom into the sectional part of the elevation file (between column lines N and L). Pick the corner as shown in Figure 3–52 to the left of column line N to start the Reference Plane. Pull the cursor straight up, so that the angle vector reads 90° and the tooltip shows Vertical. Pick a second point to create the reference plane.

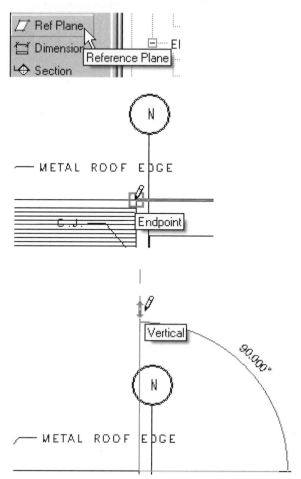

Figure 3–52 *Create a Reference Plane at the left side of the section*

6.5 Pick Modify from the Basics tab on the Design Bar. Pick the new Reference Plane, right-click, and pick Properties from the context menu. In the Element Properties dialogue, give the Reference Plane the Name **WEST SIDE OF WALKWAY**. Click OK. This plane will represent the exterior face of an upper story wall (see Figure 3–53).

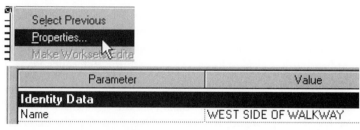

Figure 3–53 *Name the Reference Plane*

6.6 Create a second vertical Reference Plane at the right edges of the section next to column line L (see Figure 3–54). Name this plane **EAST SIDE OF WALKWAY**.

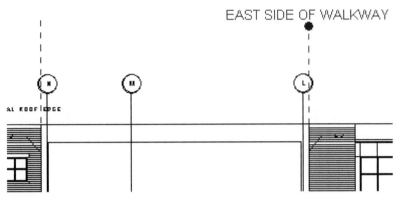

Figure 3–54 *Reference Planes at the Section*

6.7 Use Zoom to Fit to see the model. Study the building profile of the section—the first story is wider than the upper stories.

6.8 Select each Reference Plane in turn and drag its lower end to a point below the model, as you previously did with levels (see Figure 3–55.) You are preparing to create the upper story walls that the section depicts.

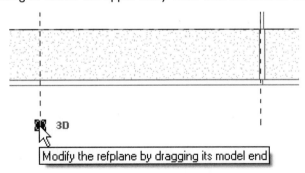

Figure 3–55 *Drag the Reference Planes so they intersect the model*

6.9 Double-click on Floor Plans>2ND FLOOR PLAN MILLER HALL in the Project Browser to open that View. Zoom In Region around the T-junction between the horizontal and vertical wings of the building. Note that the right-hand wall is aligned Interior Face with the Reference Plane, not Exterior Face.

6.10 Pick the Align tool from the Toolbar.Pick the right EAST SIDE OF WALKWAY (right-hand) Reference Plane, then pick the Exterior (right-hand) face of the right wall. The wall will shift position (see Figure 3–56). A blue lock symbol will appear at the midpoint of the wall. Toggle the symbol to locked. This will fix the wall's exterior face at the Reference Plane. If you move the plane, the wall will move with it.

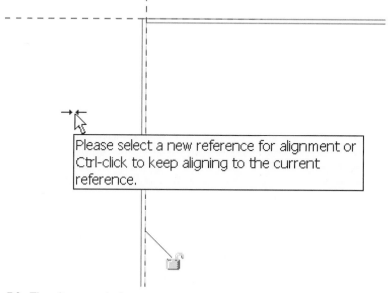

Please select a new reference for alignment or Ctrl-click to keep aligning to the current reference.

Figure 3–56 *The alignment lock*

 NOTE Revit Building's parametric capabilities allow the designer to establish a nearly unlimited set of relationships between building elements. You will explore ways to take maximum advantage of some of these possible relationships in future lessons. One of the first, most basic parametric utilities is the alignment lock. When combined with Reference Planes–Revit Building's 3D construction lines–in a carefully planned design, locking and other orientation specifications can make even late design changes straightforward to manage—closer to wholesale than retail.

6.11   Choose Modify to exit the Align command. You will now make changes in other walls. Zoom Out as necessary to see the North–South wing of the building. Pick the left-hand wall of the North–South wing (to the left of the WEST SIDE OF WALKWAY Reference Plane, just below the one you stretched). Pick the Properties icon. Set the Top Constraint value to Up to Level: 2ND FLOOR MILLER HALL. Click OK. Hit ESC twice and the wall will change color from red (highlighted) to light gray, indicating that it does not reach the current level but is visible as an underlay.

6.12   Pick the Split command from the Toolbar. The cursor changes to a scalpel icon. Pick the bottom (East–West) wall of the I-wing at the intersection with the WEST SIDE OF WALKWAY Reference Plane. Revit Building will snap to

152

the Reference Plane to make the pick easier. Pick Modify from the Basics tab of the Design Bar to terminate the command (see Figure 3–57).

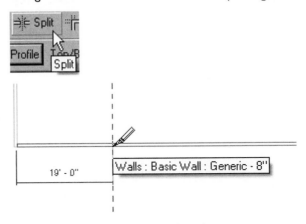

Figure 3–57 *Split a wall so that you can make other changes*

6.13 Pick the same East–West wall to the left of the Reference Plane. Only the part of the wall to the left of the Reference Plane will highlight. Right-click and pick Properties. Set the Top Constraint value to Up to Level: 2ND FLOOR MILLER HALL. Click OK. Hit ESC twice and the wall will change color from red (highlighted) to light gray, indicating it is one story in height (see Figure 3–58).

Figure 3–58 *Two walls have been re-defined, and appear gray in this 2nd Floor view.*

6.14 Pick the Wall command from the Basics tab of the Design Bar. Set the Type Selector to Generic – 8″. Set the Height to ROOF MILLER HALL as shown in Figure 3–59. You are going to create the west wall for the upper stories of the North–South wing of the building, which will become a bridgeway between the existing classroom hall and the first new building.

Figure 3–59 *Prepare to draw the wall at its face*

6.15 Put the cursor between the Reference Planes, near the bottom wall. Pull the cursor vertically to draw a wall with its exterior side located to the left. Placement and length are not critical—Revit Building allows you to sketch quickly and make precision alignments just as quickly.

6.16 Select the Align tool, pick the left (WEST SIDE) Reference Plane, then the left side of the wall; this will move the wall face to the correct position. Toggle the lock on. Select the Trim tool and make corners with the two full height walls, as shown in Figure 3–60.

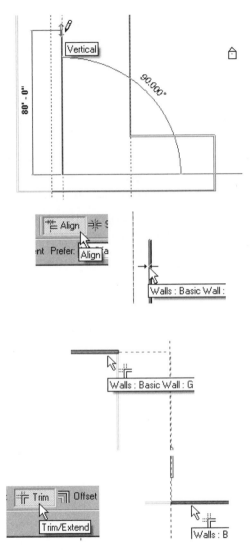

Figure 3–60 *Create a wall, align it to a Reference Plane, trim it to other walls*

DEFINE WALLS AS EXISTING-TO-REMAIN AND EXISTING-TO-BE-DEMOLISHED

6.17 Use Zoom to Fit. Pick two points outside the building outline to create a pick box that surrounds all the walls. You are selecting everything in the model. The second story walls will all highlight red, along with the South Elevation and possibly the WALKWAY Reference Planes. Pick the Filter Selection icon from the Options Bar. In the Filter dialogue that opens, clear

Elevations, Reference Planes and Views to remove them from the selection set (see Figure 3–61).

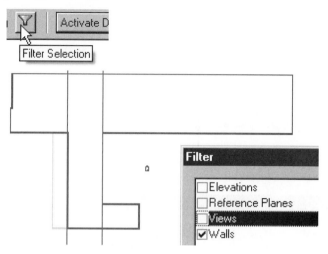

Figure 3–61 *Filter the selected objects*

Click the Properties icon. Change the value of the Phase Created Parameter to Existing for all walls. Click OK. Hit ESC twice to clear the command and selection set.

6.18 Pick one of the two gray (1st floor) walls. Hold down the CTRL key and pick the other gray wall. Pick the bottom wall. Pick the short North–South and short East–West walls that make up the ell to the right of the walkway (see Figure 3–62). Right-click and pick the Properties icon. Set the Phase Demolished Parameter to New Construction. Hit ESC twice to clear the command. The walls will display with a dashed linetype.

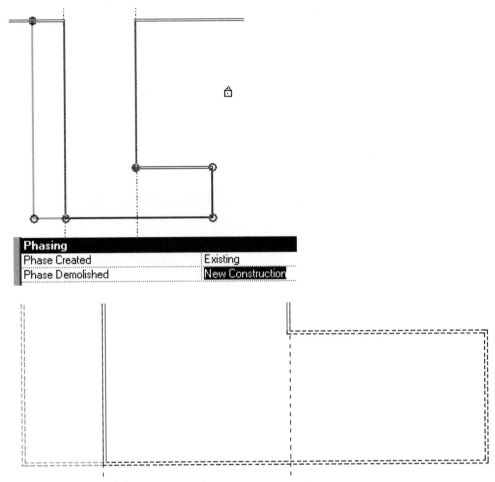

Figure 3–62 *Existing walls that will be demolished in Phase 1 construction*

Save the file.

EXERCISE 7. PLACE ROOFS ON THE BUILDING

7.1 Open the file from the previous exercise or continue working in the open project file.

7.2 Zoom to the previous close-in view of the vertical wing. Pick Roof>>Roof by Footprint from the Basics tab of the Design Bar. The Design Bar will switch to Sketch mode with drawing commands. The Pick Walls tool will be active (depressed) by default.

7.3 On the Options Bar, clear Defines Slope. This will be a flat roof. Leave the Extend into wall (to core) option unchecked. You are creating this roof for representational purposes only, and so do not need Revit Building to show it accurately in section (see Figure 3–63).

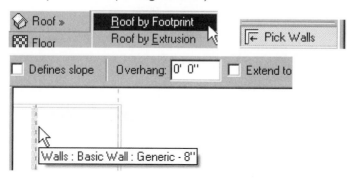

Figure 3–63 *Settings for the flat roof*

7.4   Pick the interior face of the two one-story walls. Pick the exterior face of the wall at the LEFT SIDE OF WALKWAY Reference Plane and the exterior face of the East–West wall to its left (at the top of the screen). You are placing a roof on the one-story construction only.

NOTE Revit Building will snap to either face of each wall. As you pick you create a line in the roof sketch (shown in magenta). Revit Building will display information about the sketch line—length, offset from wall face, and a double-headed orientation toggle arrow. If you mistakenly pick the wrong side of a wall, you can move the sketch line to the other side right away. You can also pick the line later to make adjustments.

7.5   Note that because the East–West wall at the top of the sketch extends past the other wall, so does the sketch line. Type **T** to activate the T hotkey. Note the Status line. It indicates the possible commands available. The only command beginning with T is trim/extend, so hit the space bar to start the command.

TIP Hotkeys and their associated keystrokes, such as arrow key cycling or space bar/enter command acceptance, should be typed within about six seconds of each other, or Revit Building clears the hotkey. For those who like to use the keyboard, Hotkey stroke combinations will be worth developing for speed. Forefinger-T>Thumb-space-bar, with just a bit of practice, can become an automatic and extremely quick way to enter the important Trim command, for example.

7.6   Use Trim to make the sketch lines meet at the corners with no gaps or overlaps, as shown in Figure 3–64. Pick Roof Properties from the Sketch tab of the Design Bar. Change the Phase Created value to Existing and the Phase Demolished value to New Construction. Click OK.

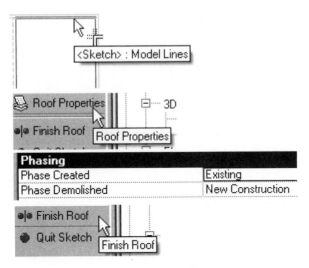

Figure 3–64 *Using Trim on sketch lines for the Roof Footprint*

7.7 Pick Finish Roof from the Sketch tab of the Design Bar. The roof will appear with a dashed linetype. (Where the roof lines overlap other lines, the dashes may not be visible.)

GET MORE INFORMATION

7.8 Double-click Elevations>South in the Project Browser to return to that view. Zoom in to the standing seam gable roof at the left side of the imported elevation drawing. Select the Tape Measure tool on the Toolbar. This will read the distance and angle between any two points.

7.9 Pick the intersection of the WEST SIDE OF WALKWAY Reference Plane and the line identified as METAL ROOF EDGE for the first point. Pull the cursor to the left, horizontally to the wall line, under the right side of the metal roof, and pick the intersection of that line and the METAL ROOF EDGE line. The Tape Measure will read 27'-0". Pick to terminate the measurement and get ready for another (see Figure 3–65).

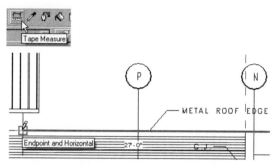

Figure 3–65 *The tape measure tool in use*

7.10 Pick two points to measure the overhang of the upper roof. The Tape Measure will read 2'-0". Pick two points to measure the height of the wall under the upper roof. The Tape Measure will read 2'1 ¾". Write down these measurements for future use. Pick Modify on the Basics tab of the Design Bar to exit the command.

7.11 Pick the WEST SIDE OF WALKWAY Reference Plane. Type **C** and hit the Space Bar to enter the Copy command. Pick any convenient point, then pull the cursor to the left until the temporary dimension reads 27'-0" (see Figure 3–66). Left-click to accept the distance.

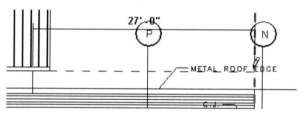

Figure 3–66 *Making a copy of the Reference Plane*

 TIP Revit Building's temporary dimensions read out in distance increments that are keyed to the zoom scale of the current view. At a close zoom, distance will display in inches, and the cursor will snap at each inch. When zoomed out at larger scale, the temporary dimension will read and snap in 4", 6", or 1' increments. You can enter a specific distance from the keyboard at any time.

7.12 Pick the new Reference Plane. Open its Properties dialogue. Give it the name EAST SIDE GABLE WALL, as shown in Figure 3–67. Click OK. You will use this plane to locate the upper walls and gable roof.

Figure 3–67 *Name the new Reference Plane*

7.13   Double-click Elevations>West in the Project Browser to open that view. From the File menu, pick File>Import/Link>DWG, DXF, DGN, SAT. Navigate to the folder with the file *Chapter 3 west elevation.dwg* and select that file. Adjust the settings in the Import/Link dialogue as before: Check Current view only; select Black and white Layer/Level Colors; and choose Manually Place/Cursor at origin.

7.14   Locate the imported symbol above the building model. You can align it exactly later. Zoom in Region around the gable roof area of the import. Use the Tape Measure tool to check the angle of the roof—pick the peak and then an eave edge. The Tape Measure will display a distance of 24′-5″ and an angle of 45° or 135°—a 12 in 12 pitch (see Figure 3–68). You will use this information in an upcoming exercise.

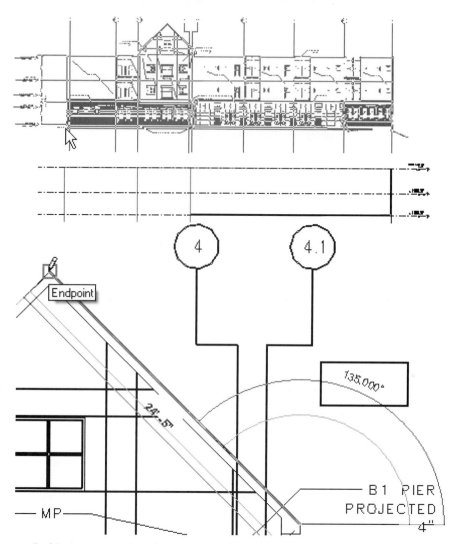

Figure 3–68 *Import another elevation, then measure the roof pitch*

CREATE TWO FLAT ROOFS—ONE TO BE DEMOLISHED

7.15   Open the ROOF MILLER HALL Floor Plan view. Zoom in to the North–South wing and the short East–West leg. Pick Roof>>Roof by Footprint from the Basics tab of the Design Bar. The Design Bar will change to Sketch mode with the Pick Walls tool active. On the Options Bar, check Extend into wall (into core). This will be a flat roof, so leave Defines slope cleared.

7.16 Pick the interior face of the left-hand wall. Pick, in turn, the interior faces of the other walls that define the two building wings. Select the Lines tool to draw a line between the endpoint of the sketch line at the top of the sketch where there is no third story wall drawn (see Figure 3–69).

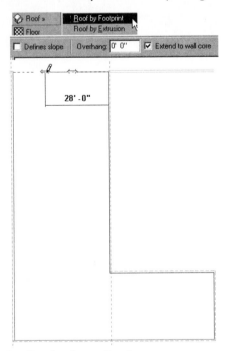

Figure 3–69 *Draw the last line for the roof outline*

7.17 Pick Roof Properties from the Sketch tab of the Design Bar. In the Element Properties dialogue, set the Phase Created value to Existing. Set the Phase Demolished value to New Construction. (This roof will be replaced by a gable to echo the roof on the main wing of the building.) Set the Base Offset From Level value to –1′. Since you are placing a generic roof of 12″ thickness, you will set it down 12″ from its Base Level so its top surface and the tops of the walls align (see Figure 3–70). Click OK.

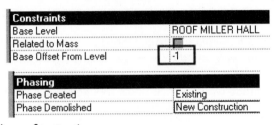

Figure 3–70 *Set the roof properties*

7.18 Pick Finish Roof on the Sketch tab of the Design Bar. Click Yes to the question about cutting the roof volume out of the walls. The roof will appear in dashed lines.

7.19 Zoom or Pan so the East–West wing of the building fills the View window. Pick Roof>>Roof by Footprint. Check Extend into wall core on the Options Bar. Accept the Pick Walls tool on the Sketch tab. Pick the interior faces of the southeast, east, north, and northwest walls of the building wing, as shown in Figure 3–71.

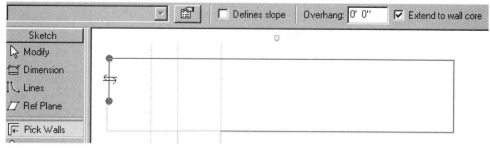

Figure 3–71 *Pick walls for the flat roof*

7.20 Select the Lines tool on the Sketch tab. Zoom In Region if necessary to draw the next lines. Check Chain on the Options Bar. Start the lines by picking the lower endpoint of the last sketch line you created. Pull the cursor horizontally right to the EAST SIDE GABLE WALL Reference Plane so that the plane highlights green, and left-click.

7.21 Pull the cursor down vertically to start a line at a 90° angle from the previous one. Revit Building will snap to either face (or the location line) of the exterior wall. Place the line endpoint at the interior face of the East–West wall. Pick the endpoint of the first sketch line you created to close the loop (see Figure 3–72). This loop does not enclose the southwest corner of the building, which you will work on in the next steps.

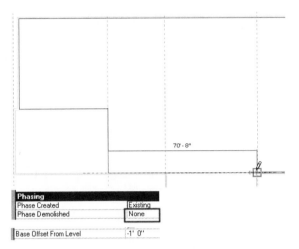

Figure 3–72 *The sketch loop and properties for the flat roof section*

7.22 Pick Roof Properties on the Sketch tab. Set the Phase Created value to Existing. Leave the Phase Demolished value at None. Leave the Base Offset From Level value at −1′ 0″. Click OK. Pick Finish Roof. Select Yes on the question.

CREATE WALLS FOR THE GABLE ROOF SECTION

7.23 Zoom in Region to the lower-left corner of the East–West wing. Pick the Wall tool from the Basics tab. Revit Building sets the Base Constraint for new walls to ROOF MILLER HALL, since the current view is at that Level. Set the Unconnected Height value for the new walls to 8′—you will adjust them to fit under the gable roof later. Set the Location Line to Finish Face Exterior. Check Chain on the Options Bar.

7.24 Pick the lower-left corner of the exterior walls to start the new walls. Pull the cursor vertically to the exterior side of the short jog in the walls below, and pick that point. Pull the cursor horizontally to the right to the Reference Plane and pick the plane when it highlights.

7.25 Pull the cursor down vertically until the exterior face of the lower East–West wall highlights, and left-click. Pull the cursor left horizontally to the starting point and pick to finish the walls (see Figure 3–73).

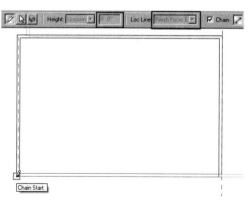

Figure 3–73 *Walls above the third floor roof align with existing walls and the reference plane*

CREATE THE FINAL EXISTING ROOF SECTION

7.26 Pick Roof>Roof by Footprint on the Basics tab. Pick Roof Properties on the Sketch tab that appears. Set the Base Offset From Level value to **2'1.75** (the distance you measured earlier). Revit Building will convert decimal inch entry to fractions. Click OK.

7.27 On the Options Bar, set the Overhang value to **2'** (the other distance you measured from the imported elevation). Check Defines Slope and place the cursor over the bottom new wall. As you move the cursor from the interior to the exterior face, Revit Building will locate a green highlighted line to show you where the sketch line will be. Select the exterior face. A magenta sketch line will appear, with a length dimension, offset dimension, double-headed control arrow to flip orientation, and slope arrow (two lines making an angle symbol). Pick the value field at the slope arrow, which reads **9"**. The field will open for editing; change the value to **12"**. Click off the field to return to the sketch (see Figure 3–74).

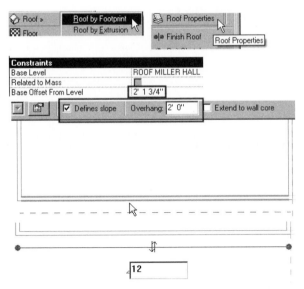

Figure 3–74 *Starting the gable roof sketch*

7.28 Pick the exterior face of the upper wall to define the other slope of the roof. Pick Properties to check the Rise/12″ value and change it if necessary. Click OK. Clear Defines slope and pick the exterior faces of the left and right walls to complete the sketching part of the roof creation (see Figure 3–75).

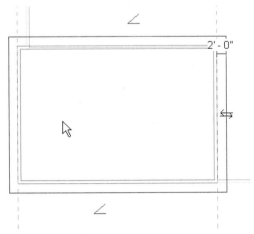

Figure 3–75 *Roof outline with offset value and slope arrows*

7.29 Pick Finish Roof. Select Yes in the question about attaching the walls to the Roof. Revit Building will trim the walls to the underside of the new Roof.

ADJUST VIEWS OF THE MODEL

7.30 The roof will display with cutlines. You will not see the ridge of the roof. In order to see the ridge line in the roof plan, you need to adjust the default settings for this view. Type **VP** to enter the View Properties dialogue. In the Element Properties dialogue for the current view, pick the Edit button for the View Range (see Figure 3–76).

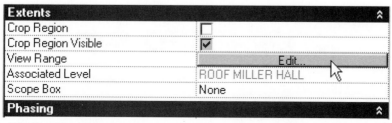

Figure 3–76 *The View Range control*

7.31 In the View Range dialogue, change the Offset value to the right of the Top parameter to **25′**. Do not change the Top parameter. Change the Cut plane Offset value to **20′** (see Figure 3–77). Pick Apply to see the changes in the roof display under the dialogue boxes. Click OK twice to exit the View Properties dialogue.

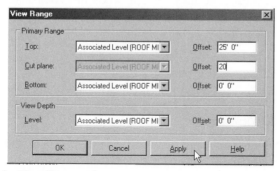

Figure 3–77 *Edit the View Range values*

7.32 Type **ZF** to fill the View window with the roof plan. Select the new roof and walls under it—pick to the left, then the right so your pick box encloses the roof. Check the Filter to make sure you have only walls and a roof. Set the Phase created to Existing (see Figure 3–78).

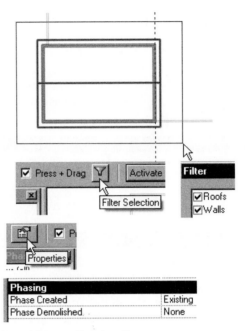

Figure 3–78 *Set the phase of the roof and walls*

7.33 Pick the 3D (Default 3D view) button on the Toolbar (see Figure 3–79). Type **Z**, then hit Spacebar to activate Zoom In Region, then pick two points around the model.

Figure 3–79 *The 3D View control*

7.34 From the View menu, pick View>Shading with Edges. Right-click and choose View Properties. The view Phase is New Construction. The parts designated as demolished in the New Construction Phase are red and transparent.

7.35 Double-click the name of 3D View 3D Ortho 1 in the Project Browser. The Phase of this view was previously set to Existing. Revit Building will display the generic walls and roofs in shades of gray and blue (see Figure 3–80).

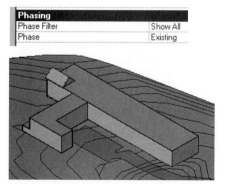
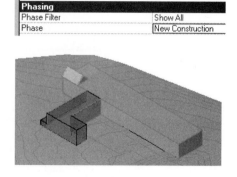

Figure 3–80 *The shaded isometric view*

7.36 Pick Window>Close Hidden Windows. Save the file.

For the purposes of our hypothetical project you now need to put wall, door, and window information on certain walls of the building so it is an accurate model of parts of the building represented in the CAD files you have been studying in Revit Building. The first stage of the design project will involve transforming the North–South wing of this building. The ground level will be removed for a pedestrian concourse, and the upper two stories will become walkways between the existing East–West wing and a new building to be located at the south end of the new concourse. The next chapter's exercises will locate the shell of this new building and those of later phases in the ambitious master plan of this little college.

Once the walls of the existing North–South wing have been accurately represented on the parts that will remain, you will design the faces of support pillars for the concourse and add a glazed roof over the top floor. The final task for the job setup will be to export an elevation of this wing in AutoCAD, to represent electronic files sent to consultants or contractors.

EXERCISE 8. DEFINE WALLS USING INFORMATION FROM IMPORTS

8.1 Continue with or open the file *Chapter 3 Phase 1.dwg*. So far you have created a site plan and a model of the existing walls and roofing. You will recall that the site plan you used to sketch the walls contained errors in the linework. Now you will make the model more accurate based on the CAD elevations that you have imported.

8.2 Open the West Elevation View. Pick the Align tool from the Toolbar. The cursor will change to the double arrow Align tool shape. Pick the left side of the building model first. It will highlight red. Put the cursor over the line representing the left building edge of the CAD import. Revit Building will snap to the linework and the line will highlight green. The Status Bar and tooltip will identify the import symbol (see Figure 3–81). Left-click to accept the selection and the import file will move over to the Reference Plane. Toggle the lock on.

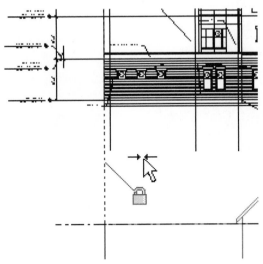

Figure 3–81 *Align the import to the model*

8.3 Zoom in Region to the upper-right corner of the CAD file so you can see the bubble identifying Column 8 and the right edge of the building. Pick Ref Plane from the Basics tab of the Design Bar. Pick the upper-right corner of the building in the CAD import. Type **ZF** to enlarge the view. Pull the cursor down at 90° past the Revit Building model to establish a Reference plane at the south side of the building wing.

8.4 Zoom in Region at the right side of the gable roof. The Reference Plane tool will still be active. Pick a point on the line representing the edge of the wall under the right side of the roof—this is also the south edge of the East–West wing of the building. Type **ZF** to zoom back out. Pull the cursor down to a point below the Revit Building model at 90° (vertical), and click to establish a vertical Reference Plane at the building edge.

8.5 Pick Modify from the Basics tab of the Design Bar. Pick the first new Reference Plane (on the right), pick Properties, and give it the Name value **SOUTH SIDE WALKWAY**. Click OK. Use the same technique to name the other new Reference Plane **SOUTH SIDE MILLER HALL**. Click OK. (See Figure 3–82.)

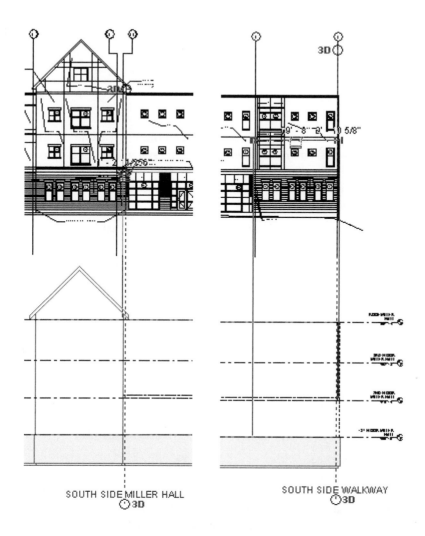

Figure 3–82 *Reference Planes in the West Elevation*

8.6 Open the 1ST FLOOR MILLER HALL Floor Plan view. Zoom in Region to the south (bottom) side of the Revit Building model. Use the Align tool to align the interior face of the south wall of the model with the SOUTH SIDE WALKWAY Reference Plane. Zoom Previous and repeat the alignment— make the two south side walls of the main wing align their south faces with the SOUTH SIDE MILLER HALL Reference Plane (see Figure 3–83). There are two unconnected walls that must be aligned separately.

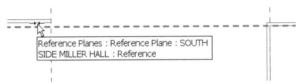

Figure 3–83 *Align the walls to the Reference Plane*

8.7 Zoom to Fit in the Floor Plan view. Open the West Elevation view. Zoom to Fit. You will simplify the view in two ways: you will turn off display of the walls to be demolished, and you will shorten the depth of the view so you will not see the wall at the far side of the building. Remember, all model views have depth.

8.8 Type **VP** to open the View Properties dialogue. Change the value of the view Phase Filter to Show Previous + New (see Figure 3–84). This will turn off the display of components marked demolished in the current phase (New Construction), while still showing components identified as Existing or New Construction.

Check the Far Clip Active box, and set the Far Clip Offset value to **140**. This value was determined by trial and error. This will stop the view short of the far (east) wall of the building wing and thus simplify the view for your work. Choose OK.

View Properties...	
Extents	
Crop Region	☐
Crop Region Visible	☑
Far Clip Active	☑
Far Clip Offset	140' 0"
Scope Box	None
Associated Datum	None
Phasing	
Phase Filter	Show Previous + New
Phase	New Construction

Figure 3–84 *Phase Filter and Far Clip for the Elevation View*

The walls will appear slightly different than before—there will be no dashed lines evident. You previously defined two west walls for the North–South extension—one from the first floor level to the second, and one from the second to the roof. The first floor wall was designated as demolished, so it has disappeared from this view.

8.9 Move the cursor over the right side of the model until the second/third story wall section highlights. Left-click to select it. The wall type will appear in the Type Selector on the Options Bar. Pick the Properties icon. In the Element

Properties dialogue, change the Base Offset value to $-1'4''$ to extend the wall below any floor structure at the second floor level. Click OK. The bottom of the wall will now appear below the 2ND FLOOR MILLER HALL Level line.

This wall, as depicted in the CAD elevation, consists of two different types: B1 and MP. You will now subdivide the single wall so you can apply alternating types to match the elevation.

CREATE REFERENCE PLANES FOR ALIGNMENT BETWEEN VIEWS

8.10 Zoom in Region to the right side of the CAD import. Create a vertical Reference Plane at the right side of the area identified as the MP type wall at column line 7. Make the Reference Plane extend below level 2ND FLOOR MILLER HALL in the model below, so it will appear in that Floor Plan view (see Figure 3–85).

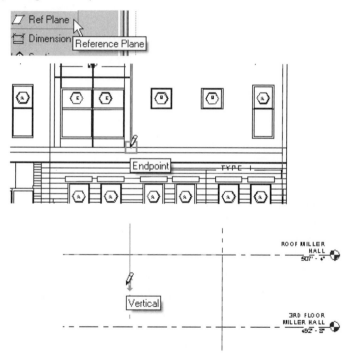

Figure 3–85 *Create a Reference Plane to mark the wall type*

8.11 Select Modify. Select the new Reference Plane. Choose Copy from the Toolbar. Select Multiple on the Options Bar. Pick the origin of the Reference Plane for the initial point, then pick the left side of the MP wall panel to place a second instance. Continue placing four more copies at the sides of the MP wall panels, for a total of six planes (see Figure 3–86).

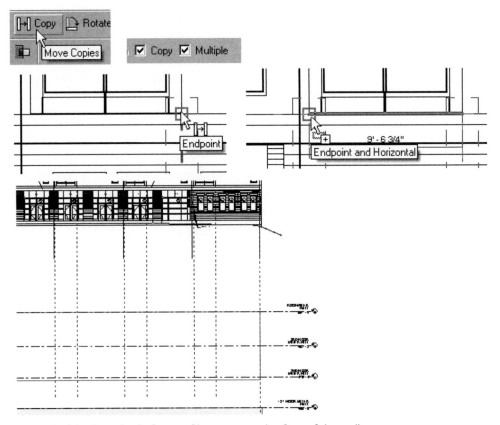

Figure 3–86 *Copy the Reference Planes across the face of the wall*

8.12 Open the 2ND FLOOR MILLER HALL Floor Plan View. Open the View Properties dialogue. Change the Phase Filter to Show Previous + New. Click OK to exit the dialogue and remove the demolished walls from the view.

8.13 Pick the Split Tool from the Toolbar. The cursor becomes a scalpel shape. Pick the west (left) wall of the extension at the intersection points of the new Reference Planes to split it into seven separate sections.

Revit Building will not snap to the Reference Planes. Create the split within about a foot of a Reference Plane, then click the temporary dimension that appears and set the dimension value to 0, as shown in Figure 3–87. Repeat for each Reference Plane, for a total of six splits in the wall.

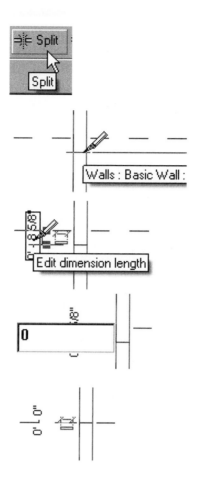

Figure 3–87 *Split the wall at each Reference Plane*

 TIP For repeated Zoom In and Zoom Out, you can use the wheel on a scroll mouse if your computer is so equipped. Rolling the wheel forward or back will Zoom In and Out, and holding the wheel down while moving the mouse will Pan back and forth.

8.14 Open the West Elevation. Zoom in Region to the left side of the CAD import. Create a horizontal Reference Plane. Use the Pick Option. Select the heavy line identified as the 2nd FLOOR TOP OF SLAB. Zoom to Fit and drag the right end of the Reference Plane beyond the CAD import. Select Modify. Select the Reference Plane. Open its properties and name the new Plane **IMPORT 2ND FLOOR T.O.S**. You will use this to find your way around the wall structure in the imported elevation (see Figure 3–88).

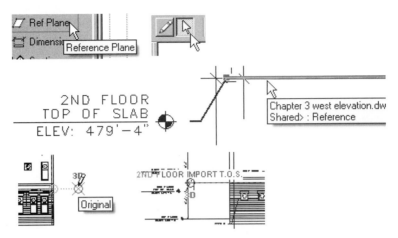

Figure 3–88 *Locate a horizontal reference plane*

IDENTIFY WALL MATERIALS

8.15 Zoom in Region to the right side of the import. Study the structure of the wall as identified by the text and leaders. Use Zoom and Pan as necessary to find identifying text.

Below the Reference Plane you just created that marks the top of the floor slab behind the wall, there is a surface marked Type 1 at the right side of the imported elevation, with metal panel/storefront window combinations between pillars dressed with brick and CMU. All of that area is to be demolished, but study its structure so that you can match it with new elements. Above the Reference Plane is a flat roof with a metal edge, and a metal panel flashing against the wall above. Above the panel, surface B1 runs to the Metal Roof Edge at the top of the building. In between the B1 sections there are areas (you identified their edges with Reference Planes) marked MP. Window assembly type E appears in the MP sections.

8.16 Study Figure 3–89 below. This is a materials specification note taken from the project that provided the elevation CAD files.

ALL AREAS DESIGNATED "MP" OR
"P" ARE METAL PANEL SYSTEM.

CMU:

TYPE 1:   4X8X16 SPLIT-FACED CMU WHITE

TYPE 2:   4X4X16 TEXTURED CMU WHITE PROJECTED 1" FROM FACE

TYPE 3:   4X8X16 TEXTURED CMU WHITE PROJECTED 1" FROM FACE

TYPE 4:   4X4X16 TEXTURED CMU BULLNOSE PROJECTED 1: FROM FACE

BRICK:

B1:   4X4X12 UTILITY BROWN 1/3 BOND

Figure 3–89 *Wall finish materials specifications*

8.17 Brick Type B1 is brown Brick. Concrete Masonry Unit (block) type 1 is 4" x 8" Split-faced Concrete Masonry Unit. MP is a Metal Panel wall system.

CHANGE WALL TYPES

8.18 Zoom to Fit in the West Elevation view. Move the cursor over the Revit Building model along the North–South wing. As the cursor moves, the sections of wall you created by using the Split tool on the original wall will highlight under the cursor. Pick the new wall segment on the far right of the model. It will highlight red, and the cursor will change shape to a double-headed Move arrow. The Type Selector will show Basic Wall: Generic – 8" and wall editing commands will appear along the Options Bar.

8.19 Use the Type Selector drop-down list to change the type of the highlighted Generic wall to Exterior – Brick on Mtl. Stud, as shown in Figure 3–90. Repeat on the other walls to define a total of four brick walls in the building wing.

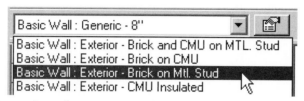

Figure 3–90 *Change the wall type on one wall*

TIP You can select more than one wall by holding down the **CTRL** key while picking wall sections as they highlight under the cursor (see Figure 3–91). Then all sections will change properties at the same time.

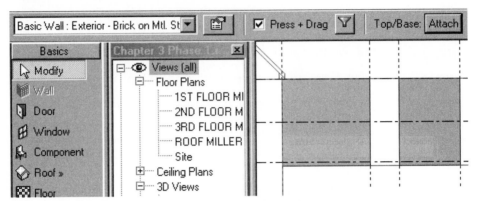

Figure 3–91 *Change the wall type on more than one wall*

8.20 Left click off the model to clear the walls you have selected. Use the same technique as the previous step to pick the three walls corresponding to the MP wall type in the imported file. When all three wall sections are highlighted red, pick the Properties icon from the Options Bar.

8.21 In the Element Properties dialogue, pick the Edit/New button. In the Type Properties dialogue that opens, change the Type to Exterior – EIFS on Mtl. Stud. Pick the Duplicate button to use that existing type as the basis for a new one you are about to name and define.

In the Name dialogue that opens, type **Exterior – MP on Mtl. Stud**, as shown in Figure 3–92. Click OK. The Type Properties dialogue will display the new name. Pick the Edit button in the Value field of the Structure Parameter.

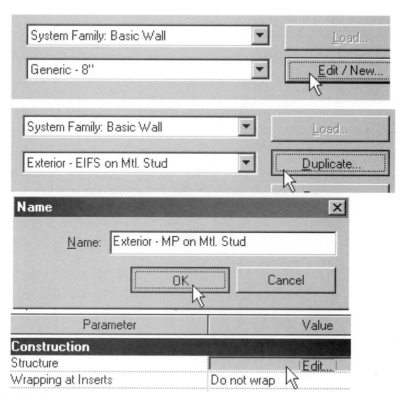

Figure 3–92 *Name and edit the new wall type*

8.22 The Edit Assembly dialogue will open. It will display the Layers of the wall structure from the Exterior side to the Interior. Study the wall structure. Pick Layer 1 to highlight it. Change the Material field to Finishes – Exterior – Metal Panel, as shown in Figure 3–93. Choose OK three times to exit the dialogues. The appearance of the walls will not change. Save the file.

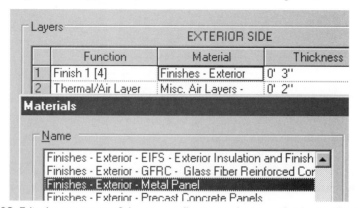

Figure 33–93 *Edit the structure of the new wall type to change a finish specification*

EXERCISE 9. ADD WINDOWS

9.1 Open or continue working in the file from the previous exercise. Make West Elevation the current view.

9.2 Zoom in to the upper-right area of the CAD file. Select the Tape Measure tool from the Toolbar (see Figure 3–94). It will read the distance and angle between any two points you then pick. Use it to determine the sizes of the Type A and Type B windows, and write down the results.

Use the Tape Measure to determine the sill height of the Type A and Type B windows—the distance between the bottom of the window unit and the IMPORT 2ND FLOOR T.O.S. Reference Plane you created in the previous exercise—and write down that information. Type A will be 2′ 8″ x 6′ 8″ with a 3′ 0″ sill. Type B will be 2′ 8″ x 2′ 8″ with a 6′ 8″ sill height.

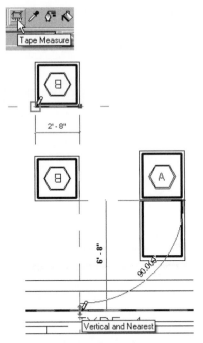

Figure 3–94 *The Tape Measure measures the windows*

9.3 Select Window from the Basics tab of the Design Bar. Since this file was started in a generic template, Revit Building will not have any special content loaded into it. The default window list will not include Type B in our imported file, fixed 32″ x 32″; pick the Properties button to begin creating a window type of the correct size, as shown in Figure 3–95.

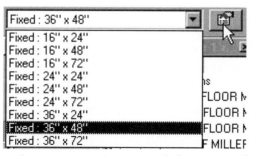

Figure 3–95 *Use Properties to start creating a new window type on the fly*

9.4 In the Element Properties dialogue, pick the Edit/New button, then the Duplicate button. In the Name dialogue, title the new fixed window size **32″ x 32″** and click OK. Make the Height value of the new window **32″**. Make the Default Sill Height **80″**. Make the Width **32″**. Revit Building will convert inch values to foot-inch (2′ 8″, 6′ 8″). Change the Type Mark value to **B** (see Figure 3–96). Click OK twice to exit the dialogue and get ready to locate a window. The cursor will change shape to show that you are inserting a component

Figure 3–96 *Properties for the Type B window*

9.5 Zoom or Pan to the lower-right corner of the model. Click to place an instance of the fixed window in the second floor section of the right-hand brick wall section. The exact location does not matter for the insertion. Place two instances of this window using the CAD file as a rough guide for locations.

Revit Building will attempt to locate the windows at the correct sill height for you, and will show temporary dimensions indicating the sill location and spacing from the center of the window to other building elements. If it's necessary to adjust the sill height of the first instance, pick the dimension tag and adjust it as shown in Figure 3–97.

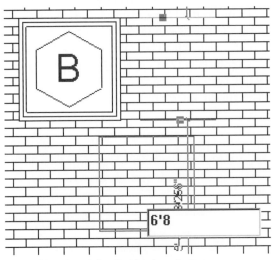

Figure 3–97 *Adjust the sill height while placing the window*

9.6 Window type A is double-hung, and not loaded into this project. You can find and use double-hung windows in the Revit Building library while still using the Window tool.

9.7 Pick Load from the Options Bar. Navigate to the *Imperial Library* folder on your system and double-click the *Windows* folder. Select the file *Double Hung.rfa* and pick Open.

9.8 Select Window from the Basics tab of the Design Bar. Revit Building will default to the Double Hung: 36″ x 48″ size window. Pick the Properties button, then the Edit/New button, and then the Duplicate button, as before, to create and name a new size. In the Name dialogue, type **32″ x 80″** and click OK. Change the Height value of this new size to **6′ 8″**, change the Default Sill Height to **3′**; and change the Width to **2′ 8″**. Make the Type Mark value **A** (see Figure 3–98). Click OK twice to exit the dialogue.

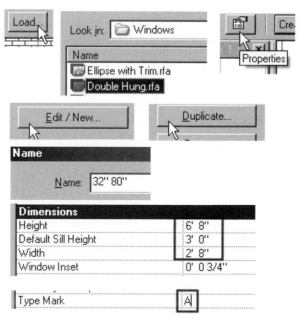

Figure 3–98 *Properties for the Type A window*

9.9 Place the new window as for the first window. Make sure the Sill Height is 3'. Snap planes will find the other windows and aid in vertical placement (see Figure 3–99).

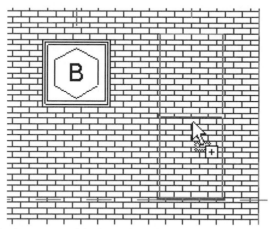

Figure 3–99 *Windows inserted but not aligned*

ALIGN AND COPY THE WINDOWS

9.10 Select Modify to terminate the window placement. Zoom to Fit. Select the Align tool. For the first pick, Zoom in and select the left side of a type A window at the right side of the CAD elevation. Pan down and select the left side of your new A window. Repeat for the two type B windows (see Figure 3–100).

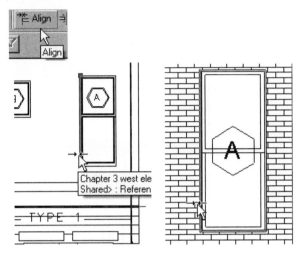

Figure 3–100 *Align windows to the CAD file*

9.11 Zoom to Fit. Pick Modify from the Basics tab of the Design Bar. Pick two points left to right around the new windows to select the three windows and their associated tags. Select Copy from the Toolbar. Check Constrain and Multiple on the Options bar. Constrain will restrict movement to horizontal or vertical planes, so you can not change the sill height by accident.

9.12 Select a point on the right hand Reference Plane you created between the two elevations. Select corresponding points to make 4 copies of the windows in the brick wall sections (see Figure 3–101).

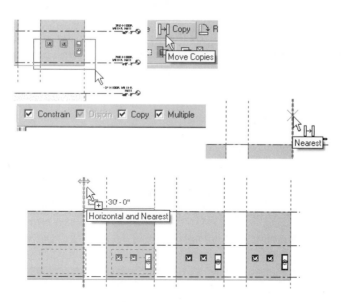

Figure 3–101 *Copy the window across the building face precisely*

9.13 There is one more type window to create in the left hand wall section to match the CAD elevation. Pick two points to the left and right of leftmost type B window. Select Copy again. For the reference picks use corners of two type B windows in the CAD file (see Figure 1-102).

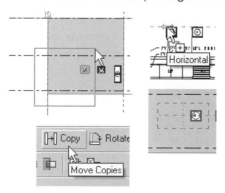

Figure 3–102 *Copy a single window using the CAD elevation for location*

9.14 Now that one level has the correct number of windows in the correct locations, you can copy them up to the next level. Select Modify. Select all the windows and window tags. Choose Copy. Choose Constrain on the Options Bar. Clear Multiple. Select a point on the 2ND FLOOR level to start the copy. Pick the level line for 2ND FLOOR MILLER HALL as the Copy start point. Pick the level line for 3RD FLOOR MILLER HALL to finish the copy. See Figure 3–103.

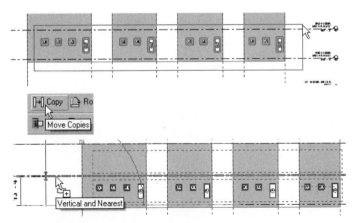

Figure 3–103 *Copy all the windows up one level*

ADD A WINDOW TYPE FROM THE REVIT BUILDING WEB SITE

Revit Building does not ship with combination window assemblies as in the Metal Panel wall sections of the CAD file. Autodesk maintain files of additional Revit Building component content on its Web site. You will need an Internet connection for this next step (or use an alternate file location specified by your instructor).

9.15 Pick Window from the Basics tab of the Design Bar. Pick Load from the Options Bar. Pick Web Library from the Open dialogue. Revit Building will open an Internet browser to the Revit Building Library index page. Pick the link for Revit Building 8.0 Library. Pick the link for Windows. Right-click the Awning-Combo 1 link and pick the Save Target As option (see Figure 3–104). Pick Save on the File Download question box that appears. If necessary, navigate to the *Windows* folder in the *Imperial Library* folder on your system. Close the browser.

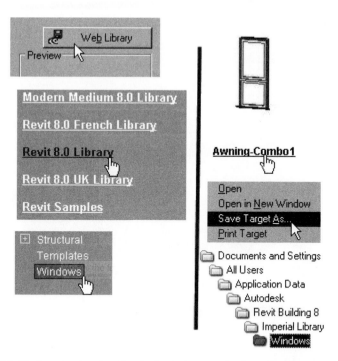

Figure 3–104 *Finding the Awning-Combo1 window on the Revit Building site*

9.16 Re-open the Window tool if necessary. Select Load. Open the Windows folder and select the new Awning-Combo1 window to load it into the project. Revit Building will list Awning-Combo1: 36″ x 48″ in the Type Selector on the Options Bar. Pick Properties. Pick Edit/New. Pick Duplicate. In the Name Box, type **48″ x 76″** and click OK.

9.17 In the Type Properties box, make the Height value **6′ 4″**. Make the Operable Unit **3′ 6″**. Make the Default Sill Height **3′**. Make the Width **4′**. Make the Type Mark **E**. Click OK (see Figure 3–105).

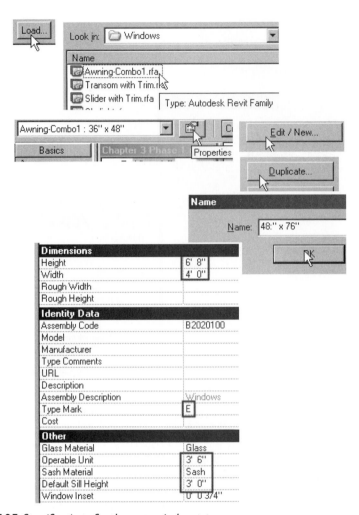

Figure 3–105 *Specifications for the new window type*

9.18 Place and align an instance of the new Window Type E in a Metal Panel wall as you did before. Select Modify to terminate the alignment tool

9.19 Select the new window and its tag. Select the Mirror command. Select the right edge of the window to create a duplicate (see Figure 3–106).

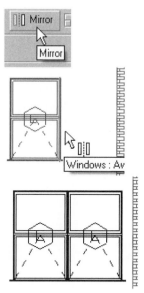

Figure 3–106 *Mirror the aligned window*

9.20 Zoom to Fit. Copy the mirrored windows up one level, then across the elevation, as before (see Figure 3–107), to create six sets total. If you copy them up first, the selection will be easier for copying across. Choose Modify to terminate the copy.

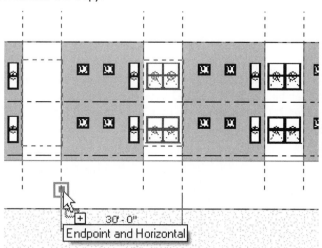

Figure 3–107 *Copy the doubled windows up, then across.*

9.21 Save the file.

EXERCISE 10. CONCEPT SKETCHES FOR THE NEW DESIGN

CREATE A NEW WALL TYPE FROM AN EXISTING ONE

10.1 Open or continue with the file from the previous exercise. Open the 1ST FLOOR MILLER HALL Floor Plan View. Pick Wall from the Basics tab of the Design Bar. Select Basic Wall: Exterior – Brick on CMU in the Type Selector. Pick Properties. In the Element Properties dialogue, pick Edit/New. In the Type Properties dialogue, pick Duplicate. In the Name box, type **Brick – CMU Column**, as shown in Figure 3–108. Click OK.

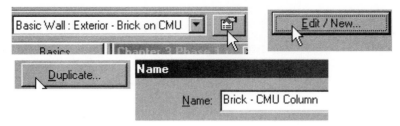

Figure 3–108 *Name the new wall type*

10.2 In the Type Properties dialogue, pick Edit to modify the Structure. In the Edit Assembly dialogue, pick <<Preview to open a view of the wall structure you are working with. In the Layers section, pick the number designator for Layer 9. The entire row will highlight black. Pick Delete. Layer 8 will then be highlighted. Delete that layer as well. The view of the wall structure will adjust.

10.3 Pick Layer 3 (insulation) so that it highlights. Click Delete. Layer 3 will now be a Membrane Layer. Click Delete. Place the cursor in the Thickness field for Layer 2. Change the value to 1″, as shown in Figure 3–109, and click OK to finish editing the wall assembly. Click OK to leave the Type Properties dialogue.

	Function	Material	Thickness
	EXTERIOR SIDE		
1	Finish 1 [4]	Masonry - Brick	0' 3 5/8"
2	Thermal/Air Layer	Misc. Air Layers -	0' 1"
3	Membrane Layer	Vapor / Moisture	0' 0"
4	**Core Boundary**	**Layers Above Wr**	0' 0"
5	Structure [1]	Masonry - Concret	0' 7 5/8"
6	**Core Boundary**	**Layers Below Wr**	0' 0"

Figure 3–109 *Working with the wall layers*

10.4 In the Element Properties dialogue, make the Top Constraint of the new wall type Up to level: 2ND FLOOR MILLER HALL. Set the Top Offset to −1′ **4″** so that the wall top will meet the upper floor wall, whose base offset you adjusted in a previous exercise. Make sure the Location Line value is Finish Face: Exterior (see Figure 3–110). Click OK.

Parameter	Value
Constraints	
Location Line	Finish Face: Exterior
Base Constraint	1ST FLOOR MILLER HALL
Base Offset	0' 0"
Base is Attached	☐
Base Extension Distance	0' 0"
Top Constraint	Up to level: 2ND FLOOR MILLER
Unconnected Height	12' 0"
Top Offset	-1' 4"

Figure 3–110 *Assign a top offset value*

10.5 Start the wall so its exterior face meets WEST SIDE WALKWAY Reference Plane. Pull the cursor up at 90° about 50'. Select the Trim command and Trim the new wall to the existing east–west wall. Remember that Revit Building can reward you for placing building elements quickly and editing them, rather than attempting to place them precisely with the cursor (see Figure 3–111).

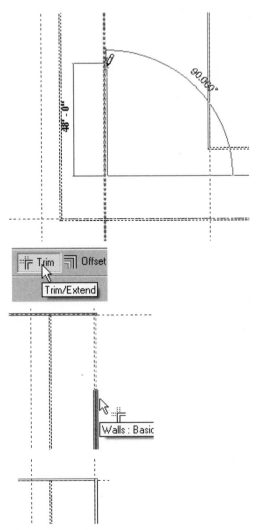

Figure 3–111 *Place and trim the wall*

10.6 Open the West Elevation View. Select the new wall. Use the control arrow that appeats at its right end to drag that end to the SOUTH SIDE WALKWAY Reference Plane as shown in Figure 3–112.

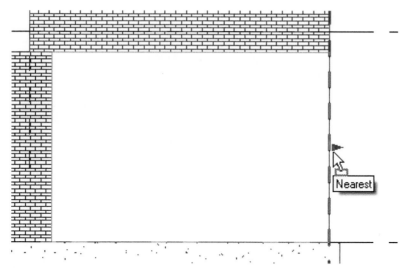

Figure 3–112 *Drag the new wall into alignment with the Reference Plane*

10.7 Select Edit Elevation Profile from the Options Bar. The surface pattern will disappear, and the edges of the wall will highlight magenta. The Design Bar will switch to Sketch mode.

10.8 Pick Lines. Pick the 3-point arc tool from the Options Bar, as shown in Figure 3–113. Create a 180° arc under the right-most brick wall section, using the intersection points of the Reference Planes and base line of the wall. The 3-point arc tool works by picking the two endpoints of the new arc, then dragging the cursor until the angle value is correct.

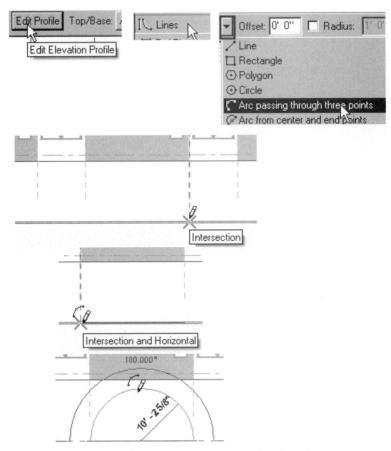

Figure 3–113 *Edit the wall profile using the 3-point arc sketch tool*

10.9 Select Modify. Pick the arc you just drew then select Copy from the Toolbar. Select Constrain and Multiple as before. Copy the arc to the left twice using the vertical Reference Planes you drew earlier, as shown in Figure 3–114.

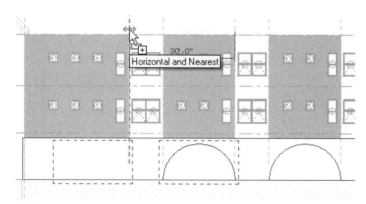

Figure 3–114 *Copy the new arc*

10.10   Select Modify to terminate the copy. Pick the Split tool from the Toolbar. Select Delete Inner Segment on the Options Bar. Split the base line of the wall profile at the intersection points with the arcs. Move from one end to the other, and the line segments under the arcs will disappear as you split them out (see Figure 3–115). Pick Finish Sketch. If a dialogue about constraints appears, click Remove Constraints.

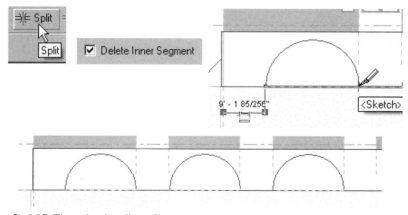

Figure 3–115 *The edited wall profile*

ADD A ROOF SKETCH

10.11   Open the ROOF MILLER HALL Floor Plan View. Select the gable roof section. Pick Copy. For the Copy start point, select the right endpoint of the ridge line. For the second point, select the midpoint of the wall at the north (upper) end of the walkway wing (see Figure 3–116).

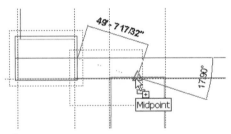

Figure 3–116 *Copy the roof*

10.12 The new roof will be highlighted. Pick Rotate from the Toolbar. A Rotation arrow symbol will appear in the center of the copied roof. Put the cursor directly over that symbol and it will highlight black and indicate that you can now move it. Move the Rotation point to the right end of the ridge line (see Figure 3–117). Pull the cursor straight down to start a rotation ray at 90°. Pick a point well away from the model so that snap points do not interfere with your cursor. Pull the cursor up to the right to create a rotation angle of 90°; when Revit Building snaps to the 90° angle, left-click.

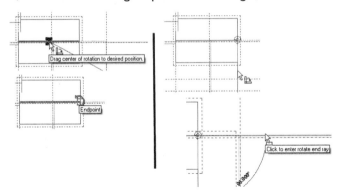

Figure 3–117 *Shift the Rotation point then rotate the sketch 90° counterclockwise*

10.13 The roof sketch will still be highlighted red. Pick Edit from the Toolbar. Pick the lower edge of the roof. Dimensions will appear. Click the value of the one defining the distance from the roof to the end wall. Type **0** in the edit field that appears, as shown in Figure 3–118. (This will extend the roof sketch down to the wall.) Click off the field to enter the value. Pick Finish Roof from the Sketch tab on the Design Bar.

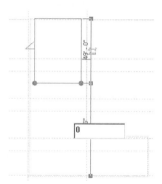

Figure 3–118 *Set the offset value for the roof sketch line to 0*

    10.14 Open the West Elevation view. Select a Reference Plane. Type **VH** to hide the category in the view. Zoom in around the right side of the model. Admire your work for a second, and then save the file.

EXERCISE 11. EXPORT YOUR WORK INTO AUTOCAD

For the purposes of our hypothetical project, as soon as you have the start on a re-design for the existing wing, you have to provide AutoCAD output to the site roadways engineer, who needs to approve the arches. File integration should be seamless in a well-run office. Revit Building makes export straightforward, with control over layers and line weights based on customizable standards.

    11.1 Open or continue working in the file from the previous exercise. Make West Elevation the current view. You are going to export this view to a DWG format file, as if you were sending it to a consultant for review.

    11.2 Pan or Zoom so the right side of the CAD file is in the left half of the View window and the right half is blank. Type **VG** to open the Visibility/Graphics dialogue. Place the cursor over the title bar of the dialogue. Left-click and hold the button down. You can now move the dialogue. Put the dialogue window on the right side of the screen so you can see both the import linework and the dialogue box Apply button, or part of it.

           The bigger your screen, the easier this will be. Few of Revit Building's dialogue boxes resize, but you can work with them as with any multi-window dialogue sequence—the active window is available to move.

    11.3 On the DWG/DXF/DGN Categories tab of the dialogue, check the + symbol next to *Chapter 3 west elevation.dwg* to expand the category. A list of the layers in the file appears. All the layers are checked to make them visible in Revit Building. Clear the A-ANNO- and A-GRID layers. Pick Apply. Lines disappear from the CAD import.

11.4 Clear each of the A-PATT layers in turn, and pick Apply after each one to see and note down the results (see Figure 3–119).

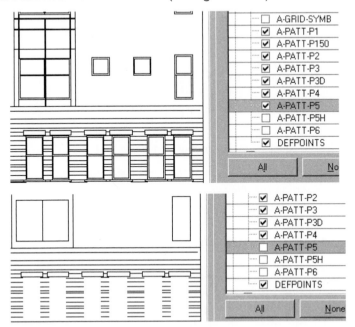

Figure 3–119 *Stepping through the CAD file layers*

The layer organization in this CAD file is not perfectly consistent, as with the lines in the site plan file you imported in the exercises at the beginning of this chapter. These files were developed from actual project files that had gone through numerous hands in different firms, and present the often casual departures from standards that anyone familiar with CAD files has seen. The layer breakdown is as follows:

A-PATT-P1	building edges
A-PATT-P150	CMU field pattern
A-PATT-P2	grade
A-PATT-P3	roof edges, wall edges, some window openings
A-PATT-P3D	foundations below grade
A-PATT-P4	sills and wall sweeps, some window openings
A-PATT-P5	window and door frames and mullions
A-PATT-P5H	door swings in elevation
A-PATT-P6	brick field pattern

11.5 Once you have examined the layers in the CAD file, Click OK to close the Visibility dialogue. Pick the CAD file. Pick Delete to remove the file from the View so it will not be part of the export.

11.6 Zoom to Fit. From the File menu, select File>Import/Export Settings>Export Layers DWG/DXF (see Figure 3–120).

Figure 3–120 *Opening the Export Settings dialogue*

11.7 Study the organization of this dialogue for a minute. The title bar holds the location of an ASCII txt file formatted to specify layer name and color ID. Revit Building comes with a dozen or more mapping files (read only by default) in a specific location. These files hold the values displayed in the dialogue panels. The left side panel lists Revit Building object Categories.

The middle panel covers Projection display (Elevations, or items not shown as cut in Plans and Sections); default layer names and color ID values come from the txt file in use.

The right panel shows layers and color IDs for Cut display. Revit Building does not make fields available where a certain condition will not apply. Area Polylines (the first Category listed) and other annotations will never have a Cut condition, for example, so the user cannot specify an inapplicable layer name.

11.8 Change the values for Revit Building object Categories in the Export Layers dialogue according to the chart that follows. See Figure 3–121 for a view of the dialogue. You will edit values for Doors, Roofs, Walls and Windows only.

Category	Projection	Layer Name Color ID
Doors	A-PATT-P5	5
Door Elevation Swing	A-PATT-P5H	5
Door Frame/Mullion	A-PATT-P5	5
Door Opening	A-PATT-P5	5
Roofs	A-PATT-P5	5
Roof Fascias	A-PATT-P3	3
Roof Gutters	A-PATT-P3	3
Walls	A-PATT-P1	1
Walls Surface Pattern	A-PATT-P6	6
Walls Wall Sweeps	A-PATT-P4	4
Walls/Exterior	A-PATTP-1	1

Windows	A-PATT-P5	5
Window Frame/Mullion	A-PATT-P5	5
Window Opening	A-PATT-P3	3
Window Sill/Head	A-PATT-P4	4
Window Trim	A-PATT-P5	5

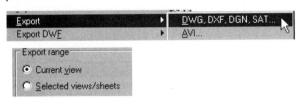

Figure 3–121 *Edits in the Export Settings dialogue*

11.9 Notice that as you apply color values to layer names, Revit Building will supply the color for you. Once a layer name has a color specified, layer names in {} parentheses will pick up layer specifications from the parent category (Window Trim from Windows, for example). When you have made the changes outlined above, pick Save As.

Save the file as ***export-layers-dwg-CHAPTER 3.txt*** in the *Program Files/ Autodesk Revit Building 6.0/Data/* folder, or as specified by your instructor.

11.10 From the File menu, pick File>Export>DWG/DXF/DGN. In the Export Dialogue, navigate to a folder where you will save an AutoCAD file. If you do not have AutoCAD 2000/2004 available but do have an earlier version, find your version in the Save as type field list. Make sure Current view is selected in the Export range section, as shown in Figure 3–122. Pick Save to create the export file.

Figure 3–122 *Export the elevation view as a DWG file*

11.11 Save the Revit Building file.

SUMMARY

Congratulations. You have created a new Revit Building file, imported files into it, and referred to the CAD content of these files while you created a site plan and building model using Revit Building elements and components. You have begun the process of detailing the shell of this sizable structure, and exported an exterior elevation view of one face of this building from Revit Building into AutoCAD. In the process you have worked with many of the basic design and editing tools within Revit Building, and some techniques that are a little more advanced than the basic level. In the next chapter you will use the model created in this chapter, plus an imported site plan sketch, to set up a multi-building, multi-file campus development project.

REVIEW QUESTIONS – CHAPTER 3

MULTIPLE CHOICE

1. To control visibility of layers in an imported CAD file

    a) use link instead of import

    b) invert the layer colors

    c) use the DWG/DWF/DGN tab in the Visibility Graphics dialogue

    d) you can't control layers in an imported file

2. Phases in Revit Building projects control

    a) the display of objects in views

    b) how many objects of a certain type the file can contain

    c) the voltage the computer uses while working in Revit Building

    d) the date display in titleblocks

3. Reference Planes

    a) can be named

    b) can be turned on and off in views

    c) are excellent alignment tools

    d) all of the above

4. Sketch mode on the Design Bar is active

    a) when creating or editing a roof

    b) when editing a wall elevation profile

c) whenever no other command is active

d) a and b, but not c

5. Revit Building's default Wall and Window families

a) can't be edited

b) are made to be edited

c) can be edited if you pay a special fee

d) fight all the time, like most families

TRUE/FALSE

6. The View Properties dialogue controls the appearance of nearly everything in Revit Building

7. Wall and Window types can not be edited during placement

8. The Align tool will snap to walls, but not to windows

9. Plan views can be duplicated in the Project Browser, but Elevations have to be copied in Plan views to create new Elevation views

10. An object's Phase Created and Phase Demolished properties work with View Phases and Phase Filters to control display of the object.

 Answers will be found on the CD.

CHAPTER 4

Design Modeling with Mass Objects

INTRODUCTION

This section continues a series of exercises using basic Revit Building techniques to create related, interconnected project files. The exercises in Chapter 3 showed how to use 2D imported CAD file information as the basis for a Revit Building model. You created the model using building elements (walls, roofs) and components (doors, windows). For your hypothetical multi-building project, the starting point is an existing building which will be modified. In this chapter's exercises you will start modeling three structures for Phases 1, 2, and 3 of this campus development.

To get started, you will import a file holding a site topography/building model similar to the one you created in Chapter 3 and a sketch of a proposed site plan into a project file, and then correlate these two sources. You will trace building footprints on top of the raster image and copy/paste those footprints into three additional empty project files. Each of those project files will become a building model, and you will work in each one using massing components to start a building. You will link those building model files back into the first file you created, which will hold the overall view of this extensive development.

OBJECTIVES

- Create mass models using a variety of techniques in three separate files
- Link Revit Building files together; share locations and coordinates
- Create perspective views of mass models for client presentation

REVIT BUILDING COMMANDS AND SKILLS

File save settings

File link

Shared levels

Coordinates—acquire; relocate project

File import attributes

Model lines—line, arc, fillet arc

Edit commands—move, copy, trim, rotate, copy to clipboard, paste

Add/Cut Mass—Extrusion, Sweep, Blend

View Camera

EXERCISE 1. IMPORT PROJECT DATA—ACQUIRE COORDINATES

1.1 Launch Revit Building. It will open to an empty project file. Select File>Save As from the File menu. On the Save As dialogue, pick the Option button. In the File Save Option dialogue, make the maximum number of backup(s) value **1**, as shown in Figure 4–1. Choose OK.

NOTE This value is set in each Revit Building file. It is **not** global. Do this step for each file you create in this and other exercises.

If you are working in a class situation, set the value as directed by your instructor. If you are working on your own on a computer that may have limited storage space, you will want to make this a habit when first saving project files.

1.2 Save the file as **Chapter 4 Project Overview.rvt**, in a location specified by your instructor if you're in a class, or in a folder on a drive with adequate space and read-write permission.

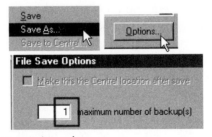

Figure 4–1 *Set the backup quanitty value*

1.3 Pick File>Import/Link>RVT from the File menu, as shown in Figure 4–2. Navigate to the folder containing *Chapter 4 Miller Hall.rvt*; select that file name so it appears in the File name field. Accept the default option to Automatically place Center-to-center (see Figure 4–3). Click Open. The file contents will not be visible in the Floor Plan: Level 1 view. You will make an adjustment to fix that in the next steps.

Figure 4–2 *Link a Revit Building file*

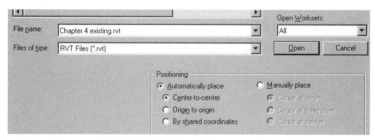

Figure 4–3 *Center-to-center positioning for the link*

1.4 Open View Elevations: East. Type **VG** to open the Visibility/Graphics Overrides dialogue for this view. Select the Linked RVT Categories tab. Click the + symbol next to the name of the linked file to expand the list of content in that file. Clear the Topography check box, as shown in Figure 4–4. Click OK. The Toposurface under the building will disappear.

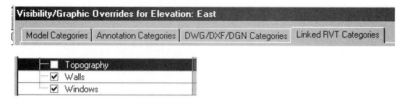

Figure 4–4 *Turn off the topography for clarity*

1.5 From the Tools menu, pick Tools>Locations and Coordinates>Acquire Coordinates, as shown in Figure 4–5. The Status Bar will read Select a linked project from which to acquire shared coordinate system. Select the elevation of the linked file. The View window will not alter its appearance.

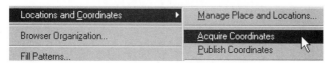

Figure 4–5 *Acquire coordinates from the link*

Revit Building does not use an "exposed" coordinate system, meaning there is no way to specify an absolute position for entities in a file independent of other entities. Revit Building does track and use the center point of each file, which changes as the building model footprint changes. Revit Building also provide a coordinate location point (unseen by the user) with each project file, so that linked files can synchronize positions. Generally the project file that contains site information becomes the basis for shared coordinates, and building models will be located according to the topography.

Since most projects are located some distance away from sea level and the absolute level will generally need to appear in each model file, the next step will be to tie the Level Elevations in this file to the shared coordinates. To do this, you will create a

Shared Level type, then Move the project up or down the required distance to pick up the base level (above or below sea level) from the site file. This sounds more complicated than it is.

    1.6 Zoom in Region to the Level lines at the bottom of the screen. These are the default Levels 1 and 2 for the current file. Pick the Level line (not the bubble) for Level 2. Pick the Elevation value text so the field opens for editing. Change the Elevation of Level 2 to **13′4**, as shown in Figure 4–6. (Revit Building will supply the inch mark.) Hit ENTER. The Level will adjust.

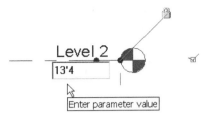

Figure 4–6 *Adjust the Elevation of Level 2*

    1.7 Pick the Level line for Level 1. Select the Properties icon from the Options Bar. Choose Edit/New in the Element Properties dialogue. Pick Duplicate in the Type Properties dialogue.

    1.8 In the Name box, type **Level – Shared**. Click OK. In the Type Properties dialogue, change the Base value to Shared, as shown in Figure 4–7. Click OK.

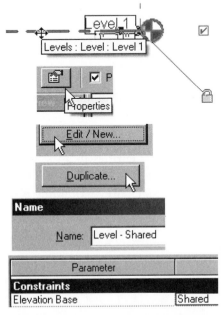

Figure 4–7 *Add a Level type named Shared, and make its base value Shared*

1.9   From the Tools menu, pick Tools>Locations>Relocate this Project, as shown in Figure 4–8. Read the information box that appears. Click OK. The cursor changes to the Move icon. Left-click anywhere off the Levels for the first Move point. Pull the cursor up at 90°. Type **466**. Hit ENTER. Type **ZF** at the keyboard, as a shortcut for Zoom to Fit. If necessary, drag the end of a Level line to align with the others.

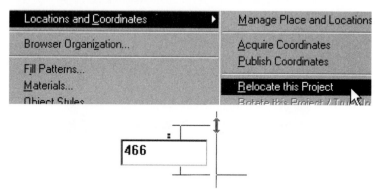

Figure 4–8 *Relocate the project to line up with the link*

1.10   Open View Floor Plans: Level 1. The topography from the linked file now appears. Zoom to Fit to see the linked model. Save the file.

EXERCISE 2. SKETCH BUILDING FOOTPRINTS FROM AN IMPORTED IMAGE

2.1   Open or continue working with the *Chapter 4 Project Overview.rvt* file from the previous exercise. Type **VV** to open the Visibility/Graphic Overrides dialogue. This is an alternate keyboard shortcut. On the Annotation Categories tab, clear Elevations. On the Linked RVT Categories tab, clear Topography. Click OK. Only the building outline will now be visible in the View window.

2.2   From the File menu, select File>Import/Link>Image, as shown in Figure 4–9. Navigate to the folder that contains the file *proposed site plan.jpg*. Select the file name. Click Open. Four blue control dots indicating the boundary of the image will appear around the cursor. Left-click anywhere in the View window. The image will appear.

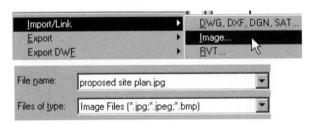

Figure 4–9 *Add a raster image to the file*

2.3 Pick the Properties icon on the Options Bar. In the Element Properties dialogue, clear Maintain aspect ratio. Change the Width value to **13′ 10″**. Change the Height value to **10′ 11″** (see Figure 4–10). Click OK.

Parameter	
Dimensions	
Width	13′ 10″
Height	10′ 11″
Maintain aspect ratio	☐

Figure 4–10 *Adjust the size of the image to fit the model*

 NOTE These values have been determined by trial and error. You will have to adjust the size of other images you may import or link into project files experimentally.

2.4 The image will change size. It will still be selected—the cursor is the double-headed double arrow, indicating that you can drag the selected item to a new position. Pick Move from the Toolbar.

2.5 Zoom in Region in the image file where the hand lettering reads **Existing Building 3 Story**. Left-click at one of the corner points of the drawn building outline, then left-click at the corresponding corner of the linked building model to move the image over the model (see Figures 4–11 and 4–12).

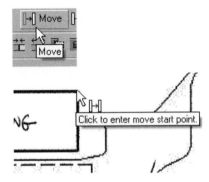

Figure 4–11 *Start moving the image*

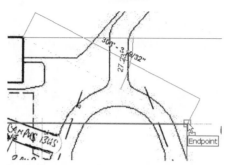

Figure 4–12 *The second pick for moving the image*

2.6 Zoom in Region to the right side of the building model. The walls of the model will show as white over the black raster image pixels. The image will still be selected. Move the image as necessary to get a careful alignment with the walls (see Figure 4–13). When you are satisfied, Zoom to Fit and study the site plan sketch.

Figure 4–13 *Careful alignment is possible*

2.7 Zoom in Region around the Phase 1 building outline. You will now start to sketch over this very rough image. Pick Lines from the Basics tab of the Design Bar. Select the Chain option from the Options Bar.

2.8 Sketch three lines of the lengths and angles shown in Figure 4–14, starting from the left end of the wall of the linked model as shown. Do not add the dimensions; they are for guidance only.

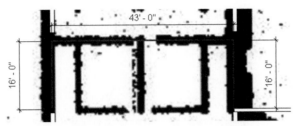

Figure 4–14 *The first sketch lines for the new building outline*

2.9 From the File menu, pick File>Manage Links. On the RVT tab, select the name of the *Chapter 4 Miller Hall.rvt* file so that it highlights (See Figure 4–15.). Select Unload. Click OK. The building model will disappear.

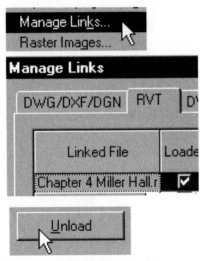

Figure 4–15 *Manage Links to unload the building model*

2.10 Continue sketching five lines around the footprint of the Phase 1 building as shown in Figure 4–16. Do not add the dimensions; they are for guidance only.

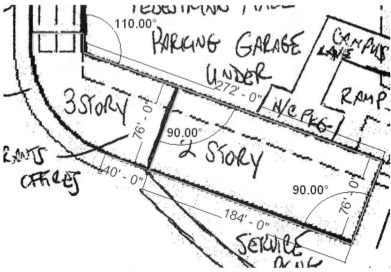

Figure 4–16 *Draw the simple building outline*

2.11   To draw the arc at the lower left of the sketch outline, pick the Fillet Arc
icon on the Options Bar, as shown in Figure 4–17. Revit Building will prompt
you to pick two lines near the endpoints where you want the fillet to be
applied. Pick the two lines shown in Figure 4–18 at the approximate locations
shown. Revit Building will start an arc and show a temporary radius
dimension.

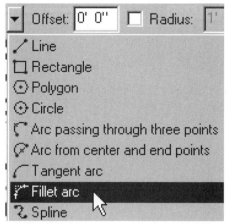

Figure 4–17 *The Fillet arc tool*

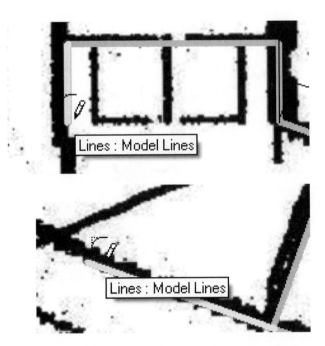

Figure 4–18 *The picks for the fillet arc; the lines have been edited for appearance*

2.12 Carefully move the cursor so the dimension reads as close to 100′ as possible, then left-click. Select the radius dimension field and edit it to **100**, then left-click off the field so it will update to 100′ - 0″ (see Figure 4–19).

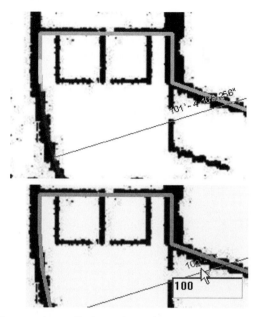

Figure 4–19 *Place the arc, then edit the radius value*

2.13 With the Lines command still active, change the Type Selector to <Overhead>, as shown in Figure 4–20. Draw the two lines shown in Figure 4–21 over the dashed lines representing a building overhang. Do not draw the dimension.

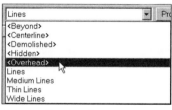

Figure 4–20 *Change the line type to Overhead*

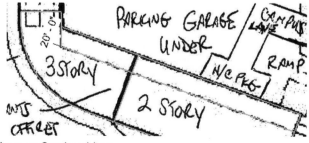

Figure 4–21 *The new Overhead lines*

2.14 Continue adding sketch lines as shown in Figure 4–22. The raster image has been removed in this figure for clarity. Do not draw the dimensions. Use a combination of Lines and <Overhead> lines as shown. Use the 3-Point Arc tool for the top and bottom lines of the outline.

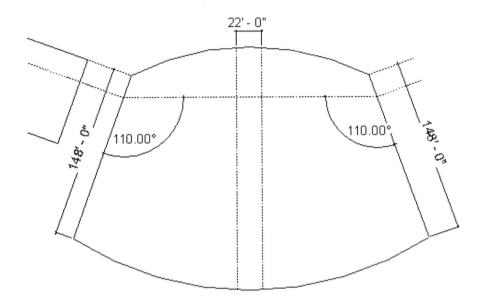

Figure 4–22 *The central building outline—the raster image has been turned off for clarity*

 TIP Try to take advantage of editing commands rather than drawing all parts of a symmetrical design. Be careful with your picks when you are zoomed in close; the raster image that provides the background will be the default pick. Learn to read the cursor (see Figure 4–23) so that you do not inadvertently move the image when trying to do something else.

Figure 4–23 *The cursor tooltip and appearance let you know when you have made a selection*

2.15 Zoom to Fit. Pick Modify from the Basics tab of the Design Bar to terminate the Lines command. Left-click over the raster image to select it. Type **VH** at the keyboard to turn the raster image category off in the view.

2.16 Zoom in Region around the sketch lines. Select the lines as shown in Figure 4–24 Select Mirror from the Toolbar. Choose the Draw option. Select the midpoint of the horizontal line in the central building outline for the first point of the mirror axis, then pull the cursor straight down (or up) to define a vertical mirror axis at the middle of the outline. Left click to mirror the selection.

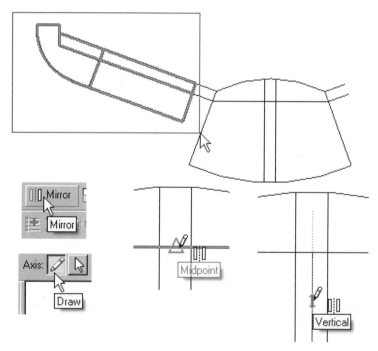

Figure 4–24 *Select the lines to mirror*

2.17 Trim the top and curved side of the right-hand building outline sketch to eliminate the gap. Trim other lines as necessary to make the sketch lines look like Figure 4–25.

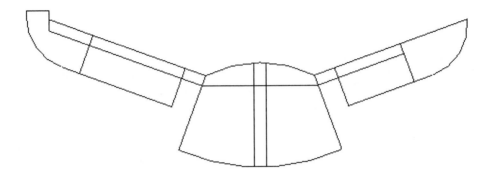

Figure 4–25 *The linework for three building footprint outlines is complete*

2.18 Zoom to Fit. Enter **VG** at the keyboard to open the Visibility Graphics dialogue. On the Annotations tab, check Raster Images and choose OK. See Figure 4–26.

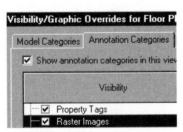

Figure 4–26 *Make the image visible to check your sketch against it*

2.19 Check your sketch for accuracy and make any necessary adjustments. Select the raster image and delete it. Zoom to Fit. Save the file.

Now that you have used the existing model file and a sketch of the proposed site design to capture information accurately, you are ready to set up files for the individual designs that will constitute this campus development. First you will move the sketched lines into separate files for each building model, and then link those model files back into the Overview file. The building models will be developed and revised extensively during the course of this (or any) project. The Overview file provides a central place for designers or project managers to check progress and design correlation.

EXERCISE 3. TRANSFER SKETCH LINES INTO NEW PROJECT FILES.

3.1 Open a new empty project file. In the new file, open any Elevation view. Set Level 2 to **13' – 4"**. Create a Shared Level type and Relocate the new project to **466'**, as you did in steps 6–9 of Exercise 1.

3.2 From the Window menu, pick *Chapter 4 Project Overview.rvt – Floor Plan: Level 1* from the list of available views in the bottom section of the menu to return to that view (see Figure 4–27). Your Project list may look different from the figure.

Figure 4–27 *Return the view to the Overview file*

3.3 With no command active, pick two points around the building outline on the left side of the View window, as shown in Figure 4–28.

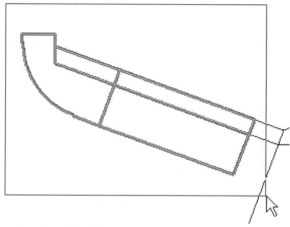

Figure 4–28 *Select the Phase 1 building outline linework*

3.4 Pick Edit>Copy to Clipboard from the Edit menu, as shown in Figure 4–29.

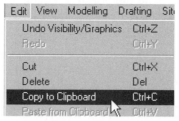

Figure 4–29 *Copy the building outline lines to the Windows Clipboard*

3.5 From the Window menu, pick *Floor Plan: Level 1* for your new project to return to that view. From the Edit menu, pick Edit>Paste Aligned>Current View, as shown in Figure 4–30.

Figure 4–30 *Edit Paste from the Clipboard*

3.6 Zoom to Fit. The linework will appear highlighted to one side of the View window. Select the pasted lines. Choose Rotate from the Toolbar. Pick a point to the right of the Rotate icon that appears in the center of the pasted content, pull the cursor up until the temporary angle dimension reads 20°, and then left-click to rotate the highlighted elements counterclockwise together (see Figure 4–31).

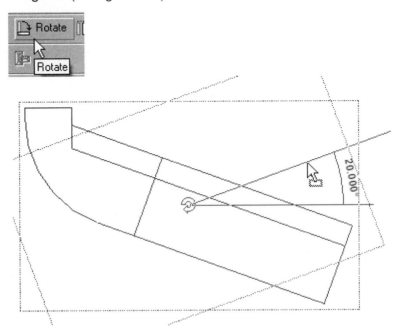

Figure 4–31 *Rotate the new lines to horizontal orientation*

3.7 Pick two points around the North and South elevations to select them, as in Figure 4–32. Select Move from the Toolbar. Pick Move start and end points to relocate the Elevations above and below the outline.

Figure 4–32 *Select two Elevations to move them*

3.8 Repeat the Move command for the East and West Elevations. Zoom to Fit. Select the elevation arrows in turn and adjust the extents line as necessary (see Figure 4–33).

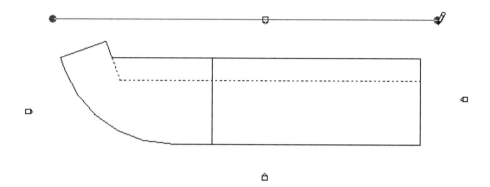

Figure 4–33 *The new project with the building outline ready to work on*

3.9 From the File menu, select File>Save As. Save the new file as *Chapter 4 Phase 1.rvt* in a location specified by your instructor. Use the Options button in the Save As dialogue to set the backup count for this new file to 1, or as specified by your instructor.

3.10 From the File menu, select File>Close.

3.11 Open a new empty project file. In the new file, open any Elevation View. Set Level 2 to **13' – 4"**. Create a Shared Level type and Relocate the new project as you did in Exercise Steps 1.6–1.9. Make the offset distance **474**, as shown in Figure 4–34.

The original topography for the referenced building model shows that there are different elevations where the concept plan sketch locates the new buildings. These steps set the new files at different absolute elevations. In the next exercise you will correlate them at correct vertical and horizontal placement in the overview file.

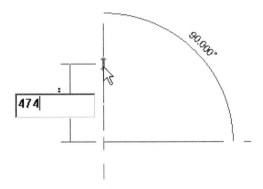

Figure 4–34 *Relocate this project to 474' above sea level*

3.12 Return to the Floor Plan: Level 1 view of the Project Overview file, as you did before. Select the right-hand building outline, as shown in Figure 4–35. Copy the selection to the Clipboard.

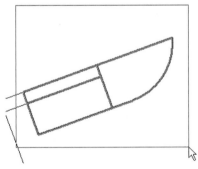

Figure 4–35 *The selection for Phase 2*

3.13 Return to the Floor Plan: Level 1 view of the new file. Paste in the contents of the Clipboard as you did before. Rotate the contents 20° counterclockwise this time. Move and adjust Elevations as necessary. Zoom to Fit (see Figure 4–36).

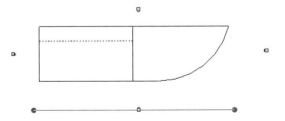

Figure 4–36 *The building outline placed horizontally and elevations adjusted*

3.14 Save the new file as *Chapter 4 Phase 2.rvt*, using the Save As, Option tool to set the backup count as before. Close the file.

3.15 Open a new empty project file. In the new file, open any Elevation View. Set Level 2 to **12′** (see Figure 4–37). Create a Shared Level type, and Relocate the new project as you did in Exercise Steps 1.6–1.9. Make the offset **470′**, as shown in Figure 4–38.

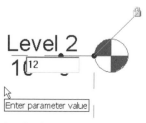

Figure 4–37 *Set Level 2 to 12′ for this building model*

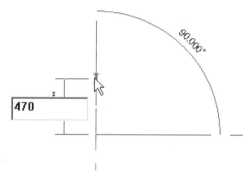

Figure 4–38 *Relocate this project to 470′*

3.16 Return to the Floor Plan: Level 1 view of the Project Overview file, as you did before. Select the middle building outline, as shown in Figure 4–39. Copy the selection to the Clipboard.

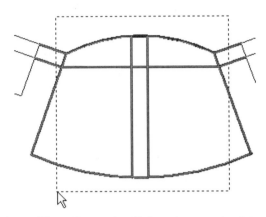

Figure 4–39 *Copy the middle outline to the Clipboard—note the right-to-left selection window*

3.17 Return to the Floor Plan: Level I view of the new file. Paste in the contents of the Clipboard using the Edit>Paste Aligned>Current View option, as before. Move and adjust Elevations as necessary. Zoom to Fit (see Figure 4–40).

Revit Building allows the user to copy and paste from file to file, as you have just done, and also allows copy/paste operations within a project, with alignment as an option. A particular arrangement of walls or furniture components can easily be copied from floor to floor of a multi-level project in exact alignment. Revit Building's capacity to align objects with each other or with Reference Planes is sophisticated and extremely useful for the busy designer.

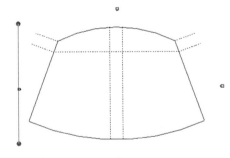

Figure 4–40 *The new file with Elevations in place*

3.18 Save the new file as *Chapter 4 Phase 3.rvt*, using the Save As Option to set the backup count to I. Close the file.

EXERCISE 4. LINK THE NEW PROJECT PHASE FILES TO THE OVERVIEW FILE

4.1   Open or continue working with the *Chapter 4 Project Overview* file. Zoom to Fit in the Floor Plan: Level 1 view.

You have located the three building outlines in plan, and now you will prepare the overview file so that they all appear at the correct elevations.

4.2   Open Elevations: East view. Use Zoom in Region around the Elevation bubbles at the right side of the View window. Pick the text label identifying Level 1. Pick the same text again to open the label for editing. Rename the Level **1ST FLOOR EXISTING**. Click Yes to rename the corresponding Views.

4.3   Repeat for Level 2, naming it **2ND FLOOR EXISTING** (see Figure 4–41). Zoom To Fit.

2ND FLOOR
EXISTING
13' - 4"

1ST FLOOR
EXISTING
466' - 0"

Figure 4–41 *Levels in the current file renamed*

4.4   Pick Level from the Basics tab of the Design Bar. Create a level **4' 0"** above 1ST FLOOR EXISTING. Click to the left of the screen, then pull the cursor to the right. Revit Building will snap to an alignment above 1ST FLOOR EXISTING.

4.5   Click the name field and rename the first new Level (named Level 3 by default) at the 4" offset 1ST FLOOR PHASE 3. Click Yes to change View names. The level tool will still be active

4.6   Create a Level **4' – 0"** above the new one (8' – 0" above 1ST FLOOR EXISTING). The name field will default to 1ST FLOOR PHASE 4. Rename the second new Level 1ST FLOOR PHASE 2. Click Yes to change View names. Select Modify.

4.7 Select the Level line for 1ST FLOOR PHASE 2. Use the Type Selector on the Options Bar to change the Level type to Level – Shared. The Elevation will read 470' – 0". Repeat for 1ST FLOOR PHASE 2 (see Figure 4–42).

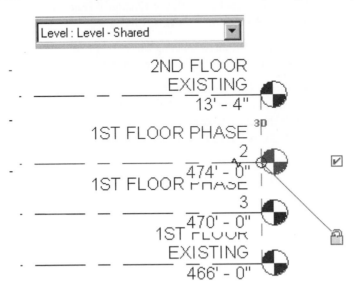

Figure 4–42 *New Levels shared*

4.8 Open the Site view. Turn off visibility of the Elevations to simplify the screen.

4.9 From the File menu, select File>Import/Link>RVT, as shown in Figure 4–43. In the Look in: section of the Add Link dialogue, navigate to the folder that contains the file *Chapter 4 Phase 1.rvt* that you created earlier. Select that file name so that it appears in the File Name field in the Add Link dialogue box. In the Positioning section of the dialogue, select Manually Place>Cursor at origin. Click Open.

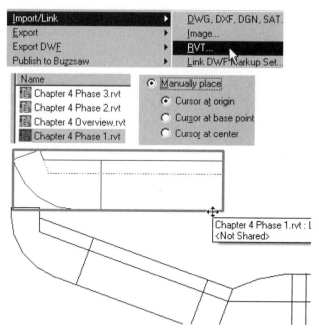

Figure 4–43 *Link a Revit Building file into the current file*

4.10 Linework from the linked file will appear in the View window. The cursor
tooltip and Status Bar will read Left-Click to place Import Instance. Place the
linked instance to the left of the View window near the linework you pasted
into the link file.

4.11 Repeat the Add Link process for files *Chapter 4 Phase 2.rvt* and *Chapter 4
Phase 3.rvt*. Place the Phase 2 file to the right of the View window. In the Add
Link dialogue for the Phase 3 file, select Automatically Place/Center-to-center
in the Positioning section of the dialogue. Zoom to Fit.

4.12 Select the Phase 3 link Instance. It will highlight red and a boundary will
appear around it. The cursor will become the double-headed edit arrow.
Move the linked file using appropriate pick points so that it sits directly on
top of the linework in the current file. See Figure 4–44.

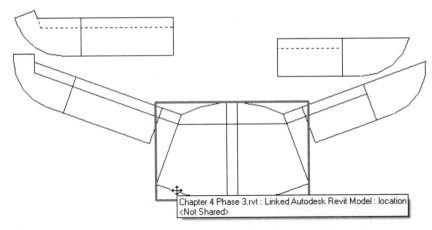

Figure 4–44 *Move the center link so it sits on top of its originating linework*

4.13 Select the Phase I link INSTANCE. Move it so that the upper-left corner of the link sits directly on the corresponding corner of the linework in the current file (see Figure 4–45). Choose Rotate from the Toolbar. The Rotation icon will appear in the center of the link Instance.

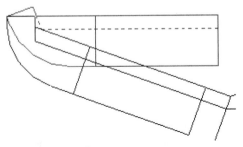

Figure 4–45 *Move the link so corners align*

4.14 Place the cursor over the icon so that it turns black. Left-click and hold the button down. Drag the Rotation icon to the corner you used for your alignment point. Pull the cursor vertically and pick to start a rotation angle vector. Pull the cursor down to the right (clockwise), so that the angular listening dimension reads 20° (see Figure 4–46). Left-click to enter the angle. You can also enter **20** at the keyboard to define the angle numerically.

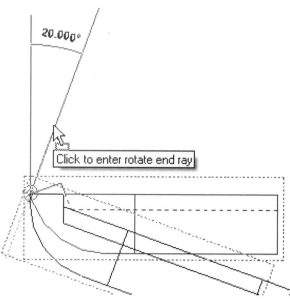

Figure 4–46 *Rotate the link after you relocate the Rotation center*

4.15 Select the Phase 2 import Instance. Repeat Move and Rotate (20°
counterclockwise) so that the import sits directly on top of the linework.
Zoom to Fit.

4.16 Choose two points around the edges of the View window to select
everything visible. Use the Filter button on the Toolbar to clear the imports
as shown in Figure 4–47. Click OK. Delete the lines from the project. The
screen will not change in appearance, but as you move your cursor over the
visible entities, it will only read the import instances.

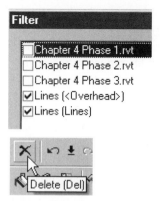

Figure 4–47 *Filter out links from the selection before deleting lines*

4.17 Type **VV** to activate the Visibility/Graphic Overrides dialogue. On the
Annotations tab, check Elevations, as shown in Figure 4–48. Click OK. Move

the now visible Elevations out around the imported file instances as you have done previously.

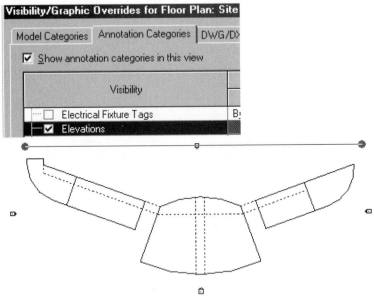

Figure 4–48 *Select Elevations in the Visibility control dialogue*

4.18 Open the View Elevations: South. Zoom in Region to the Level Elevation bubbles in the middle of the screen. As you move the cursor over Level lines, Revit Building will let you know that some belong in the current file and some are in linked files. Level 1 for the Phase 1 file is at 466′ elevation. It does not need to be moved.

4.19 Pick the Align tool from the Toolbar. Pick 1ST FLOOR EXISTING (470′ - 0′ elevation) as the reference to align to. Carefully select Level 1 (470′ – 0″) in the *Chapter 4 Phase 3.rvt* Linked Revit Building Model with the second pick, so that the Phase 3 file moves to the correct Elevation relative to the current file. (See Figure 4–49.)

Figure 4–49 *Align the link Level to the host file Level*

4.20 The Align tool will still be active. Pick 1ST FLOOR PHASE 2 (474′ – 0″ elevation) as the Reference. Pick Level 1 (474′ – 0″) in the *Chapter 4 Phase 2.rvt* linked model to move that file to the correct Elevation.

4.21   Open the View Floor Plans: 1ST FLOOR EXISTING. The Phase 2 file is no longer visible. That instance now sits above the default cut plane for this Floor Plan View.

4.22   Open the 1ST FLOOR PHASE 2 and 1ST FLOOR PHASE 3 Floor Plan Views in turn to note the differences in default visibility.

4.23   Open the View Floor Plans: 1ST FLOOR EXISTING. Type **VP** to open the View Properties dialogue. Pick the Edit button in the View Range Value field. Use the drop-down list in the Top field to select 2ND FLOOR EXISTING. Make the Offset from that Level **0′ − 0″**, and make the Cut Plane offset **9′ − 0″**. Click Apply. Phase 2 will now appear on the right side of the screen. Click OK. Click OK again to exit the View Properties dialogue. See Figure 4–50.

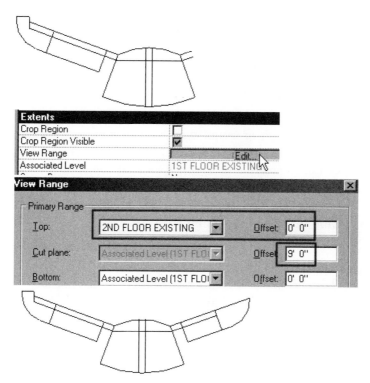

Figure 4–50 *Change the Plan View Properties to see links at different elevations*

4.24   From the File menu, select File>Manage Links. On the RVT tab in the Manage Links dialogue, choose *Chapter 4 Miller Hall.rvt* so that it highlights. Select Reload. Click OK. Zoom to Fit. The building outline will appear at the upper left. (See Figure 4–51.)

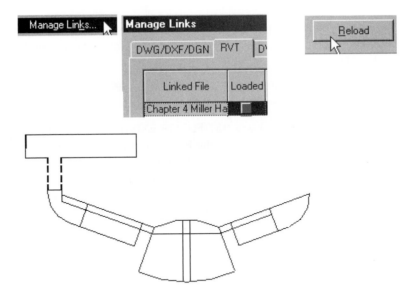

Figure 4–51 *The Overview file with all links loaded*

4.25 Save the file. Close the file.

This file is now set up. As the designs of the other project building models develop and evolve, a project manager or senior designer can monitor overall progress. In the next exercise you will start mass models in the three associated project files, then return to this file to create a perspective view of the combined facades for a hypothetical client presentation.

EXERCISE 5. CREATE A BUILDING MODEL USING MASSING— EXTRUSIONS

5.1 Launch Revit Building if it is not running. Open file *Chapter 4 Phase 1.rvt*. Open View Elevations: East. Select Level from the Basics tab of the Design Bar. Add a Level at **13′ – 4″** above Level 2. Add a level at **14′ – 4″** above the new Level 3.

5.2 Rename Level 1 1ST FLOOR PHASE 1. Rename Level 2 **2ND FLOOR PHASE 1**. Rename Level 3 **3RD FLOOR PHASE 1**. Rename Level 4 **ROOF PHASE 1**. Click Yes to rename the corresponding Views in each case (see Figure 4–52).

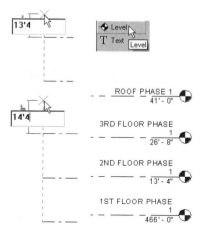

Figure 4–52 *New and renamed Levels*

5.3 Open the Site View. If the Design Bar does not display a Massing tab head in the stack at its bottom, put the cursor over the Design Bar and right-click to bring up the list of available tabs. Check Massing, as shown in Figure 4–53. That tab will open up in the Design Bar.

Figure 4–53 *Toggle on the Massing tab*

5.4 Pick Create Mass from the Massing tab of the Design Bar, as shown in Figure 4–54. A notice about visibility settings will display. Check the box next to Don't show this message again and choose OK. In the Name dialogue, Enter **Phase I Shell** and click OK. The Design Bar will switch to Mass mode. The linework will display as halftone gray.

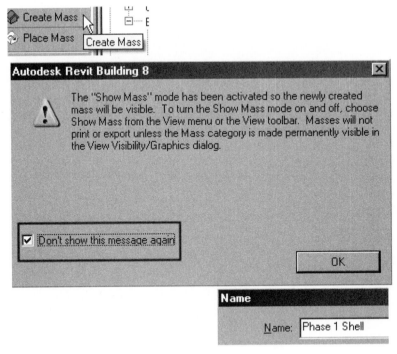

Figure 4–54 *Create and name a mass*

5.5 Choose Solid Form>Solid Extrusion from the Mass bar. It will switch to Sketch mode, with the Lines tool active. The cursor will change to a pencil outline. The Options Bar will show linework controls and a Depth control default of 20′ – 0″.

5.6 Set the Depth value for this extrusion to **28′**. Select the Pick icon (arrowhead) on the Options Bar. Line Options will disappear from the Options Bar once you make that selection. The cursor shows the selection arrow icon. Select the lines that make up right half of the footprint of the building. Each will highlight magenta as it is selected, so that you can start to modify the profile sketch you are creating (see Figure 4–55). The top line will extend the length of the outline. Choose Trim and trim lines as necessary. When you have a complete outline selected and highlighted, click Finish Sketch on the Sketch tab of the Design Bar.

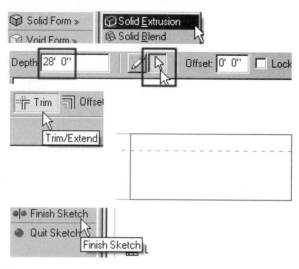

Figure 4–55 *Select the outline for the Extrusion sketch—trim to make a simple rectangle*

5.7 Repeat the Solid Form>Solid Extrusion command. Set the Depth value for this extrusion to **41'**. Select Pick on the Options Bar. Start by picking the line in the middle of the outline, then the lines and arc to its left (see Figure 4–56).

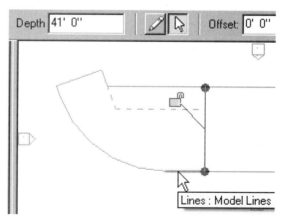

Figure 4–56 *The first picks for the second extrusion*

5.8 Use the Trim command to trim back the upper and lower outline lines as necessary to make a complete outline with no overlapping lines (see Figure 4–57). Select Finish Sketch from the Sketch tab on the Design Bar.

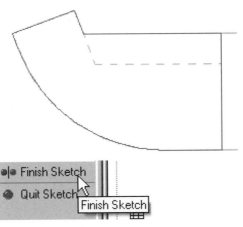

Figure 4–57 *The complete, trimmed outline for the second extrusion*

5.9 The lines will disappear under the masses. Enter **WF** at the keyboard to shift to Wireframe display mode.

5.10 Pick Void Form>Void Extrusion from the Mass tab of the Design Bar. Set the Depth value of the Cut to **12′**. Select the Pick icon on the Options Bar.

5.11 Pick the two Overhead lines inside the building outline as shown in Figure 4–58. Pick the top and right edge lines of the outline. Trim as necessary to make a complete outline. Select Finish Sketch from the Sketch tab on the Design Bar. Select Finish Mass from the Mass tab.

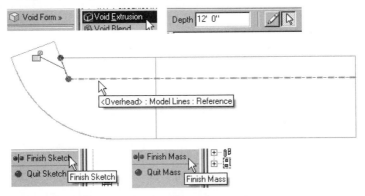

Figure 4–58 *Create a Void Extrusion. It will cut the previous masses to create a building overhang.*

5.12 Type **HL** to return the current View to Hidden Line mode.

5.13 Open Elevation view East. Select a Level line and drag the bubbles to the right of the building outline, as shown in Figure 4–59. Drag the left ends of the levels close to the outline as well. Zoom to Fit.

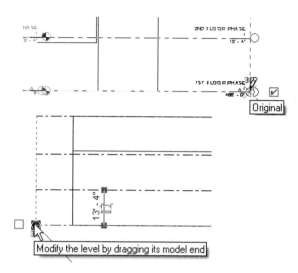

Figure 4–59 *Drag Level lines close around the building shell*

Extrusions can be created in other directions and other views than simply vertically in plan. You will now create an extrusion in this elevation.

5.14   Select Create Mass from the Design Bar. Name the new mass **Atrium Roof** and pick OK.

5.15   Select Solid Form>Solid Extrusion. Select Pick a Plane in the dialogue, and then choose OK. Place the cursor over the left side of the building outline and press the TAB key until the outline highlights as shown in Figure 4–60 and then left-click

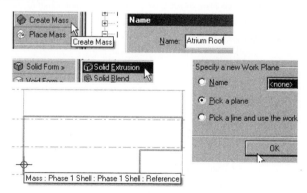

Figure 4–60 *Start an Extrusion. You need to specify a Work Plane in Elevation views.*

5.16   Draw a line starting in the middle of the line that is just above the 3RD FLOOR level (the roof of the two-story section). Make the line 12′ long vertically. Draw two lines to the left and right at 30° below vertical. You do not need to carry them to the roof line.

5.17 Draw a line along the roof line. Use Trim to make a complete triangle outline. Delete the first sketch line. Select Finish Sketch. See Figure 4–61.

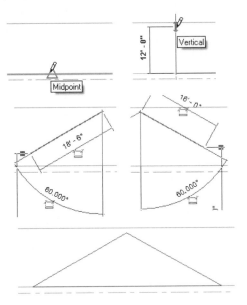

Figure 4–61 *The triangular sketch for the roof extrusion*

5.18 Open the Site View. The Extrusion has been created, but it extends away from the building rather than over the lower half. Select the Extrusion and its control arrows activate. Select the right arrow and drag it to the left until it snaps to the left side face. The depth indicator will read −184′ − 0″. See Figure 4–62. Select Finish Mass.

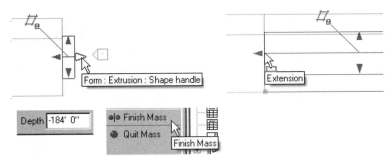

Figure 4–62 *Adjust the Extrusion proportions using the Shape Handle*

5.19 Select the Default 3D View icon on the Toolbar, as shown in Figure 4–63. Revit Building will open an SE isometric view of the model and add a View named {3D} under 3D Views in the Project Browser. The void will be visible under the other extrusions.

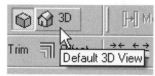

Figure 4–63 *The Default 3D View icon*

5.20 Left-click anywhere off the model in the View window to make sure the view is active. Select View>Orient>Northeast from the View menu, as shown in Figure 4–64.

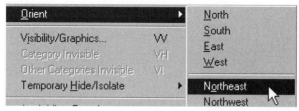

Figure 4–64 *Orient the 3D View to the Northeast*

5.21 Select View>Orient>Save View from the View menu, as shown in Figure 4–65. Enter **Northeast** in the Name box and choose OK.

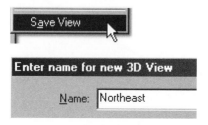

Figure 4–65 *Save the view orientatio under a name*

5.22 Pick Shading with Edges from the View Control Bar. Revit Building will display the mass, as shown in Figure 4–66. Select the View Mass toggle. The mass will disappear. Select the lines and delete them. Toggle the View Mass control back on.

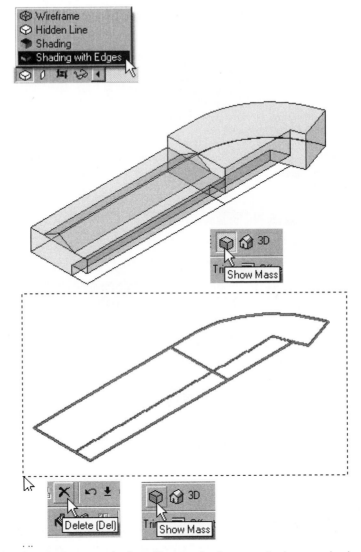

Figure 4–66 *Toggle the mass display off, erase the lines, toggle the mass back on*

TRANSFORM THE MASS INTO BUILDING ELEMENTS

5.23 The Massing Tab contains tools to create building elements—walls, floors, roofs and curtain systems—directly from mass faces. See Figure 4–67.

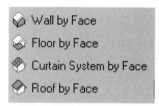

Figure 4–67 *Tools for creating building elements from masses*

5.24 Select the mass object as shown in Figure 4–68. Do not select the atrium roof. Select Floor Area Faces from the Options Bar. In the dialogue that opens, select 1ST FLOOR and 2ND FLOOR. Pick OK. The mass now contains faces at the selected levels.

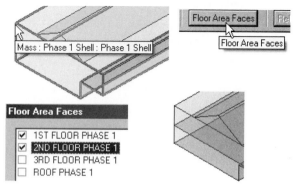

Figure 4–68 *Creating Floor Area Faces*

5.25 Select Floor by Face on the Massing tab. Select the two floor faces. Select Create Floors from the Options Bar. See Figure 4–69.

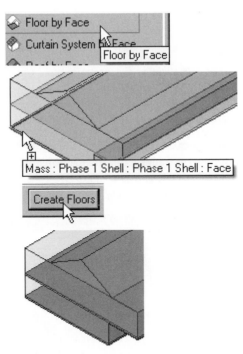

Figure 4–69 *Create two floors using the Floor by Face tool*

5.26 Select Roof by Face. Select the two flat roof faces. Choose Create Roof from the Options Bar.

5.27 Select Curtain System by Face. Select the slanted faces and the end face of the triangular extrusion. Select Create System on the Options Bar. See Figure 4–70.

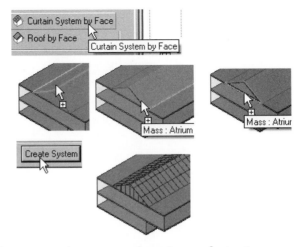

Figure 4–70 *Create a curtain system on the atrium roof extrusion*

5.28 Select Wall by Face. Set the Type Selector to Exterior: Brick on CMU. Pick vertical faces as shown in Figure 4–71 to create walls.

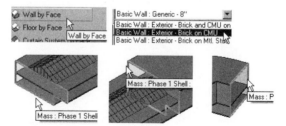

Figure 4–71 *Create Brick on CMU Walls by Face*

5.29 Set the Type Selector to Curtain Wall: Storefront. Pick Faces as shown in Figure 4–72 to create glass curtain walls with predefined panel sizes. You will define your own curtain wall type in a later exercise.

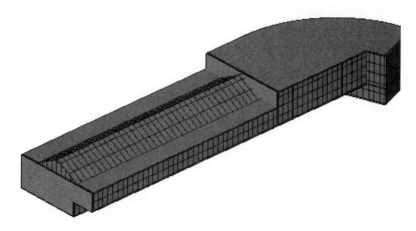

Figure 4–72 *Curtain Walls by Face*

5.30 Open the default 3D view. Create Walls by Face using Curtain Wall: Storefront along the south and west sides of the building. See Figure 4–73.

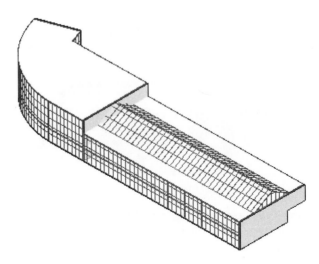

Figure 4–73 *Curtain walls on the curved side*

    5.31 Return to View Northeast. Toggle the View Mass control off. Zoom to Fit. Save the file. Close the file.

EXERCISE 6. CREATE AN EXTRUDED MODEL WITH SWEEPS AND REVOLVES

    6.1 Open the file *Chapter 4 Phase 2.rvt*. Make the Basics tab on the Design Bar active. Open the West Elevation View. Add 5 Levels (3 through 7) at a standard offset distance of **13′ – 4″** from the Level below. Add a Level **14′ – 4″** above Level 7. Rename Levels 1 through 7 to **1ST FLOOR PHASE 2** through **7TH FLOOR PHASE 2**. Rename level 8 to **ROOF PHASE 2**. Click Yes each time to rename the corresponding Views (see Figure 4–74).

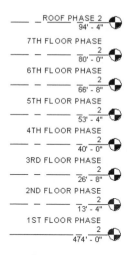

Figure 4–74 *Levels in the Phase 2 file*

6.2 Open the Site view. Select the Massing tab head on the Design Bar to activate that tab. Select Create Mass from the Massing tab of the Design Bar. Name the Mass **Phase 2 Block**. Click OK.

6.3 Choose Solid Form>Solid Extrusion. Set the Depth Value for the first extrusion to **28′**. Select the Pick icon on the Options Bar. Pick the outer lines and center line of the footprint. Trim to the rectangular shape shown in Figure 4–75. Select Finish Sketch.

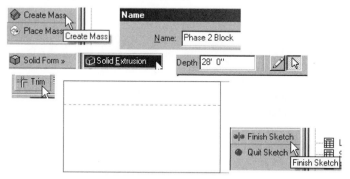

Figure 4–75 *The outline for the first extrusion*

6.4 Type **WF** to set the View mode to Wireframe. Select Void Form>Void Extrusion on the Massing tab of the Design Bar. Set the Depth to **12′**. Select Pick on the Options Bar. Pick the overhead line, top and side lines of the rectangle. Trim as shown in Figure 4–76. Select Finish Sketch. Select Finish Mass. Type **HL** to return to Hidden Line display.

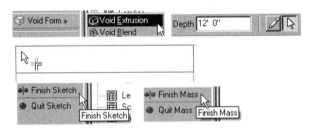

Figure 4–76 *Define a void and fnish the mass*

6.5 Select Create Mass. Name the Mass Phase 2 Tower. Select Solid Form> Solid Extrusion. Set the Depth to **94′–4″** (the height of the Roof level). Select Pick on the Options Bar. Choose the vertical line and then the lines and fillet arc to the right of that line to complete the outline. Trim lines as necessary (see Figure 4–77). Select Finish Sketch.

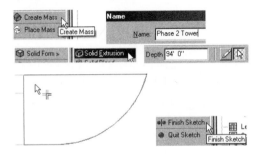

Figure 4–77 *The outline for the second extrusion*

6.6 Open the default 3D view. The tower mass has been defined. You will now create setbacks at each floor level to create a slimmer profile for the tower. Select Void Form>Void Sweep on the Massing tab of the Design Bar. The Design Bar changes to Sketch mode.

6.7 A Sweep consists of a Profile moved along a Path. Select Pick Path. The Design Bar changes to Pick Path mode, with Pick selected. Choose the line segment as shown in Figure 4–78; a profile locator will appear. Choose the arc to the right of your first line selection to define the rest of the path. Choose Finish Path.

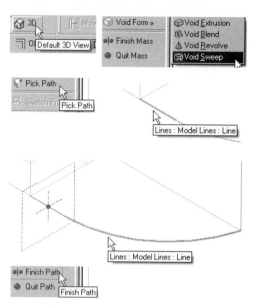

Figure 4–78 *Void sweep—pick the path segments carefully*

6.8 Select Sketch Profile on the Sketch tab, as shown in Figure 4–79. Open the West Elevation View. Choose Lines from the Sketch tab. Select the Chain option. Start a profile sketch at the right side of the building outline at the top of the Block mass, just above Level 3. Make the first line 12' long to the left as shown. Create a profile outline, as shown in Figure 4–80. Do not draw the dimensions; they are for your guidance only.

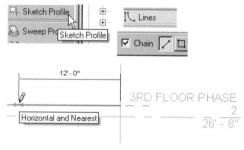

Figure 4–79 *Start a sketch profile on the 3RD FLOOR level*

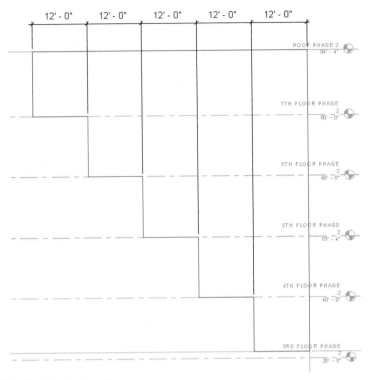

| 12' - 0" | 12' - 0" | 12' - 0" | 12' - 0" | 12' - 0" |

ROOF PHASE 2
94' - 4"

7TH FLOOR PHASE 2
80' - 0"

6TH FLOOR PHASE 2
66' - 8"

5TH FLOOR PHASE 2
53' - 4"

4TH FLOOR PHASE 2
40' - 0"

3RD FLOOR PHASE 2
26' - 8"

Figure 4–80 *The finished profile*

You will see by a study of the figure that the profile consists of a series of 12′ setbacks moving from right to left, starting at the height of the lower extrusion and using the project Levels above that. Close the profile at the roof level and along the right side.

6.9 When the Sweep Profile is a complete loop, select Finish Profile. Choose Finish Sweep. Open the default 3D view to check the results as shown in Figure 4–81. Choose Finish Mass.

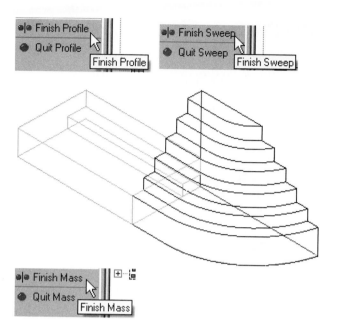

Figure 4–81 *Finish the profile, finish the sweep, finish the mass*

6.10   Select View>Orient>Northwest from the Menu Bar. Select
        View>Orient>Save View. Name the view Northwest and choose OK (see
        Figure 4–82).

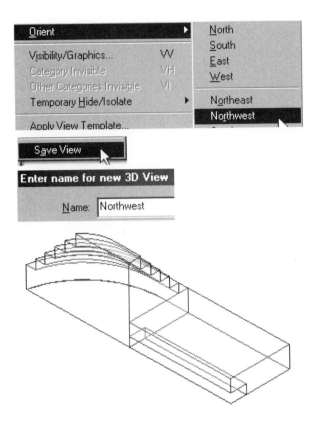

Figure 4–82 *Northwest view of the stepped extrusion*

6.11 Toggle the View Mass control off. Select and delete the lines. Toggle the View Mass control on.

6.12 Select Create Mass from the Massing tab on the Design Bar. Name the mass **External Elevator**. Click OK. Select Solid Form>Solid Revolve. See Figure 4–83. The Design Bar shifts to Sketch mode.

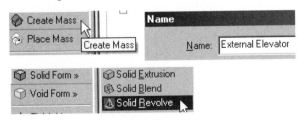

Figure 4–83 *Create a mass using a solid revolve*

6.13 A Revolve consists of a profile spun around an axis. Since you are working in a 3D view, you will first define the work plane, then locate the axis, and then draw a profile as before.

6.14 Select Set Work Plane. Select Pick a Plane in the dialogue and choose OK. Put the cursor over the edge of the tower profile and TAB until the outline of the tower highlights, as shown in Figure 4–84. Left-click to select the profile and define the plane of the working face.

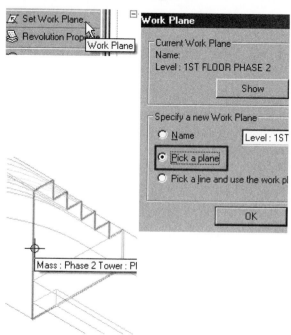

Figure 4–84 *Set the work plane*

6.15 Choose Axis from the Sketch tab. Draw a vertical line from the roofline of the Block mass to the Roof of the Tower mass, as shown in Figure 4–85. You will adjust it precisely in the next step.

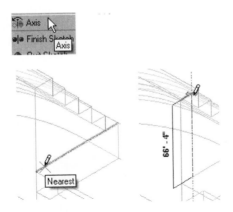

Figure 4–85 *Draw a vertical axis on the work plane*

6.16 Open view West Elevation. Select Dimension from the Sketch Bar. Place the dimension between the axis line and the left side of the tower. Select the axis. Change the dimension value to **5′** as shown in Figure 4–86. Check to make sure that the axis line terminates at the upper edges of the two masses.

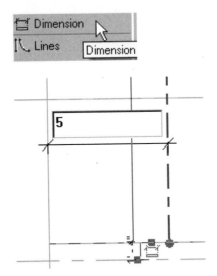

Figure 4–86 *Locate the axis with a dimension*

6.17 Choose Lines from the Sketch tab to start drawing the profile. Draw a line starting at the midpoint of the axis. Make the line 3′6″ long horizontal to the left.

6.18 Select the 3-point arc tool from the Lines options on the Options Bar. Start the arc at the left side of the tower by the lower end of the axis. The second point will be the upper left end of the tower, and the middle of the arc will be the endpoint of the horizontal sketch line, as shown in Figure 4–87.

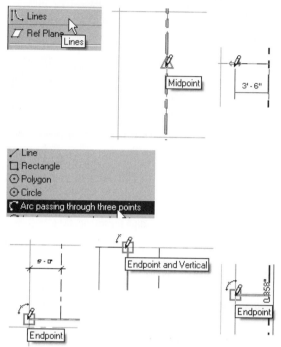

Figure 4–87 *Picks for a 3-point arc*

6.19 Select the straight line tool, select the Chain option, and finish the profile as shown in Figure 4–88. The right side of the profile sketch is on top of the axis. Delete the horizontal line. Select Revolution Properties on the Sketch tab. Change the End Angle value to −180; this will make the revolution travel in a half circle around the axis towards you. This value was determined by trial and error. Select Finish Sketch.

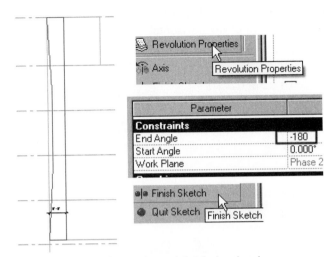

Figure 4–88 *Finish the profile, set the angle, and finish the sketch*

6.20 Open 3D view Northwest to check your revolved mass. Select Finish Mass.

6.21 You will now create floors, walls, roofs and curtain systems as you did in the previous exercise. Select the tower. Create Floor Area Faces at floor levels 1 through 7. Select the block. Create Floor Area Faces at floor levels 1 and 2.

6.22 Select Floor by Face. Select all the floor faces and create floors. Set the view display to Shaded with Edges, as shown in Figure 4–89.

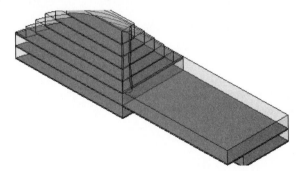

Figure 4–89 *New floors*

6.23 Select Roof by Face. Select the roof of the block. Select Create Roof. Carefully select roof faces on the tower, as shown in Figure 4–90. If you have difficulty selecting horizontal faces in this view, open the default 3D (Southeast) view to continue the command. Place roofs on all the exterior horizontal segments of the tower.

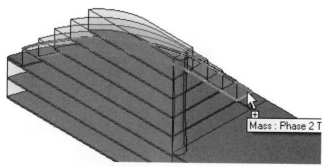

Figure 4–90 *Create roofs on the block and tower*

6.24 Select Wall by Face. Place instances of wall type Exterior: Brick on CMU and Curtain Wall: Storefront as shown in Figure 4–91. Use the default 3D view to place curtain walls on the setbacks.

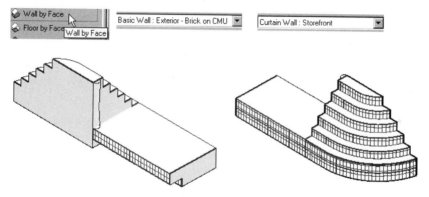

Figure 4–91 *Place walls by face. Shading has been removed for clarity*

6.25 Open the Northwest view. Select Curtain System by Face. The default curtain grid spacing will not fit on the elevator tower profile. Select the Properties icon. Select Edit/New. Select Duplicate. Name the new Curtain System Type **2' x 2'**. Choose OK. Set its grid values to **2'** each way. Choose OK twice to finish the settings. Select the face and top of the elevator mass. Select Create System. See Figure 4–92.

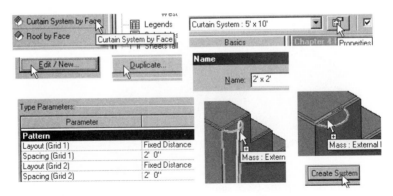

Figure 4–92 *Define a curtain system and apply it to the elevator mass*

    6.26 Zoom to Fit in the view. Toggle the View Mass control off.

    6.27 Save the file. Close the file.

EXERCISE 7. CREATE A MASS MODEL USING BLENDS

    7.1 Open the file *Chapter 4 Phase 3.rvt*. Open View Elevations: East. Choose Level from the Massing tab. Add a Level at **12′ – 0″** above level 2. Add a level at **16′ – 0″** above the new Level 3.

    7.2 Rename the Levels **1ST FLOOR PHASE 3**, **2ND FLOOR PHASE 3**, **3RD FLOOR PHASE 3**, and **ROOF PHASE 3**. Click Yes to accept View name changes (see Figure 4–93).

ROOF PHASE 3
40'- 0"

3RD FLOOR PHASE 3
24'- 0"

2ND FLOOR PHASE 3
12'- 0"

1ST FLOOR PHASE 3
470'- 0"

Figure 4–93 *Levels in the Phase 3 file*

    7.3 Open the Site Floor Plan View. Select Create Mass from the Massing tab. Name the new mass Phase 3 Shell. Click OK. Select Solid Form>Solid Blend (see Figure 4–94). The Design Bar will change to Sketch mode.

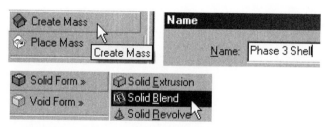

Figure 4–94 *The blend choice for a Mass component*

A blended Mass in Revit Building uses two profiles, one each for the top and bottom, and also creates a volume between the two. You will start by sketching the bottom profile, then the top.

7.4 Set the Depth value to **40'**. Select the Pick icon and pick the sides of the building outline. Select the Draw icon and draw horizontal lines between the endpoints of the magenta lines you created by your picks (see Figure 4–95).

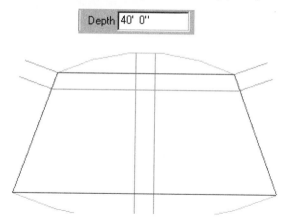

Figure 4–95 *The outline for the blend bottom*

7.5 Pick Edit Top in the Sketch tab. Select the Pick icon and pick the outline of the building footprint (see Figure 4–96). Select Finish Sketch.

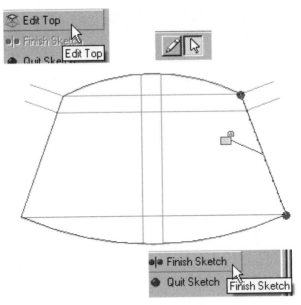

Figure 4–96 *Edit the top and pick the profile*

7.6 Choose the Default 3D View icon. Set the display controls to Shaded with Edges. See Figure 4–97.

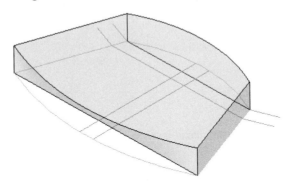

Figure 4–97 *View the Blend*

7.7 Open the Site View. Type **WF** to see the model lines. Select Void Mass>Void Extrusion on the Massing tab.

7.8 Set the Cut Depth to **23′**. Select the Pick icon on the Options Bar. Pick the Overhead line near the top of the figure. Pick the arc at the top of the building outline. Select the Draw icon and trace lines connecting the endpoints of the magenta lines created by your previous picks (see Figure 4–98).

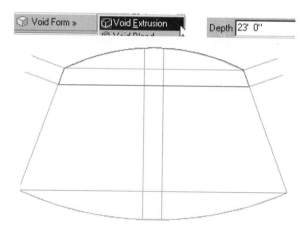

Figure 4–98 *Outline for the Extrusion cut*

   7.9 Choose Finish Sketch. Examine the results in the 3D View. Return to the Site View.

7.10 Select Void Mass>Void Blend from the Massing tab.

7.11 Set the Depth to **40′**. Select the Pick icon from the Options Bar. Pick the two Overhead lines running vertically down the middle of the outline. Pick the upper and lower horizontal lines as with the previous blend. Trim the lines to complete the outline of the base (see Figure 4–99). Select Edit Top.

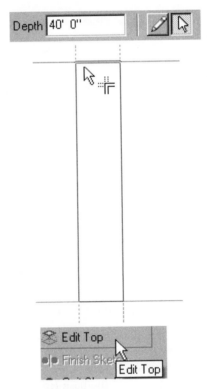

Figure 4–99 *Outline for the blend cut bottom*

7.12 Select Pick. Make the Offset value **4′**. Pick the two Overhead lines so that the offset lines appear to the outside of each (wider). Pick the upper and lower arcs so that the offset lines appear to the inside. Trim the arcs back to the vertical lines (see Figure 4–100). Select Finish Sketch. Select Finish Mass. Examine the results in the 3D view, as shown in Figure 4–101.

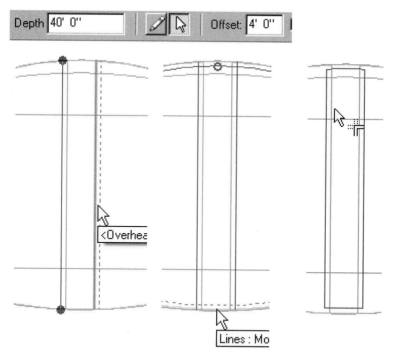

Figure 4–100 *Outline with offsets for the blend top*

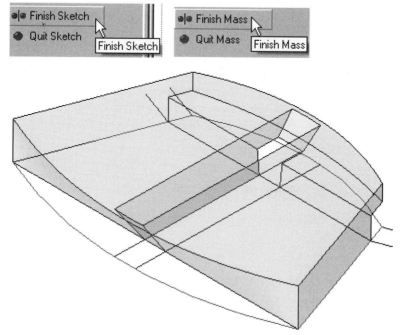

Figure 4–101 *Results of the blend cut*

7.13 Open the ROOF PHASE 3 Floor Plan View. Make the Basics tab active. Select Roof>>Roof by Footprint.

7.14 Choose Lines on the Sketch tab. Select Pick from the Options Bar. Check Defines Slope. Carefully select the outer sides of the blend cut. Select Draw from the Options Bar. Clear Defines Slope. Select the upper and lower ends of the blend cut (see Figure 4–102). Select Finish Roof. The roof will appear cut because of the view settings.

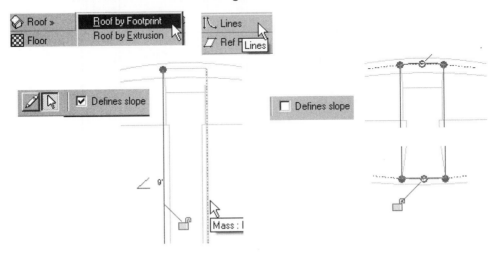

Figure 4–102 *Profile for the atrium roof*

7.15 Open the 3D View to see the roof you just created.

7.16 Open the 2ND FLOOR PHASE 3 Floor Plan View. Select Roof>>Roof by Footprint. Choose Lines from the Sketch tab. Select the Pick icon from the Options Bar. Put a check next to Defines Slope. Pick the two Overhead lines at the upper-left of the building outline.

7.17 Clear Defines Slope. Select Draw. Draw two lines between the endpoints of the lines you just created (see Figure 4–103). Select Finish Roof.

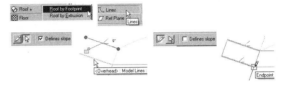

Figure 4–103 *Profile for the passageway roof*

7.18 Select Roof Properties from the Sketch tab. Set the Base Offset from Level value to **8'**; this will align the roof with the height of the void cut you made earlier. Select Finish Roof. See Figure 4–104.

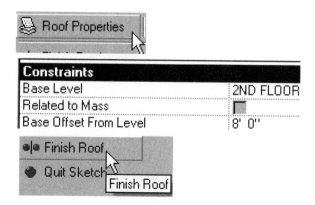

Figure 4–104 *Set the height offset and finish the roof*

7.19 Open the Site view. Select the new roof you just created. Choose Mirror from the Toolbar. Select the Draw icon. Pick the midpoint of the building front, as shown in Figure 4–105. Pull the cursor straight up or down to create a vertical mirror axis. Left click to create the copy on the right side of the building.

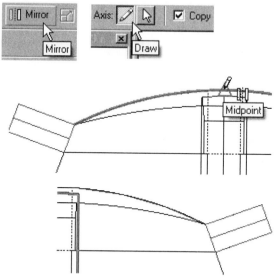

Figure 4–105 *Mirror the left hand roof to the right side.*

7.20 Open the 3D view. Toggle the View Mass control off. Select and delete the lines. Toggle the View Mass control on.

TIP Select everything In the view and use the Filter Selection tool from the Options Bar. Clear Roofs from the list in the Select dialogue, and what will remain in your selection set will be the two types of lines. The Filter Selection tool speeds up selections.

7.21 Select the three roofs. Use the Type Selector to change them to Sloped Glazing. They will change appearance, as shown in Figure 4–106.

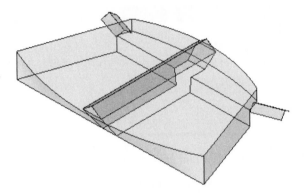

Figure 4–106 *The new roofs are defined as glass, but without any grids yet*

7.22 Select the shell mass. From the Options Bar select Floor Area Faces. Choose Floors 2 and 3 in the dialogue. Select OK. Select Floor by Face from the Massing tab. Select the two floor faces. Choose Create Floor. See Figure 4–107.

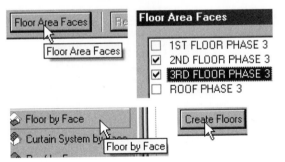

Figure 4–107 *Create floors in the mass*

7.23 Select Roof by Face. Select the flat roof of the mass shell. Select Create Roof.

7.24 Select Curtain System by Face. Select the arc blend (south) face of the building. Select Create System. See Figure 4–108.

Figure 4–108 *Create a curtain system on the south wall*

> 7.25 Select Wall by Face. Set the Type Selector to Brick on CMU. Select the right face to create a wall. See Figure 4–109.

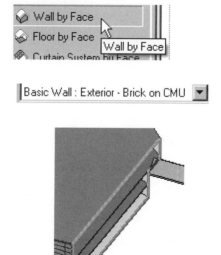

Figure 4–109 *Create a wall on the right side*

> 7.26 Orient the 3D view to the Northwest. Place curtain systems on the blend faces (there are two, so pick carefully. Place a Brick on CMU wall on the right side. Toggle View Mass off. See Figure 4–110.

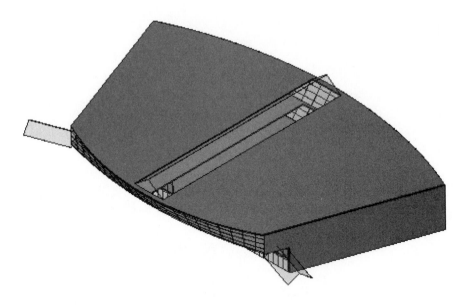

Figure 4–110 *The completed model*

    7.27 Save the file. Close the file.

VIEW THE NEW MODELS IN THE OVERVIEW FILE

    7.28 Open the file *Chapter 4 Project Overview.rvt.* Open the Default 3D View. Orient the view to the Northwest and examine the model. If your mouse is equipped with a scroll wheel, hold down the wheel and the SHIFT key together and you can spin the view to other angles.

    7.29 From the menu bar select View>Orient>Save View. Name the View **Northeast.** Set the view display mode to Shading with Edges. See Figure 4–111.

Figure 4–111 *Shaded view of the linked models*

    7.30 Save the file. Close the file.

> **SUMMARY**
>
> You have seen in this series of exercises how Revit Building accepts information from other sources, including CAD files, images, and Revit Building files. You used simple modeling techniques based on Massing to create building shells for a multi-building project, and linked them into a central file so as to view them all at once.

REVIEW QUESTIONS – CHAPTER 4

MULTIPLE CHOICE

1. Typed shortcuts that deal with Views include

   a) VV, VG, VP

   b) WF, HL, SD

   c) F8

   d) all of the above

2. To create a Blended Massing Form you define

   a) a Top Profile

   b) a Bottom Profile

   c) both a and b

   d) both a and b plus a Profile Angle

3. Sharing Coordinates by Acquire or Publish between files

   a) allows synchronized elevations

   b) requires Administrator privileges on the network

   c) takes too much system memory to be useful

   d) all of the above

4. A Shared Level

   a) is easily created from a default Level by changing the Project Base for the Level

   b) takes its elevation information from a file with Shared coordinates

   c) always has to be a different color from regular Levels

   d) a and b, but not c

5. Edit/Copy and Edit/Paste

a) are only for use between files, not within a project file

b) can only be used on door and window objects, not model lines

c) can take advantage of Paste Aligned

d) can only be used once per file

TRUE/FALSE

6. Erasing a linked Revit Building file does not remove the link from the host

7. Linked images can't be deleted

8. The Solid Form and Void Form types are Extrude, Revolve, Sweep, and Blend

9. Sketches for Roofs and Massing always have to be drawn—there is no option to Pick lines or edges

10. Model Line types include Medium, Thin, and Wide lines, plus Overhead, Demolished, and Hidden

 Answers will be found on the CD.

CHAPTER 5

Subdivide the Design: Worksets and Phases

INTRODUCTION

We have repeatedly mentioned and concentrated on the teamwork aspects of design projects. The ability to share information directly from a building design model with collaborators, consultants, clients, and officials will only grow in importance for designers of buildings. In Chapter 3 you examined how Revit Building imports from and exports to CAD files. In Chapter 4 you explored how to link Revit Building model files together in order to coordinate multi-building site developments.

In this chapter you shall look at Revit Building's Worksets and Phasing mechanisms. Worksets are designed to allow real-time collaborative team design. A Workset is simply a named, segregated collection of elements in a model file. When a project model is divided according to Worksets, more than one user can work simultaneously on different parts of the model. All Worksets are part of the same building model file, so when design developments are published from a local copy to the Central File and reloaded by other users, changes to the model are propagated and coordinated tightly.

Worksets are a professional toolset, and their proper and efficient use takes planning and adherence to common-sense best practices. There isn't enough room in this book to discuss advanced uses of Worksets (editing at risk, rolling back changes, or restricting references). Worksets provide the ability to open only part of a building file's contents. Many Revit Building users divide nearly all their projects—even those that only one person will work on—into Worksets according to a standard setup, to take advantage of the reduced load on their computer system resources.

Phases are Revit Building's way of putting depiction of time span into a building model. All building projects have at least two implicit phases (existing conditions and new construction); all remodeling projects have at least three phases (existing, demolition, and new); many projects have four or more phases (existing, demolition, temporary construction, removal of temporary work, new construction phase 1, and so forth). Revit Building allows the user to define any number of phases for purposes of viewing a project at any stage along its projected timeline, which can be of significant value in a collaborative setting.

Phases and Worksets are not designed explicitly for exploring alternative options or variations on a design. The Design Options feature in Revit Building 8.0 will be explored in Chapter 10. Both phases and Worksets can be utilized in linked files, allowing for specific views across many models in a multi-building project.

OBJECTIVES

- Activate, create, and use Worksets in a model file
- Examine Worksets in linked files
- Set-up and use Phases in a model file
- Examine phases in linked files

REVIT BUILDING COMMANDS AND SKILLS

Worksets—rename, new

Save to Central

Save Local

Create walls from mass elements

Create sloping floor

Split walls

Convert to Curtain wall

Add and Edit Curtain Grid

Manage Links

Phasing—create Phases

View Properties for Phases

Add Components in Phases

EXERCISE 1. ACTIVATE AND CREATE WORKSETS IN A PROJECT FILE

1.1 Launch Revit Building. Open the file *Chapter 5 start.rvt*. It will open to a floor plan view that shows exterior walls, interior walls, and a collection of toposurfaces. In your hypothetical project the design for the building and site has thus far been prepared by a single person working in the file. You will now prepare the file so that a team will be able to work in the file simultaneously.

1.2 From the File menu, pick File>Worksets, as shown in Figure 5–1. The Project Sharing – Worksets dialogue will appear, which names the default Worksets that will be created automatically. Click OK to accept the default Workset names.

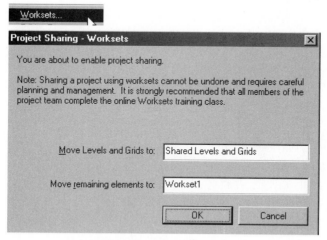

Figure 5–1 *Activating Worksets from the file menu*

1.3 The Worksets dialogue appears. It shows the active Workset (Workset1 by default) and the properties of the User-Created Worksets now in the file. They are all Opened and Editable. Your Revit Building username will appear next to the Workset name in the Owner column (see Figure 5–2).

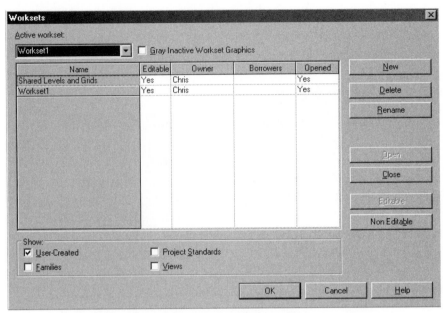

Figure 5–2 *The Worksets dialogue with default Worksets*

1.4 Take a moment to study the Worksets that Revit Building creates, other than the User-Created ones. Clear User-Created in the Show section of the Worksets dialogue, and check Families. Note the Revit Building object categories that have been moved into Worksets, as shown in Figure 5–3.

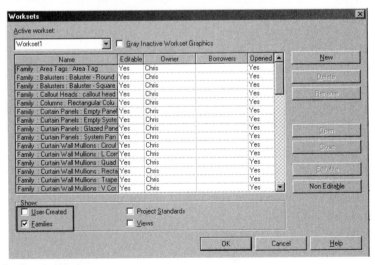

Figure 5–3 *Family Worksets*

1.5 Clear Families, and check Project Standards and Views, in turn, to display the relevant Worksets, as shown in figures 5–4 and 5–5.

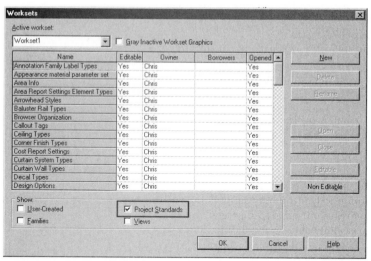

Figure 5–4 *Project Standards Worksets*

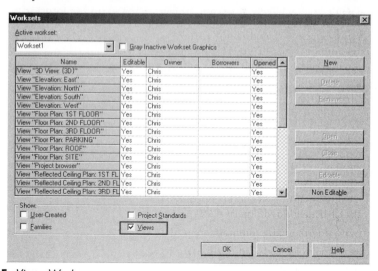

Figure 5–5 *Views Worksets*

1.6 Clear Views and check User-Created. Select Workset1 in the Name field of the dialogue, and pick Rename. In the Rename dialogue, type **Exterior** in the New name field (see Figure 5–6). Click OK.

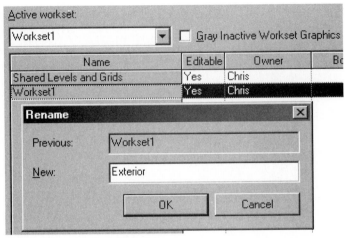

Figure 5–6 *Rename Workset1 to Exterior*

1.7 Select New. Enter the name **Interior** (see Figure 5–7). Note that Visible by default in all views is checked—this is appropriate for interior walls. Click OK.

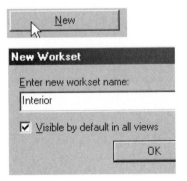

Figure 5–7 *A new Workset named Interior*

1.8 Select New. Enter the name **Site**. Clear Visible by default in all views, as shown in Figure 5–8. Click OK. Select New. Enter the name **Furniture**. Clear Visible by default in all views. Click OK. Click OK in the Worksets dialogue to close it.

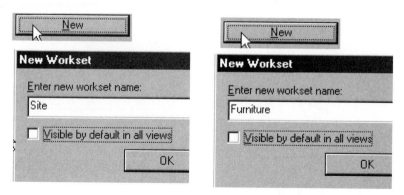

Figure 5–8 *The Site and Furniture Worksets will not be visible in all views*

1.9 Place the cursor anywhere over the Toolbar. Right-click to bring up the list of available toolbars. Check Worksets. The Worksets toolbar will appear, with Exterior in the drop-down field, indicating that Exterior is the current Workset. Any elements created now will appear in the Exterior Workset (see Figure 5–9).

The Pick selection on the Options Bar now shows Editable Only, which is checked by default. With this option checked, you will not be able to select elements that are in a non-editable Workset.

Figure 5–9 *Turn on the Worksets toolbar*

PLACE THE MODEL CONTENTS INTO WORKSETS

1.10 Zoom to Fit. Pick two points to select the entire visible model. Select the Filter icon. In the Filter dialogue, pick Check None, then check Topography, as shown in Figure 5–10. Click OK.

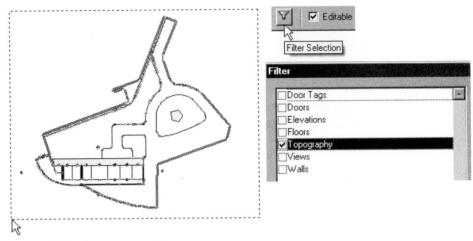

Figure 5–10 *Filter the selection*

1.11 Pick the Properties icon. In the Element Properties dialogue, change the Workset Value to Site, as shown in Figure 5–11. Click OK. The highlighted topography will disappear, as the Site Workset is not visible in all views by default.

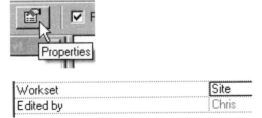

Figure 5–11 *Change the selected topography elements to the Site Workset*

1.12 Zoom to Fit again. Select all interior doors and walls, as shown in Figure 5–13. Pick the filter icon and clear Door Tags. Select the Properties icon. Change the Workset Value to Interior, as shown in Figure 5–12. The doors and walls will not disappear, as the Interior Workset is visible in all views by default.

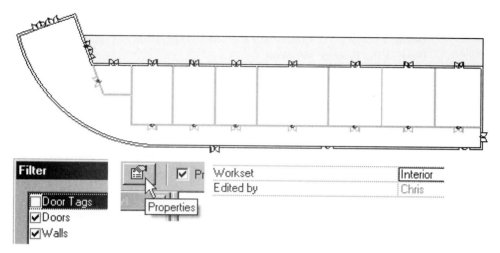

Figure 5–12 *Select interior doors and walls in the 1st FLOOR view and place the selection in the Interior Workset*

1.13 Left-click to the right (interior) side of the curved wall, hold down the left button, pull the cursor to the left (exterior) side of the wall, and release the button, to select the wall and also the floor that touches it. Use the Filter to clear Walls. Pick the Properties icon and place the floor into the Interior Workset. See Figure 5–13.

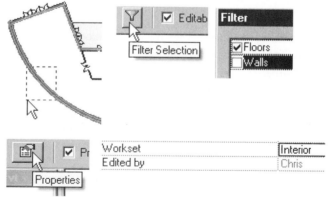

Figure 5–13 *Select the wall and floor, filter out the wall, and place the floor in the Interior Workset*

 NOTE The floor in the current floor plan view is selectable, but since no pattern has been applied it's difficult to know if a floor has been created as an object in this view. Depending on the complexity of a plan, it can also be tricky to select the desired object. Using right–left picks to create a Crossing selection window at wall-floor joins is a quick way to "fish" for floor edges and other hard-to-select objects.

1.14 Open the 2ND FLOOR plan view. Select all interior doors and windows, as shown in Figure 5–14 (filter out the Door Tags, as in Step 1.13), and place them in the Interior Workset.

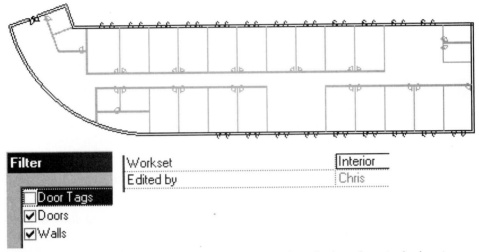

Figure 5–14 *The 2nd FLOOR interior walls and doors selected; place them in the Interior Workset*

1.15 Select the floor as you did in step 1.13. Pick the Properties icon and place the floor into the Interior Workset.

You are removing Door Tags from our selections for the Interior Workset because Tags are automatically placed in a Workset by Family type (as are all annotations).

1.16 Open the PARKING plan view. Pick two points around the model to select everything visible. Only the Elevations and the green trapezoidal shape (sketch lines for the Parking Garage) will highlight red. Use the Filter to clear Elevations and Views, leaving Lines (Lines) checked. Click OK to exit the Filter. Hold down the CTRL key and select the two lines to the right of the trapezoid. These are also sketch lines to use in creating a ramp (see Figure 5–15).

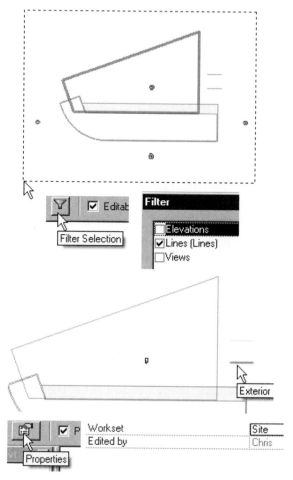

Figure 5–15 *Pick the two short lines one by one with the CTRL key held down*

1.17 Select the Properties icon. Place the lines in the Site Workset. Click OK. The lines will disappear, as the Site Workset is not visible in all views by default. Zoom to Fit.

1.18 Open the SITE plan view. Type **VV** to enter the Visibility/Graphic Overrides dialogue. There is now a Worksets tab. Make that tab active. Clear Furniture and Interior. Check Site (see Figure 5–16). Click OK. Topography is now visible around the building outline.

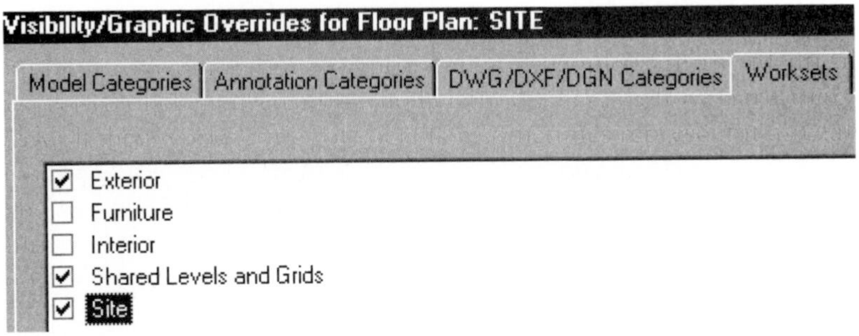

Figure 5–16 *Change the Workset visibility in the SITE plan view.*

1.19 Open the PARKING view. Type **VV** to enter the Visibility/Graphic Overrides dialogue. Repeat the previous checks on the Worksets tab to make the Site Workset visible, and Interior not visible, in this view.

1.20 Open the 1ST FLOOR view. Zoom to Fit.

EXERCISE 2. CREATE A CENTRAL FILE

2.1 Pick File>Save As to save the project as a Central File.

The Central File is created automatically the first time a file is saved after Worksets are enabled. Location of this file is important—all team members who will use the file should have access to it, using the same network path.

2.2 In the Save As dialogue, select Options. In the File Save Options dialogue, notice that the option to Make this the Central location after save is checked and grayed out—this will automatically happen the first time, and then become optional for later saves. Change the number of backups to 1, or as specified by your instructor (see Figure 5–17). Click OK.

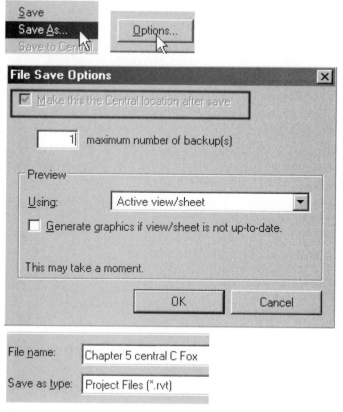

Figure 5–17 *The backup and Make this Central options*

2.3 In the Save As dialogue, give the file you are about to save the name *Chapter 5 central.rvt,* and save it in a location specified by your instructor. If you are working this exercise on your own, or outside a network environment, save the Central File on your hard drive. Click Save.

INSTRUCTOR'S NOTE If at all possible have the students save their central files to a shared network location with a drive path that will be the same for all. If your network security policy mandates private folders for student work, create a temporary public folder for the duration of this exercise. Have the students save the Chapter 5 central file with their first initial and last name as part of the file name, as in *Chapter 5 central J Lopez.rvt*. *This will facilitate sharing of central files in the second part of the exercise. The central file storage folder will need to be on a drive with adequate space (see the next exercise step).*

2.4 Open the File Manager or Windows Explorer (My Computer) on your computer and navigate to the folder where you just saved the Central File.

Note that Revit Building has created a folder named *Chapter 5 central_backup*. Open that folder. There will be over 150 folders cryptically named, files to hold workset information, and one file named *Chapter 5 central 0001.bak*. More bak files will be created as the Central File is updated, up to the number of copies you entered in the Save As Options dialogue earlier (see Figure 5–18). Revit Building will not create *Chapter 5 central.000x.rvt* files for the Central File as it does for non–Central Files.

 CAUTION If you ever move a Central File after creating it, the backup folder and its contents should be moved with it. Make no changes to the contents of the backup folder at any time.

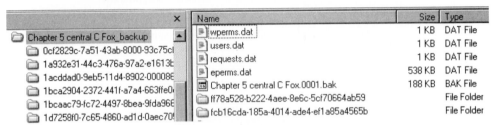

Figure 5–18 *A typical central_backup folder and contents*

2.5 Close the File Manager without making any changes to it or in the *Chapter 5 central_backup* folder.

2.6 Before closing the new Central File, release the Worksets so they will be accessible to other users who will use this file to create local files to develop the design. Pick the Worksets icon on the toolbar you made visible earlier. The Worksets dialogue will open. Select the top Workset name field, hold down the left mouse button, and drag down to select all the Worksets, as shown in Figure 5–19. Pick Non-Editable. Your name will be removed from the Owner fields and the Editable values become No. Select Close. Click OK.

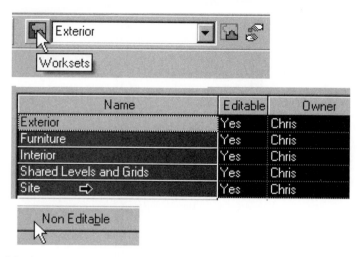

Figure 5–19 *Relinquish Workset editability in the Central File*

2.7 Pick File>Close from the File menu.

CREATING AND USING LOCAL FILES

At this point you are about to create a copy of the Central File for your local use. You will need to check out at least one Workset to make edits. To publish your design changes so others can see them, you should save work frequently (every 30 minutes is recommended) and save centrally—a different step—every one to two hours. After your last central save, remember to release Worksets as you just did, then save locally and close the file.

In the next part of the exercise you will create three local files and work in each as a separate user, accessing the same project Central File. If you are working individually without a network you can open a separate session of Revit Building for each file (depending on system resources you may want to limit yourself to two open sessions) to see the changes propagate.

2.8 Pick File>Open from the File menu. In the File Open dialogue, use the Open Worksets drop-down list and select Specify. Select the *Chapter 5 central.rvt* file you created (see Figure 5–20).

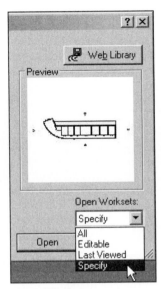

Figure 5–20 *The choices for opening Worksets*

2.9 In the Opening Worksets dialogue, you will choose which Worksets to make open (make visible) in the copy of the Central File you are about to create. Since you did not close Worksets when closing the file, all are open by default. In this file you will work on the site, so select the Furniture and Interior Worksets (as shown in Figure 5–21) and pick Close. Click OK. You can also close Worksets at any time after the file is opened.

Name	Editable	Owner	Borrowers	Opened
Exterior	No			Yes
Furniture	No			Yes
Interior	No			Yes
Shared Levels and Grids	No			Yes
Site	No			Yes

Figure 5–21 *Select the Worksets you do not need to see*

2.10 The *Chapter 5 central.rvt* file opens. No interior walls or doors are visible. Select File>Save As from the File menu. Click Options. Note that the option to Make this the Central location after save is now available. Make sure it is not checked (see Figure 5–22). Click OK to close the options dialogue. Save the file as *Chapter 5 site.rvt* on your local drive. Click Save.

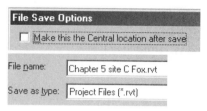

Figure 5–22 *Do not make this the Central File*

NOTE Once you have saved the file as a central copy, Revit Building activates the icon for the Save to Central function on the Toolbar (see Figure 5–23). Set the save reminder—Settings>Options—as appropriate so you can keep your work safe and your design changes published to the Central File regularly. Remember—Revit Building **does not** have an autosave.

This file is now your local copy to work in. It will maintain a connection to the Central File so that certain functions and conditions (Workset editability) are constantly monitored, while other changes update when local files are saved and Worksets reloaded.

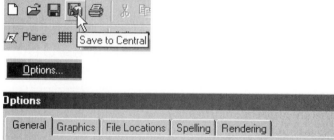

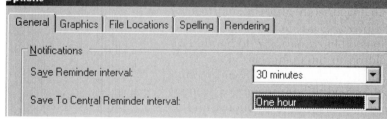

Figure 5–23 *The Save to Central icon appears in Workset-activated files*

2.11   Pick the Worksets icon on the Toolbar. Certain Worksets are open (visible), but none have been made editable (checked out of the Central File). Select the Site Workset. Pick Editable. Your name appears as the Owner, as shown in Figure 5–24. Use the Active Workset drop-down list to make Site the active Workset. Click OK.

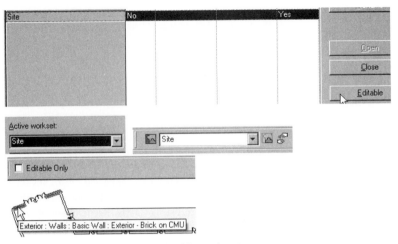

Figure 5–24 *Make the Site Workset editable and active*

2.12 The Worksets Toolbar shows Site as the active Workset. Clear Editable Only on the Options Bar to enable you to select elements not in the editable Worksets. Otherwise walls will not pre-highlight under the cursor or allow you to select/identify them.

2.13 Select a wall. The tooltip will identify its type. Pick the Properties icon. All the properties are visible, but if you try to modify any of them an error message appears (see Figure 5–25). Change the wall type to Exterior – CMU Insulated. Choose OK. Select Cancel to close the warning.

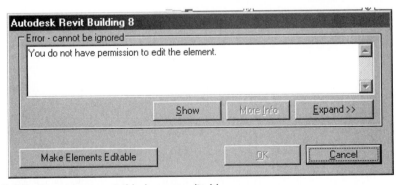

Figure 5–25 *Properties are visible but not editable*

EXERCISE 3. WORKING SIMULTANEOUSLY

If you are working this exercise on your own, at this point you will open a second session of Revit Building and run the second session under a different username. If you are working in a class situation or with a partner, skip to Step 3.3.

3.1 Minimize but do not end the current session of Revit Building. Start another session using the desktop icon or Start menu. Close the file that opens. From the Settings menu, pick Settings>Options, as shown in Figure 5–26.

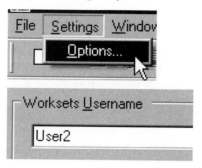

Figure 5–26 *Reach Revit Building's Options with no project open; change your Worksets Username*

3.2 On the General tab of the Options dialogue, enter **User2** in the Worksets Username value field. Click OK.

The Workset Username value field will hold the last entered value. After completing this exercise and closing all open Revit Building sessions, change the Worksets Username back to your Windows Login name to coordinate Revit Building with your usual login for future work.

CREATE A SECOND LOCAL FILE

If you are working with a partner in this part of the exercise, you will both have created a Central File on the network. Choose one of those Central Files to use for the rest of this exercise. The open local file from the selected Central File will stay open—that user will now be called User1, and the user who creates a second local file in the next steps will be called User2.

USER2:

3.3 Close any open Revit Building files. Open the Central File *Chapter 5 central.rvt* (open the file created by your partner for this exercise if you're working with a partner). Accept the Last Viewed option for Visible Worksets. Select File>Save As from the File menu.

3.4 In the Save As dialogue, select Options and make sure that Make this the Central Location after save is **not** checked. Save the file on your local hard drive as *Chapter 5 exterior.rvt.*

You are now working in a local file. Revit Building places a Save to Central icon on the Toolbar so design changes can be published to the Central File and shared with your partner.

3.5 Place the Worksets icon on the Toolbar as before. Open the Worksets dialogue. Pick Exterior in the Name field to highlight it. Left-click in the Editable column; change the value from No to Yes. Your username (which will be different from User2 if you are working the exercise with a partner) will appear in the Owner column (see Figure 5–27). Click OK.

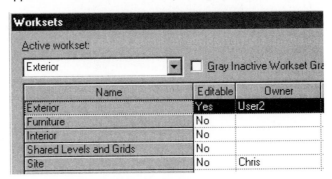

Figure 5–27 *Making the Exterior Workset Editable*

USER1:

3.6 You are still working in the file *Chapter 5 site.rvt.* Open the Worksets dialogue. The Exterior Workset will show that User2 has checked it out. The Editable status of that Workset in your file is No. Try to change the status of the Exterior Workset to Yes—Revit Building will display a warning that the Workset is being edited by another user, as shown in Figure 5–28. Click OK to close the warning. Click Cancel to exit the Worksets dialogue.

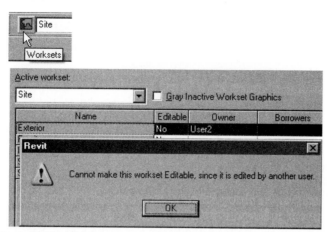

Figure 5–28 *Revit Building knows this Workset is checked out*

 NOTE FOR INSTRUCTORS The next steps in this exercise for both User1 and User2 introduce new skills other than Worksets. We recommend that you have users complete both sections of this exercise. Once both users have completed their relative exercises 4 or 5, have them return to step 3.3, open the copy of Chapter 5 central.rvt NOT used the first time through, and work the exercise steps (site or exterior) that they did not complete the first time.

USER2:

3.7 Move to Step 5.1 to work by yourself—you will continue to work in the file *Chapter 5 exterior.rvt.*

3.8 When your partner notifies you of an editing request, go to Step 4.30.

USER1:

Start Exercise 4:

EXERCISE 4. CREATE WALLS

4.1 Open the PARKING plan view. This view shows the level above as the Underlay, so you see the walls and topography at ground level. Pick Wall from the Basics tab of the Design Bar.

4.2 Set the Type value of the selected walls to **Foundation – 12″ Concrete**. Foundations walls are created from the top down. Set the Base Constraint to Level **1ST FLOOR**; leave the Base Offset of 0′ 0″. Set the Top Constraint to Unconnected, and the Unconnected Height to 11′ 0″ (see Figure 5–29). Click OK.

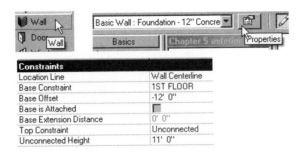

Figure 5–29 *Create Foundation walls*

4.3 Check the Chain Option on the Options Bar and trace over the green model lines to create four walls. Hit ESC twice to terminate the Wall tool.

4.4 Pick the Floor icon from the Basics tab of the Design Bar. The Design Bar will change to Sketch mode. Pick Walls will be selected by default on the Design Bar. Clear Extend into wall (into core) on the Options Bar. Select the walls you just created. Select Floor Properties from the Design Bar. Make the Floor Type **Generic – 12"**; make the Level value **PARKING** with a **0′ 0″** Height Offset. See Figure 5–30.

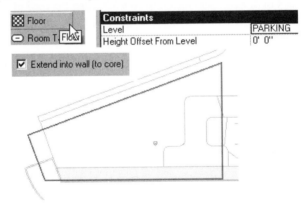

Figure 5–30 *Floor Properties for the Parking level*

4.5 Click OK in the Element Properties dialogue. Pick Finish Sketch on the Design Bar.

CREATE A SLOPED FLOOR USING THE SLOPE ARROW

4.6 Zoom in to the right side of the new walls. Pick the Floor tool again. The Design Bar will change to Sketch mode. Select Lines on the Sketch tab of the Design Bar. Select the Pick icon (arrow) on the Options Bar.

4.7 Two Model lines have previously been placed in the Central File to indicate edges of a ramp from the driveway on the 1ST FLOOR level down to the PARKING level. Select those two lines to start the outline of the ramp.

4.8 Select Draw on the Options Bar. Draw two vertical lines to create left and right edges for the floor outline. Trim the four new lines to complete a rectangular outline, as shown in Figure 5–31.

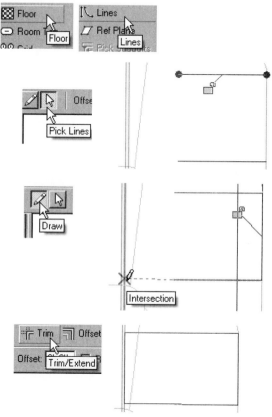

Figure 5–31 *Pick two lines then draw two lines and trim to finish the sketch*

4.9 Pick Slope Arrow from the Sketch tab of the Design Bar. Pick the midpoint of the left side of the outline as the start point. Pull the cursor horizontally to the right to define the Slope Arrow, as shown in Figure 5–32. Pick the midpoint of the right side of the floor outline for the end point of the arrow.

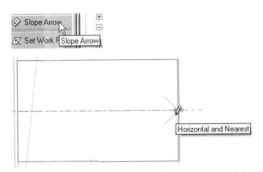

Figure 5–32 *Draw the slope arrow from the left-side line to the right-side line*

4.10 Select the new Slope Arrow. Pick the Properties icon from Options Bar. Set the Specify value to Level at Tail. Set the Level at Tail value as **PARKING**, with a Height Offset of **0′ 0″**. Make the Level at Head value **1ST FLOOR** (see Figure 5–38). Click OK. Pick Floor Properties. Make the Type value **Concrete walkway 4″**. (See Figure 5–33). Click OK. Click Finish Sketch. The ramp will display according to the cut line for the view.

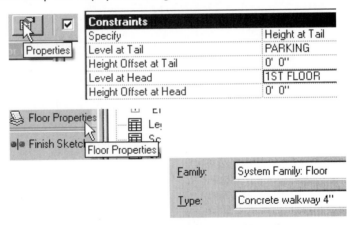

Figure 5–33 *Properties for the head and tail of the ramp Slope Arrow*

4.11 Select the Split tool from the Tool Bar. Check Delete Inner Segment from the Options Bar. Split the wall that crosses the left end of the ramp at the edges of the ramp, as shown in Figure 5–34.

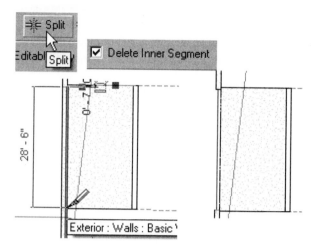

Figure 5–34 *Splitting the wall to remove a section*

CREATE SLOPED WALLS AND AN OPENING FOR THE RAMP

4.12 Select Wall from the Basics tab of the Design Bar. From the Type Selector drop-down list, select **Basic Wall: Retaining – 12″ Concrete**. Make the Base Constraint **PARKING**. Set the Height value to **Up to level: 1ST FLOOR**, as shown in Figure 5–35. Create two horizontal walls **47′** long along the sides of the ramp. Be sure the walls are on the ramp, not beside it.

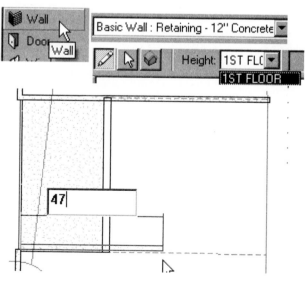

Figure 5–35 *Draw walls on the ramp*

4.13 Select Modify. Pick the two new walls, then select Attach from the Options Bar, as shown in Figure 5–36. Choose Base. Pick the Ramp. The walls will change appearance and end at the cut line, consistent with the ramp.

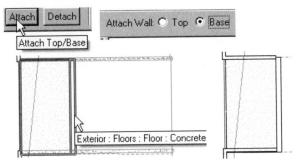

Figure 5–36 *Attach the bottom of the new walls*

4.14 Zoom to Fit in the PARKING plan view. Open the Elevations: South view. Type **VG**. On the Model Categories tab, make sure Topography is cleared. On the Worksets tab, make sure that Exterior, Shared levels and Grids, and Site are checked. Click OK.

4.15 Zoom in Region to the right side of the view. Examine the ramp and side walls you have just created (see Figure 5–37).

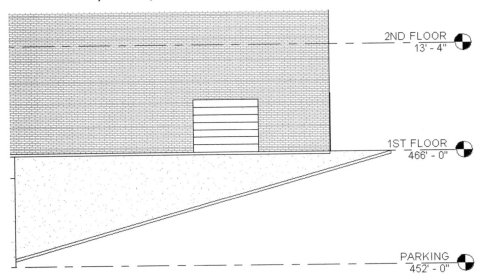

Figure 5–37 *A side view of the walls and ramp*

4.16 Open the Floor Plan view SITE. Type **VG**. On the Model Categories tab of the Visibility dialogue, make sure that Topography is checked. On the

Worksets tab, make sure that Exterior, Shared levels and Grids, and Site are checked. Click OK.

4.17 Pick two points right to left around the topography. Filter to remove Topography from the selection, leaving two types of lines. Choose Delete to remove those lines and make the next steps easier. See Figure 5–38

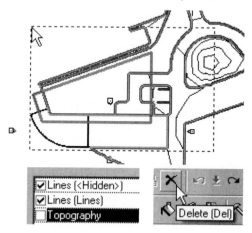

Figure 5–38 *Remove two sketch ines in the parking lot area*

4.18 Make the Site tab on the Design Bar active. Pick Split Surface from the Site tab on the Design Bar.

4.19 Select the L-shaped Toposurface that defines the area between the main drive and the shuttle bus loop/handicapped parking access (see Figure 5–39).

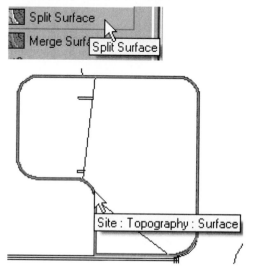

Figure 5–39 *The topography to edit*

4.20 The Site tab will change to Sketch mode. Type **WF** to enter Wireframe View mode. If a warning about View Worksets appears, choose OK for temporary changes to the view.

4.21 Select the Rectangle tool on the Options Bar, as shown in Figure 5–40. Pick two corners at the edges of the new ramp side walls. Select Finish Sketch.

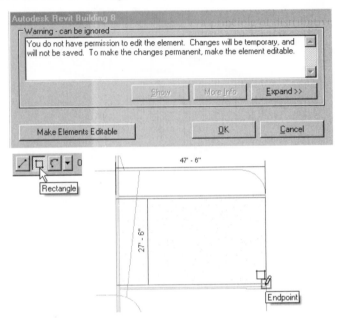

Figure 5–40 *Adjust the view to wireframe, then sketch a rectangle over the ramp and walls*

4.22 Type **HL** to return to Hidden Line mode. Select the new rectangular Toposurface you split off from the existing one. Delete it. The ramp and walls will appear, as shown in Figure 5–41.

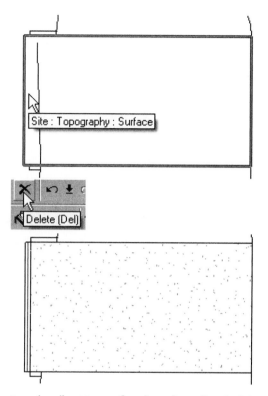

Figure 5–41 *The ramp and walls appear after the split surface is deleted*

SHARING WORKSETS: EDITING REQUESTS

4.23 Open the Floor Plan View 1ST FLOOR. Type **VP**. In the View Properties dialogue, make the Underlay value **PARKING**. Pick Edit in the Visibility field. In the dialogue that opens, choose Make Elements Editable. On the Model Categories tab of the Visibility dialogue, check **Topography**. On the Worksets Tab, check **Site**. See Figure 5–42. Click OK twice.

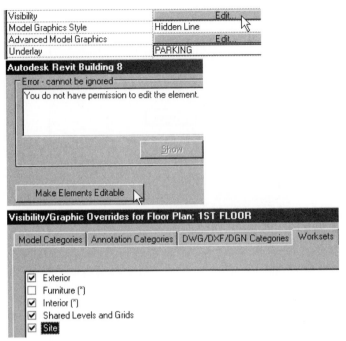

Figure 5–42 *Change the Underlay in this view and make the Site workset not visible by default) visible*

4.24 Zoom in Region to the lower-right corner of the now-visible parking garage walls. In the Worksets dialogue, make Exterior the current Workset, even though it is not editable in this local file (see Figure 5–43).

Figure 5–43 *Make the Exterior Workset the active one*

4.25 Make the Modeling tab active on the Design Bar. You are going to create an opening in the walkway for a future stair down to the parking garage. Pick the Opening tool. In the Options dialogue that opens, check Pick a roof, floor, or ceiling and cut vertically, as shown in Figure 5–44. Click OK.

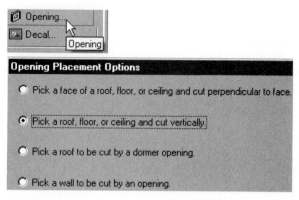

Figure 5–44 *Use the Opening tool to cut a hole in the walkway, which is a floor*

4.26 The cursor will change shape. Pick the exterior walkway. The cursor will change shape again, and the Design Bar will switch to Sketch mode. Lines will be the default option.

4.27 Select the rectangle option from the Options Bar. Sketch a 6' x 10' rectangle as shown in Figure 5–45, starting at the intersection of the walkway edge and the wall below.

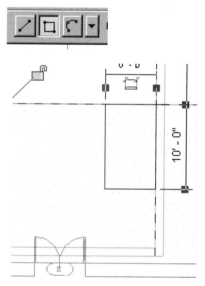

Figure 5–45 *Sketch the cut in the floor*

4.28 Pick Finish Sketch from the Design Bar. An error message will appear. Click Make Elements Editable. A question box will appear. Click Yes to request permission from your partner to make a change in the Exterior Workset in this exercise (see Figure 5–46).

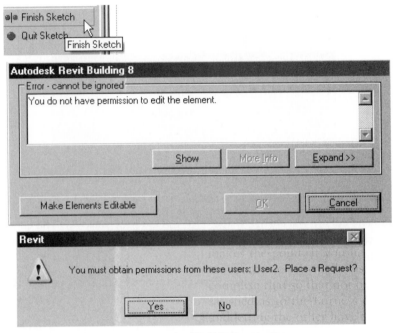

Figure 5–46 *Request to change a non-editable Workset entity*

4.29 Revit Building will display a confirmation dialogue with options to continue work or to check for a grant to your editing request. If you are working this exercise alone, move to the second session of Revit Building that has the file *chapter 5 exterior.rvt* open. If you are working with a partner, notify her or him that you have reached this point in your exercise.

USER2:

4.30 You are working in the file *chapter 5 exterior.rvt*. When your partner notifies you (or if working alone, when you have made the editing request in Step 4.29), pick the Editing Requests icon from the Worksets toolbar (see Figure 5–47).

Figure 5–47 *Open the Editing Request dialogue*

4.31 The Workset Editing Requests dialogue will open. You can expand the list to see who is requesting what change in the Worksets you control from your local file. Select the one from your partner (or yourself) and click Grant (see Figure 5–48).

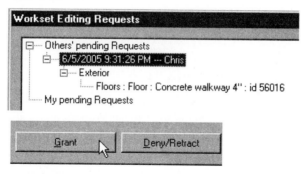

Figure 5–48 *Grant the request*

4.32 Click Close and continue with your work in Exercise 5.

USER1:

4.33 Once you (or your partner) has completed Steps 4.30 to 4.32, pick Check Now in the Check Editability Grants dialogue. Click OK in the confirmation dialogue (see Figure 5–49). Revit Building will create the opening in the walkway.

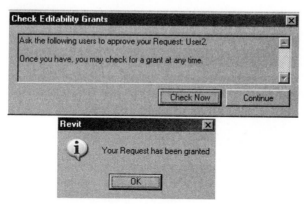

Figure 5–49 *Check the request and finish the sketch*

You could have picked Continue in the Grant Check dialogue and continued with other work in your project file, rather than requesting an immediate grant. In a real-world situation, the demands of work flow, and a possibly far-flung design team, may very well mean that requests accumulate while other work goes on.

4.34 Open the Worksets dialogue. Note that your name now appears as a Borrower(s) of the Exterior Workset (see Figure 5–50). Click Cancel to exit the dialogue.

Name	Editable	Owner	Borrowers	Opened
Exterior	No	User2	Chris	Yes
Furniture	No			No
Interior	No			No
Shared Levels and Grids	No			Yes
Site	Yes	Chris		Yes

Active workset:
[Not Editable] Exterior ▼ ☐ Gray Inactive Workset Graphics

Figure 5–50 *You are now a borrower of the Exterior Workset in this local file*

4.35 Open the default 3D view. Type **VG**. On the Model Categories tab of the Visibility dialogue, check **Topography**. On the Worksets Tab, check **Site**. If a warning appears, select Make Editable. Click OK. From the View menu, pick View>Orient>Northeast.

4.36 From the View menu, pick View>Shading with Edges (see Figure 5–51 for the results).

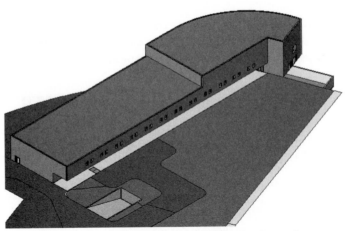

Figure 5–51 *A shaded NE view of the ramp and stair cut for the parking garage in place*

4.37 Pick the Save to Central icon from the Toolbar. In the Save to Central dialogue, check all the options to relinquish element editability after the save. Check the option to save the local file. In the Comment field, type **Site plan with underground parking area and ramp**. Click OK.

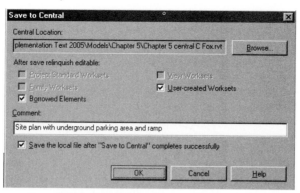

Figure 5–52 *Save to Central with a comment*

4.38 Close the File. Move to Step 5.1 unless you have already worked exercise 5.

USER2:

If User1 has Saved to Central before you get to this step, Revit Building will ask you to reload Worksets before continuing work. Certain views may therefore look different from the illustrations in this chapter.

EXERCISE 5. CREATE AND EDIT A CURVED CURTAIN WALL

You are working in file *Chapter 5 exterior.rvt*. Open the Floor Plans: 1ST FLOOR view. Zoom in Region to the left side of the model.

5.1 Pick the Worksets icon on the Toolbar. Close Site. Make Exterior and Interior Editable. Keep Exterior the Active Workset (see Figure 5–53). Click OK.

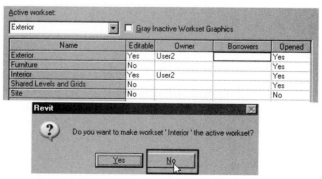

Figure 5–53 *Make Worksets editable*

5.2 Activate the Split tool. Split the curved wall on the left side of the building, as shown in Figure 5–54. The exact location is not critical; wall lengths will be adjusted in the next few steps.

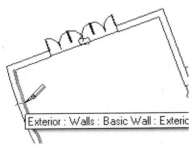

Figure 5–54 *Split the wall*

5.3 Pick Modify from the Basics tab of the Design Bar. Pick the short segment you just created. The wall will highlight red, and blue control dots will appear. Select the lower control dot and drag it up to separate the wall ends. Left-click when the temporary arc dimension reads close to **5°** (see Figure 5–55).

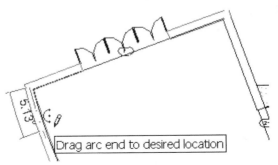

Figure 5–55 *Adjust the position of the wall end*

5.4 If a warning appears about walls overlapping, click OK. Drag the end of the longer wall down past the end of the short one.

5.5 Pick the other curved wall segment. Pick Activate Dimensions from the Toolbar. Select the angular dimension text so that the text field opens for input. Type **70** in the field and hit ENTER, as shown in Figure 5–56. The wall length will adjust.

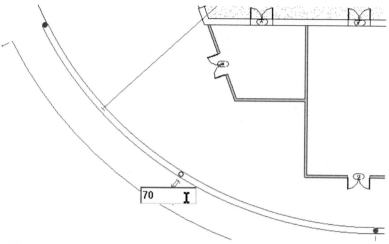

Figure 5–56 *Adust the wall arc angle value*

5.6 Select the short wall segment you previously created. Drag its lower end so it aligns with the upper end of the 70° curved wall segment. The wall centerline will highlight with a dashed green line when the alignment is correct (see Figure 5–57).

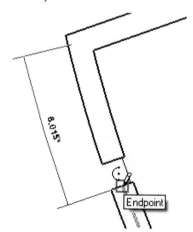

Figure 5–57 *Pull the short wall end down to meet the other one*

5.7 Open the Default 3D view. From the View menu, pick View>Orient>Southwest. Zoom in Region at the curved wall section

5.8 Select the Split tool. As you move the cursor over the building exterior, the two curved wall segments will highlight in turn. When the larger segment highlights, move the cursor up the wall, as in Figure 5–58. Pick a point near but below the roof to split the wall into two segments vertically.

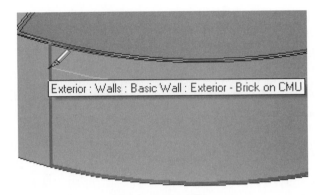

Figure 5–58 *Using the Split on a vertical wall*

5.9 Select Modify from the Basics tab of the Design Bar. Pick the large curved wall at the floor line. Choose the Properties icon. Make the Unconnected Height value **40**. Revit Building will adjust the field value to 40' 0". Click OK.

5.10 With the wall still selected, use the drop-down list in the Type Selector to select Curtain Wall: Curtain Wall 1 (the bottom item on the list). The wall displays as a single straight (not curved) transparent panel. Make the Modeling tab of the Design Bar active. Pick Curtain Grid from the Modeling tab, as shown in Figure 5–59.

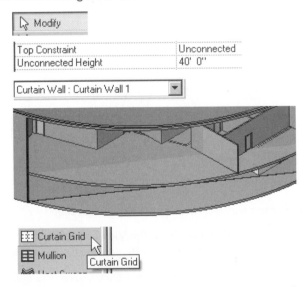

Figure 5–59 *Change the wall to a curtain wall, and pick the curtain grid to to subdivide it*

5.11 Set the view display mode to Hidden Line to make the next steps easier to read. Choose OK in the dialogue, as in Figure 5–60. The curtain grid tool will stay active.

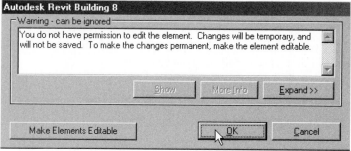

Figure 5–60 *Change the view for easier work*

5.12 Place the cursor to the left end of the curtain wall and start moving it to the right. Angular dimensions will appear. When the angle values read **5.00°/ 65.00°**, left-click to place a curtain grid.

 NOTE The temporary dimension precision level is keyed to the Zoom scale. Use the mouse scroll wheel, if so equipped, to move the zoom in or out so that the angular dimensions read in whole degrees. You can do this while using the grid placement tool.

5.13 Continue pulling the cursor to the right and place a curtain grid element every 5° around the arc. Revit Building will snap weakly to whole degree increments (see Figure 5–61).

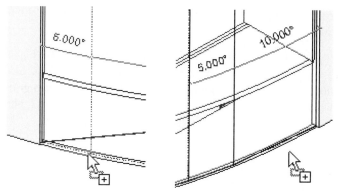

Figure 5–61 *Adding curtain grids at 5°*

5.14 Repeat the process, placing vertical curtain grids at **2.50°** arc increments between the 5° grid elements. Revit Building will snap weakly to the midpoints and one-third divisions of the grid panels.

5.15 Move the cursor to the left side of the curtain wall. Move it up the left side and Revit Building will show vertical dimensions. Place horizontal curtain grids at **8'**, **16'**, **24'**, and **32'** up the wall (8' up from the bottom or previous grid), as shown in Figure 5–62.

 TIP If you locate a grid line in the wrong place, you can select its vertical dimension value and edit it while remaining in the command.

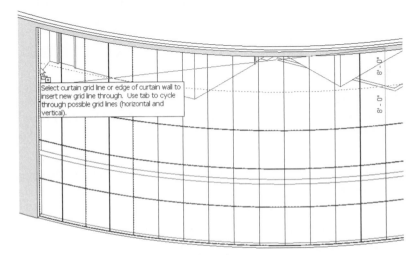

Figure 5–62 *Adding horizontal curtain grids at 8' increments*

5.16 Open the South Elevation view. Zoom in Region to the right side of the curtain wall.

5.17 Select the first (right-most) vertical curtain grid so that it highlights with a dashed green line. The Options Bar will change its options. Select Add or Remove Segments. Pick the lowest segment of the curtain grid. It will highlight with a dashed green line, as shown in Figure 5–63. Hit ESC twice to terminate the command. The grid segment will disappear.

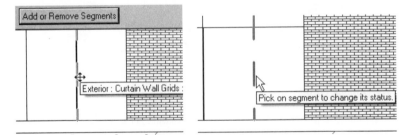

Figure 5–63 *Removing a curtain grid segment*

5.18 Select Curtain Grid from the Modeling tab of the Design Bar. Choose One Segment from the Options Bar. Place a horizontal curtain grid at **7'** elevation in the open grid you just created, as shown in Figure 5–64.

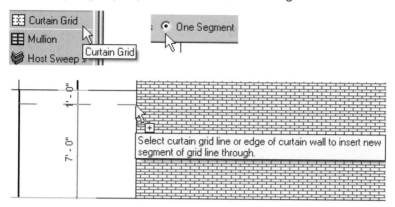

Figure 5–64 *Adding a curtain grid segment*

5.19 Zoom to Fit. Pick Mullion from the Modeling tab of the Design Bar. From the Options Bar, choose All Empty Segments. Accept the default Mullion type in the Type Selector. Put the cursor over the curtain wall and the whole wall will highlight. Left-click on the wall. Revit Building will create mullions throughout the curtain wall. See Figure 5–65.

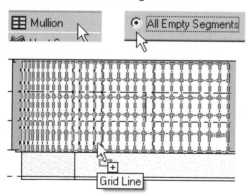

Figure 5–65 *Create mullionis on the entire curtain wall*

5.20 Expand the Families section of the Project Browser. Expand the Curtain Panels Section under Families. Note that it contains four types of panels: Empty Panel, Empty System Panel, Glazed Panel, and System Panel. You need a door to go in a curtain wall panel.

5.21 From the Modeling tab, select Door. On the Options Bar, choose Load from Library. Navigate to the *Content/Imperial Library/Doors* folder. Select *Curtain Wall-Store Front-Dbl.rfa* (see Figure 5–66). Click Open.

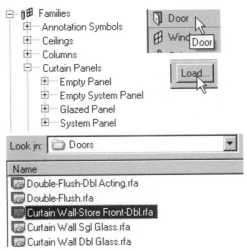

Figure 5–66 *The door type in the Imperial Library*

5.22 Revit Building will attempt to place a door, but not the type you just loaded, and the new door type will not be in the Type Selector list. Click ESC. The Families/Curtain Panels section of the Browser will now show **Curtain Wall-Store Front-Dbl** (see Figure 5–67). The Door family has been loaded into the file, but does not work with the Door tool. There is another way to specify the correct door type in the curtain wall panel.

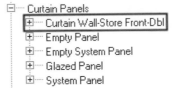

Figure 5–67 *The new curtain panel type in the Project Browser*

5.23 Zoom in Region around the open panel you created earlier. Place the cursor over a mullion at the edge of the panel. Hit TAB until the four sides of the panel pre-highlight and the tooltip identifies the System Panel (see Figure 5–68). Left-click to select it.

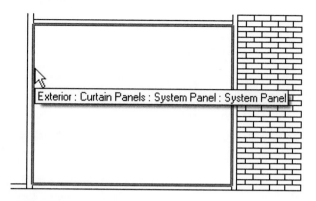

Figure 5–68 *The System Panel selected*

5.24 From the Type Selector drop-down list, choose Curtain Wall-Store Front-Double Door. Revit Building will fit a double door into the panel, as shown in Figure 5–69.

Figure 5–69 *The panel type set to double door*

5.25 Open the 3D View. Zoom to Fit. Open the Worksets dialogue. Open the Site workset. Select OK. From the View control bar, change the display to Shading with Edges. See Figure 5–70.

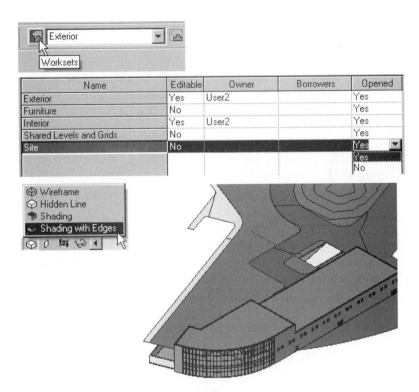

Figure 5–70 *A shaded view of the curtain wall*

5.26 Pick Save to Central from the Toolbar. Check User-created Worksets and Family Worksets to relinquish the Worksets for editing by others. Check the option to save the local file automatically. In the Comment field, type: **Building shell with curtain wall**, as shown in Figure 5–71. Click OK. If you have not yet worked Exercise 4, move to Step 4.1.

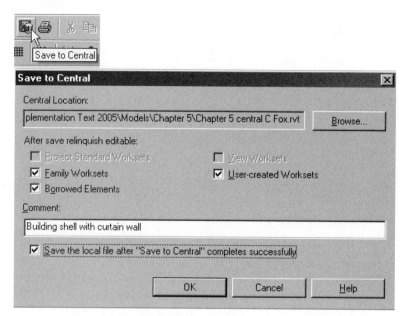

Figure 5–71 *Save Central with a comment*

This ends the tandem-user section of exercises in this chapter. For the rest of the exercise steps there will be no usernames specified. Close all open files. If you have worked the tandem-user exercises by yourself, close one session of Revit Building and use the menu to choose Settings>Options to set the Worksets Username to its previous value.

EXERCISE 6. WORKSETS IN A LINKED FILE

6.1 Open a new Revit Building project. Select File>Worksets from the File menu to activate Project Sharing. Click OK in the Project Sharing dialogue.

6.2 In the Worksets dialogue, click New. Name the new Workset **Links**. Clear Visible by default in all views. Click OK. In the question box that appears, click No, as shown in Figure 5–72. Workset1 will be the active Workset.

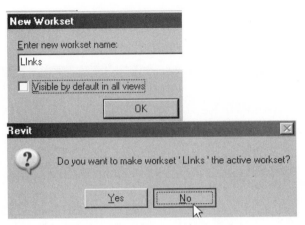

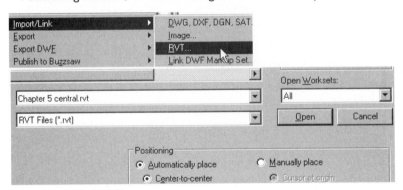

Figure 5–72 *Create a Links Workset, but do not make it active*

6.3 From the File menu, select File>Import/Link>RVT. Navigate to the folder that holds *Chapter 5 central.rvt*. Select the file so that it highlights and appears in the File Name field. In the Open Worksets field, select All. Accept the Positioning defaults, as shown in Figure 5–73. Click Open.

Figure 5–73 *Open a linked file*

6.4 Zoom to Fit. Type **VG**. In the Visibility/Graphic Overrides dialogue, check the Worksets tab. Note that Links is not checked. Leave it for now. On the Linked RVT Categories tab, expand the list under the linked file name. Note that all model categories in the linked file are checked. Click OK.

 NOTE The Site Workset in the linked Central File was not set to display in all views by default. At this time there is no way for Revit Building to override that condition in a linked file, so the topography of the link will not display.

6.5 Select the linked file. Pick Copy from the toolbar. Make a copy 100′ down (–90° vertically) on the screen, as shown in Figure 5–74.

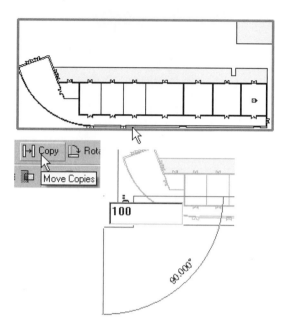

Figure 5–74 *Copy the link*

6.6 From the File menu, select File>Manage Links. On the RVT tab, highlight the name of the linked file. Click Unload. Place the cursor on the title bar of the dialogue, left-click, and drag the dialogue to the right of the drawing window so that you can see what is under it. Both instances of the linked file have disappeared from the screen.

6.7 Choose Reload. Both link instances will reappear. Select Reload From. See Figure 5–75.

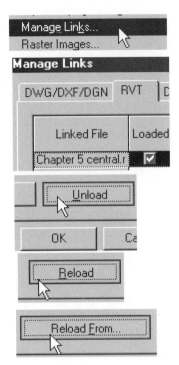

Figure 5–75 *Unload, Reload and Reload From are controls for links*

    6.8 In the Add Link dialogue, make sure that the same file name appears in the File name field and that the Open Worksets value is Specify. Click Open. In the Linking Worksets dialogue, select all but Exterior, as shown in Figure 5–76. Click Close.

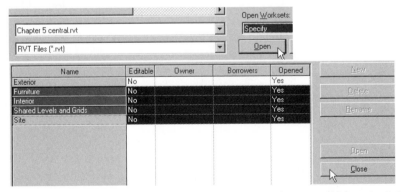

Figure 5–76 *Select all but the Exterior Workset to close (make not visible) in the link*

    Limiting the visibility of Workset elements in linked files reduces the amount of system memory Revit Building has to use.

6.9   Click OK. The linked files will appear without the interior walls and doors visible, as shown in Figure 5–77. Click OK to exit the Manage Links dialogue.

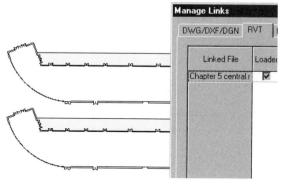

Figure 5–77 *Two instances of a linked file—the Interior Workset is not open*

6.10   Pick the lower of the two instances of the linked file. Select Properties. In the Properties dialogue, change the Workset value to Links, as shown in Figure 5–78. Click OK. The instance will disappear.

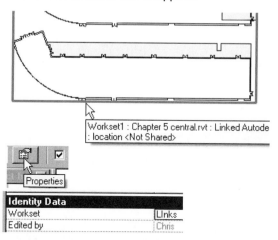

Figure 5–78 *Place one linked file instance in the Links Workset*

6.11   Open the Level 2 Floor Plan view. In this view the linked file instance in Workset1 is visible. The instance in Links Workset is not. Type **VG**. In the Visibility/Graphic Overrides dialogue, open the Worksets tab. Check Links. Click OK. The second link instance appears. See Figure 5–79.

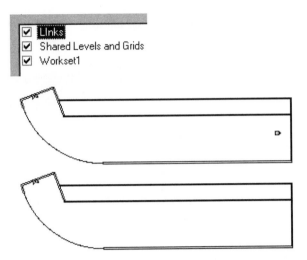

Figure 5–79 *Turn on the Links Workset in the 2nd floor plan*

6.12 Open Level I Floor Plan. Only one link instance is visible.

6.13 Close the file without saving.

This is an example of a way to organize linked files in a campus development project file using Worksets. If your file contains models of three or four housing types, for instance, using a Workset specifically for links allows you to control visibility of more than one link at a time, rather than unloading or reloading files in the Manage Links dialogue.

Revit Building gives the user a number of different ways to control the visibility of project elements or entire models, as you have just seen. The Visibility/Graphics Overrides dialogue allows you to adjust the visibility of elements in a view by object category in the project file, by Workset, and (within limits) by category within a linked file.

EXERCISE 7. PHASING

7.1 Open the file *Chapter 5 central.rvt*. Pick File>Save As from the File menu. Save the file in a location specified by your instructor under the name *Chapter 5 interior.rvt*. In the Save As>Options, make sure that Make this the Central location after save is cleared, so that you create a local file. See Figure 5–80.

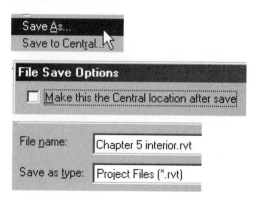

Figure 5–80 *Save a new local copy of the central file to work on the interior*

7.2 Open the 2ND FLOOR plan view. Type **VV** to open the Visibility/Graphic Overrides dialogue. On the Worksets tab, check Furniture and Interior. Clear Shared Levels and Grids. On the Annotation Categories tab, clear Door Tags (see Figure 5–81). Click OK. Select Make Elements Editable in the Warning box to save the view properties changes.

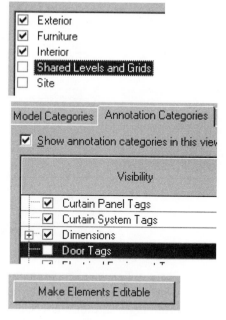

Figure 5–81 *Editing Visibility—turn off Door Tags*

7.3 Pick the Worksets icon on the Toolbar to open the Worksets dialogue. Make Furniture and Interior open and editable. Close Site and Shared levels and Grids. Click OK. Make Interior the active Workset. See Figure 5–82.

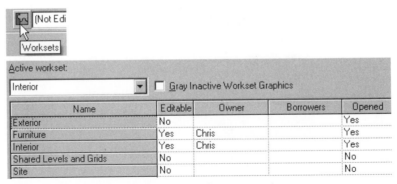

Figure 5–82 *Workset settings for interior and furniture work*

7.4 From the Menu bar, pick Settings>Phases, as shown in Figure 5–83. On the Project Phases tab of the Phasing dialogue, two default Phases will display: (1) Existing and (2) New Construction. Select the Name field for Phase 2 and change it to **Building Shell**. Choose the After button, in the Insert section of the dialog box, to add a new line. Revit Building will create Phase 1 on line 3.

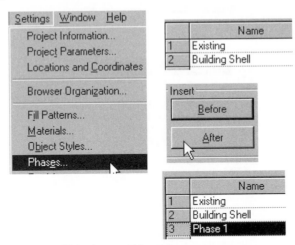

Figure 5–83 *Rename a default phase, add another*

7.5 Make the Name for Phase 1 **Temporary Classrooms**. In the Description field, type **2nd floor used for classes during Miller Hall renovations**. Select Insert After button again. In the Name field for Phase 1 (on line 4), type **Offices**. Type the following in the Description field: **2nd floor offices and labs** (see Figure 5–84).

Project Phases	Phase Filters	Graphic Overrides		
		PAST		
	Name	Description		
1	Existing			
2	Building Shell			
3	Temporary Classrooms	2nd flor used for classes during Miller Hall renovations		
4	Offices	2nd floor offices and labs		

Figure 5–84 *One Phase renamed, two new ones added with descriptions*

7.6 Select the Phase Filters tab. Study the default Phase display settings for a moment.

7.7 Select the Graphic Overrides tab and study it so that you gain an understanding of the display characteristics for the different Phase conditions.

7.8 Click OK to exit the dialogue. Select Make Elements Editable in the Warning box that appears (see Figure 5–84).

7.9 Choose two points on either side of the screen to select everything visible in a crossing pick box. Note that the exterior walls and elevations do not highlight, since the pick box setting with Worksets activated includes an option for Editable Only, which is checked by default.

7.10 Select Filter. Clear Floors. Click OK. The selection set will include only interior walls and doors. Choose the Properties icon. Note that the Phase Created value shows Building Shell, the edited name for Phase 2. Change that value to **Temporary Classrooms**, as shown in Figure 5–85. Click OK.

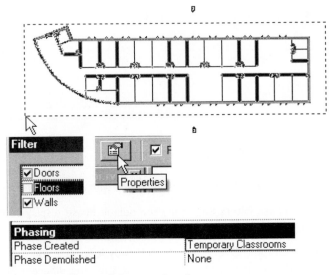

Figure 5–85 *Change the Phase of doors and walls*

7.11 The walls and doors will disappear. Type **VP** to enter the View Properties dialogue. Change the Phase Value for the View to Temporary Classrooms, as shown in Figure 5–86. Click OK. The walls and doors appear.

Phasing	
Phase Filter	Show All
Phase	Temporary Classrooms

Figure 5–86 *Change the Phase of the view*

7.12 Pick walls and doors as shown in Figure 5–87. The selected items are light gray in this image. Select the Properties icon. Change the Phase Demolished value to Offices. Click OK.

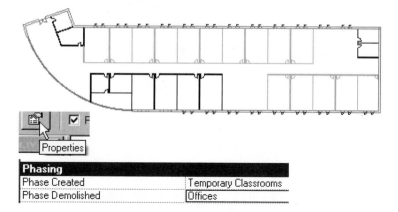

Figure 5–87 *Select walls and doors to designate as temporary*

7.13 Right-click on 2ND FLOOR in the Project Browser. Select Rename. Name the view **2ND FLOOR CLASSROOMS**. Click No in the rename level query.

7.14 Right-click on 2ND FLOOR CLASSROOMS in the Project Browser. Choose Duplicate. Rename the new view **2ND FLOOR OFFICES**. Type **VP**. Change the Phase of the View to Offices. Change the Phase Filter to Show Previous + New, as shown in Figure 5–88. Click OK. The walls that you designated to be removed in the Office construction Phase disappear. The Walls constructed in the previous phase (Temporary Classrooms) that remain in the phase for this view appear halftone.

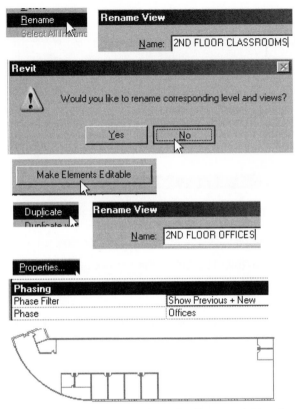

Figure 5–88 *Rename a view, duplicate it and set the Phase and Filter for the new view*

7.15 Create 20 Office spaces each 11′ wide, as shown in figures 5–89 and 5–90. Do not draw the dimensions. Make the walls Generic 6″ and the doors Single Flush 36″ x 84″. Ignore wall-window conflict warnings.

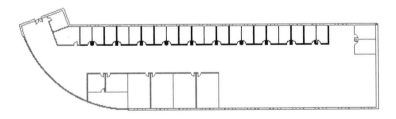

Figure 5–89 *New offices on the second floor*

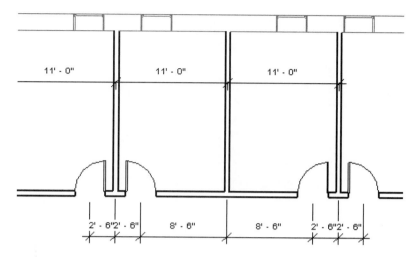

Figure 5–90 *Typical spacing for doors*

 TIP Try using the array tool to avoid drawing 20 copies of walls and doors. Clear Group and Associate and use an 11′ interval distance.

PLACE, EDIT, AND GROUP OFFICE FURNITURE

7.16 Use the Workset tool on the Toolbar to make Furniture the Current Workset.

7.17 Select Component from the Basics tab on the Design Bar. Choose Load. Navigate to the *Content/Imperial Library/Furniture System* folder and select *Work Station Cubicle.rfa* (See Figure 5–91). Click Open.

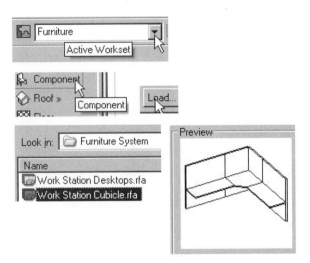

Figure 5–91 *Find a Work Station Cubicle component to insert*

7.18 Locate an instance of the workstation as shown in Figure 5–92. Do not draw the dimensions.

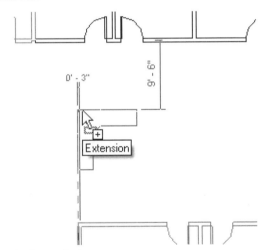

Figure 5–92 *Locate the first cubicle*

7.19 Choose Modify from the Design Bar to terminate component placement. Select the new workstation. Pick Array from the Toolbar. Make a linear array of the workstation as shown in Figure 5–93 (24 copies x 8'-2" move distance). Clear Group and Associate on the Options Bar. Choose a convenient point to start the distance, and pull the cursor to the right until the temporary dimension reads 8'-2", then left-click to accept the distance. Revit will snap to points on the cubicle. You can also enter **8'2** at the keyboard.

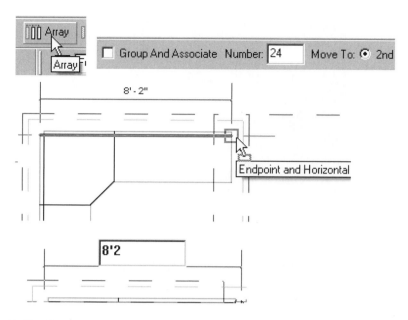

Figure 5–93 *Getting ready to array the workstation*

7.20 Zoom to the right side of the model. Copy the right most cubicle 25' down (–90°). The exact location is not critical.

7.21 The new component will still be highlighted. Select the Properties icon from the Options Bar. Pick Edit/New in the Type Properties dialogue, choose Duplicate In the Name box, type **108″ x 108″** in the Name box and click OK.

7.22 In the Type Properties dialogue, change the Panel1 Length and Panel2 Length values to **6**—Revit Building will translate the value to 6' 0″, as shown in Figure 5–94. Click OK twice to exit the edit dialogues. The cubicle will grow.

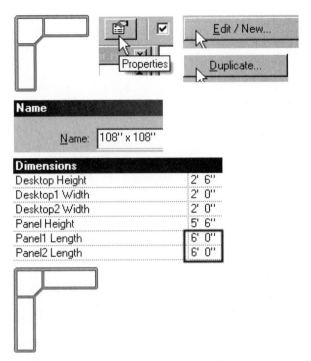

Figure 5–94 *Adjust the panel length in the new larger cubicle type*

7.23 Select the cubicle. Mirror it along one edge. Mirror the two instances to make a four-sided cubicle arrangement as shown in Figure 5–95.

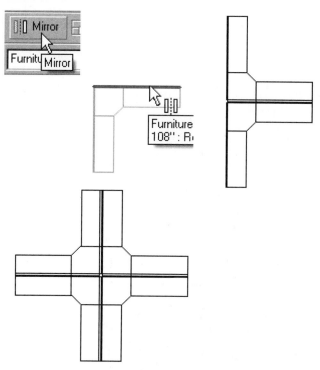

Figure 5–95 *The large cubicles arranged back to back to back*

7.24 Select all four cubicles. Choose Group from the Toolbar. Revit Building will make a Group of the four cubicles (which are, themselves, Groups). Expand the Project Brower display under Groups>Model. Select the name of Group 1, right click and choose Rename. Change the name of the Group to **Quad Workstation 108″** as shown in Figure 5–96.

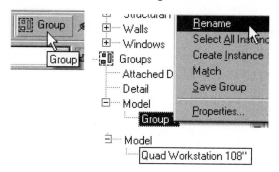

Figure 5–96 *Rename the Group*

CAD Manager Note: Groups are a very useful tool in Revit Building. When combined with Worksets they introduce a level of complexity that needs to be carefully managed. Groups can be created with elements from different Worksets; this can create headaches later, as designs develop and teams interact.

7.25 Select the new group. Select Rotate from the Toolbar. Accept the default location of the rotation icon and rotate the Group 45° (see Figure 5–97).

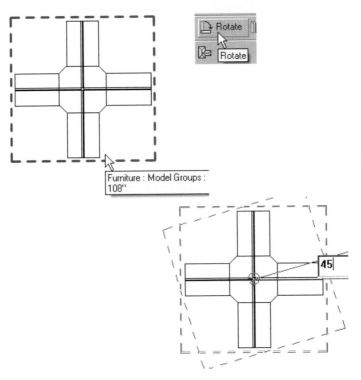

Figure 5–97 *Rotate the workstation*

7.26 Pick Copy. Clear Constrain on the Options Bar. Check Multiple. Copy the Group to the left and right at a distance interval of **25′** four times, as shown in Figure 5–98. When the copies have been created, hit ESC twice to terminate the Copy tool.

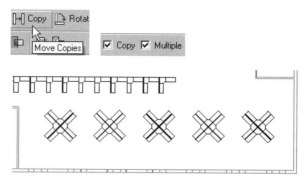

Figure 5–98 *Workstations in place in the Furniture Plan*

7.27 Zoom to Fit. Save to Central. Relinquish Family Worksets and User-Created Worksets. For the Comment field, type **Interior design with two phases on 2nd floor**. Save the local file. See Figure 5–99. Close the file.

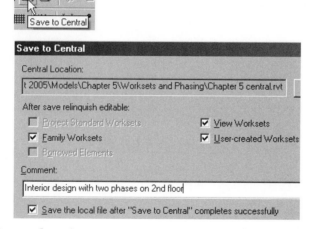

Figure 5–99 *Save to Central*

PHASES AND FILE LINKING

7.28 Open a new file. Enable Worksets in the new file. Do not create any new Worksets. Click OK.

7.29 From the File menu, select File>Import/Link>RVT. Find the *Chapter 5 interior.rvt* file you just created and link it, accepting the defaults. Zoom to Fit.

7.30 From the Settings menu, select Settings>Phases. Note on the Project Phases tab that linking a file does not make the Phases in that file accessible in the host file.

7.31 Check the Project Browser. There are no Floor Plans corresponding to the Phased Floor Plans you created in the previous file.

7.32 Select the Worksets Icon. Note that Worksets in a linked file are not available in the host file.

7.33 Close the file without saving it.

SUMMARY

This chapter's exercises have introduced you to the uses of Worksets and Phasing, two important mechanisms in Revit Building for organizing a building model project file. Worksets divide a project's contents according to whatever system you decide works best for you. The Save to Central/Save Local function allows numerous users to work on a project simultaneously and coordinate their efforts. Many firms find that using Worksets in projects without multiple users is worthwhile for the control it gives designers over component visibility and organization.

Phases provide a way to show changes in a project over time. Phases, together with their associated Filters and Graphic Overrides, give complete control over view visibility characteristics—new work, old work, temporary work and future work can all be displayed or hidden according to your preferences.

REVIEW QUESTIONS – CHAPTER 5

MULTIPLE CHOICE

1. A Workset can be

    a) Open but not Editable

    b) Editable but not Open

    c) Active but not Editable

    d) all of the above

2. To select elements in a Workset-enabled file,

    a) at least one Workset must be editable

    b) Editable Only must be cleared on the Options Bar

    c) either a or b

    d) both a and b

3. Properties of Slope Arrows for floors include

a) Width and Rotation

b) Height at Tail, Height at Head, Offsets

c) Thickness

d) all of the above

4. When placing grid lines in a curtain wall,

    a) you can set vertical and horizontal lines in any order

    b) you must work right to left for vertical lines

    c) you must work top–down for horizontal lines

    d) Revit Building disables snaps

5. To place a door in a curtain wall,

    a) Create a hole in the curtain wall and place a standard door

    b) Create a section of standard wall inside the curtain wall

    c) Load the curtain wall door family you want to use, and edit the curtain wall panel to the correct door type

    d) You can't place doors in curtain walls

TRUE/FALSE

6. In order to make a change in a model object that is part of a Workset already being edited by another user, you must make an Editing Request of that user.

7. The user needs to create Worksets for annotation types, as Revit Building does not create those Worksets automatically.

8. When a Workset-enabled file is linked into a host file, Revit Building displays only the Worksets from the link that were originally set to display in all views.

9. When you Attach the bottom of a wall to a roof or floor below the wall, Revit Building will split the wall at the floor line and you have to delete the extra section by hand.

10. Revit Building will let you create as many Phases in a project as you want.

 Answers will be found on the CD.

CHAPTER 6

Managing Annotation

INTRODUCTION

So far you have been working on the design tools integrated into Revit Building. In previous chapters you have massed buildings, and added walls, doors, and windows. You have linked models and sites, and shared models with others. The main purpose of building a model is to communicate to the construction team what the design team wants to build. Traditionally, design communication has been in the form of construction documents: printed collections of plans, sections, elevations, and specifications. This may not be the only way to deliver or communicate design intent in the future, but documents of a certain widely accepted appearance are today's standard and will not soon be completely replaced. If a wonderful model of a wonderful building cannot be understood by those who have to approve, pay for, or build it, the building won't be built. Proper annotation is essential to effective design practice.

Revit Building is comprehensive building information modeling software for the full spectrum of architectural practice, so you shall now explore its tools for documentation of your designs. This chapter will contain several sections to illustrate different types of standard documentation and the management of each technique. In the following exercises we will demonstrate techniques for adding, editing, and managing annotation. In the interest of keeping the exercises moving, we will focus on one area of our hypothetical project. Please note that many of these exercises are organized to show you techniques for setting up and managing annotations, and therefore a certain similarity will become evident. In a working environment, once annotations have been set up in a project, the settings can be saved in company standard templates and therefore will not have to be repeated.

For the sake of simplicity, while previous chapters included worksets (a one-way process, and a normal step in Revit Building design team environment), the model file that starts this chapter has been created without worksets.

OBJECTIVES

- Create, insert, manage, and modify tags—Room, door, window, sections, elevation, and grids

- Insert, manage, and modify annotation—Notes and dimensions
- Transition entire set from Schematic annotation style to Design Development annotation style

REVIT BUILDING COMMANDS AND SKILLS

Creation and customization of:

Room Tags, Section Indicators, Elevation Symbols, Titleblocks, Dimensions, Units, Font Styles, Notes, and Grids

Working with Phases

View Templates

Project Browser as a Tool

Views and Sheets

View Titles

Tag All Not Tagged

One of the great benefits of using Revit Building is its coordination of all objects in the model, including tagging. Because the tag is merely a view or "looking glass" into one specific piece of information about the object it annotates, the coordination of that information is handled by the object and the computer. Very little user input is necessary, which eliminates opportunities for user error. There are tags for just about any Revit Building object you can think of, and customizing them to match company standards is a simple task.

 NOTE In an effort to demonstrate the management of annotation and its appearance in different phases of design, we will assume that you are in our project's Schematic Design phase during the first portion of the exercises. Annotation during this phase will be loose in appearance. You will be using the Comic Sans MS font style (standard on Windows PCs) to illustrate this. Midway through the chapter you will switch to a Design Development phase appearance with more formal "hard-line" tags and annotations. You will make this change in a few short steps.

In many cases you will want your annotations to convey a different image at different times. You may want to show or not show certain information at different stages of design. The best way to do that is through the customization of the tag that reports that information. In the following exercise you will create some custom tags for a Schematic Design look.

EXERCISE 1. CUSTOMIZING TAGS: ROOM TAGS

1.1 Launch Revit Building. From the File pulldown menu, select File>Open. In the dialogue that opens, navigate to the *Library/Chapter 6* folder. Select *Room Tag With Area.rfa* from the *Library* folder (see Figure 6–1). Select File>Save As from the File menu. Save the file as **Room Tag With Number - Area – SD.rfa** (the SD in this filename stands for Schematic Design).

Room name

$$\boxed{101}$$

150 SF

Figure 6–1 *The Room tag file before changes*

1.2 Select and delete the lines that form the box around the room number.

 NOTE Pre-highlighting one of the lines and selecting TAB *will select all the lines in the box*

1.3 Select the Room Name label and pick the Properties button in the Options toolbar. Select Edit/New and change the Text Font to Comic Sans MS. Check the Bold, Italic, and Underline checkboxes, as shown in Figure 6–2. Click OK twice to see the results.

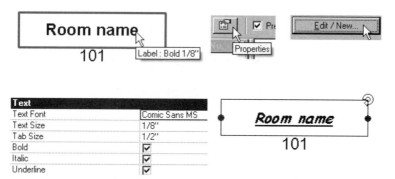

Figure 6–2 *Label font control dialogue*

1.4 Repeat the steps above with the Room Number label. Change the Text Font in the Room Number label to Comic Sans MS, and select the Italic checkbox. Click OK twice to see the change.

NOTE Both the Room Number and Room Area label text fonts are changed. This is because they are both driven by the same Type Parameters.

1.5 Type **VG** to open the View Graphics dialogue. Select the Annotation Categories tab and check the Reference Planes checkbox. Click OK.

1.6 The intersection of the reference planes represents the insertion point of the tag. Move the Room Number and Room Area labels closer together so they look more like a single text item than individual lines (see Figure 6–3). Save the file. Close the file.

Room name
101
150 SF

Figure 6–3 *Room tag with room number and area set up for Schematic Design annotation*

 NOTE Use the new "nudge tool" to move the text closer together by tapping the arrow keys once you have selected the text. You will note that the closer you are zoomed in, the less the text actually moves. The move distance is controlled by the snap distance, which is zoom-dependant

ROOM TAGS

1.7 Open the file *Chapter 6 start.rvt*. It will open to the 1ST FLOOR – FLOOR PLAN view that shows exterior and interior walls. Save the file as **Chapter 6 annotation.rvt** in a location determined by your instructor.

1.8 From the File menu, select File>Load From Library>Load Family. Navigate to find the *Room Tag With Number - Area – SD.rfa* you just created and click Open. This room tag will already be loaded in the *Chapter 6 start* file; click Yes to overwrite it with your newer customized version.

TIP If you have the family file open in the family editor, you can select Load into Projects from the Design Bar. This will load the family directly into your current project (or ask you which open files to load it into). You can open a family file directly from a Revit Building project by selecting the family and picking the Edit Family button on the options bar. See Figure 6–4.

Figure 6–4 *Load into Projects and Edit Family icons*

1.9 From the Basics Design Bar, select the Room Tag tool. Make sure Room Tag with Number - Area – SD is the active tag. Place a room tag in all of the rooms that do not already have one, plus the open curved area at the left of the plan. Select Modify to terminate placement. See Figure 6–5.

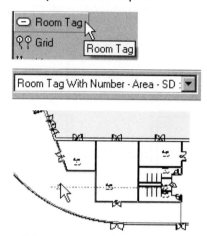

Figure 6–5 *Place instances of the new room tag*

1.10 Zoom in Region to the far left of the plan. Select the Room Tag in the large curved area, click on the Room label text, and it will become editable. Change the name to **Lobby**. Do the same for the Number label and change it to **101**. Continue from left to right and name the rooms as follows: **102 – Office, 103 – Office, 104 – Storage, 105 – Men, 106 – Women, 151 – Janitor** (as shown in Figure 6–6), and **152 – Elevator**. Actual room names and numbers are not critical.

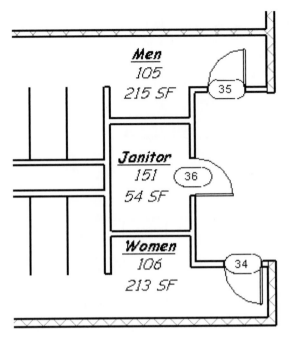

Figure 6–6 *Room tags with room numbers and area*

SECTION INDICATORS

1.11 From the Basics Design Bar, choose the Section tool. In the Type Selector, pick Section: Building Section from the list, and add three sections as shown in Figure 6–7.

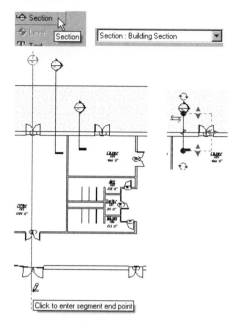

Figure 6–7 *Section indicators—controls allow you to flip the section head, split the line, and cycle through alternate head/tail symbols*

1.12 Select the far right section indicator and pick the Properties icon from the Option toolbar. Pick Edit/New in the Element Properties dialogue, and then select Duplicate. In the Name field, type **Wall Section** as shown in Figure 6–8, and click OK. Change the Section Tag value to Detail View 1 and click OK twice. The new section will now have an open section head.

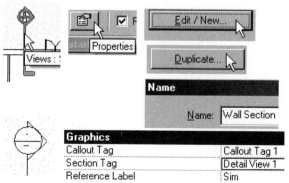

Figure 6–8 *Changing the section head properties for a detail (wall) section*

1.13 You will now have two section types to choose from when adding a section, plus a Detail View, as shown in Figure 6–9.

340

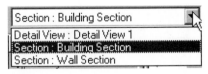

Figure 6–9 *Section type selection*

Many offices differentiate wall sections and building sections with different section heads: thus the solid fill and no fill section heads. Additionally, many experienced Revit Building users find that using a section to study a model and a different one for documentation purposes is useful. Some offices develop a temporary section that is graphically different to indicate to other users that it is permissible to move, edit, or even delete it without consequences to the documentation set.

1.14 Select the middle section. Cycle the head as shown in Figure 6–10 so that the section shows as a double tail section indicator. Select the Split Segment tool from the Options Bar, split the section, and draw the lower portion into the restrooms as shown.

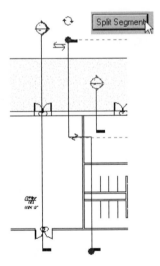

Figure 6–10 *Building section, temporary section, and wall section indicators*

ELEVATION INDICATORS

1.15 From the View Design Bar tab (right-click over the Design Bar and check View from the list if necessary), select the Elevation tool. Place an elevation indicator on the bottom of the women's room, as shown in Figure 6–11. Hit ESC to terminate the tool.

TIP Revit Building will snap the point of the elevation to the nearest wall. You can also use **TAB** to cycle through pointer locations before placing the elevation indicator.

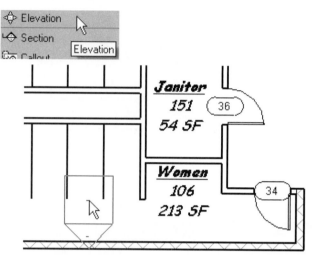

Figure 6–11 *Interior elevation indicator*

1.16 You will now create a Schematic Design style elevation marker to go with the Schematic Design room tags you created before. From the Menu Bar, pick Settings>View Tags>Elevation Tags. There are two default types, square and circular. Select Duplicate. Type **Interior Elevation – SD** in the Name dialogue and click OK. Change the Width, Shape, and Text Position values as shown in Figure 6–12. Place a check in the Filled value and change the Text Font to Comic Sans. Click OK twice to close the dialogue box.

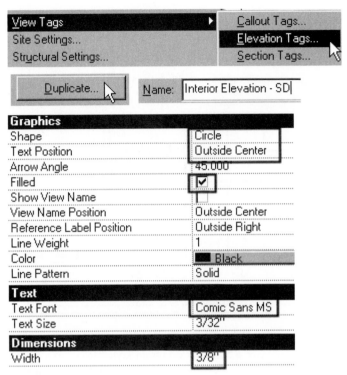

Figure 6–12 *SD elevation tag properties*

1.17 Now you apply the new Elevation Tag type to a new Elevation Type.

1.18 Select the square portion of the elevation. Pick the Properties icon from the Options toolbar. Pick Edit/New from the Element Properties dialogue. Choose Duplicate from the Type Properties dialogue. Type **Interior Elevation – SD** in the Name dialogue and click OK. Change the Elevation Tag value to Interior Elevation – SD as shown in Figure 6–13. Click OK twice to close the dialogue box.

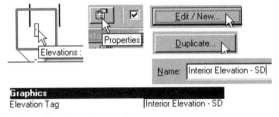

Figure 6–13 *Interior elevation type properties*

1.19 Select the elevation marker (now a circle); checkboxes will appear at the quadrants. Check the three boxes that are not checked (see Figure 6–14).

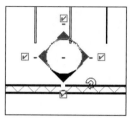

Figure 6–14 *Checkboxes at the quadrants indicate whether an elevation has been generated or not*

 NOTE Elevation views will be added to your Project Browser (see Figure 6–15).

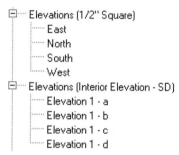

Figure 6–15 *Elevation Views are automatically added when you check the associated checkboxes*

CREATE AN ELEVATION VIEW TEMPLATE

1.20 Double-click on any one of the arrow portions of the interior elevation indicator. This will open the associated interior elevation. Type **VP** to open the Elevation View Element Properties dialogue. Change the View Scale to 3/8″ = 1′ – 0″ and click OK.

1.21 Choose the View pulldown and select Save As View Template. Type **Interior Elevation – 3/8″** and click OK. In the View Templates dialogue that opens, change the Detail Level value to Fine. Click OK (see Figure 6–16).

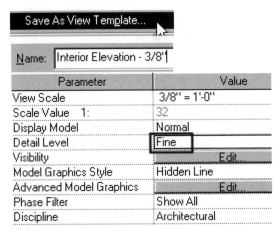

Figure 6–16 *Save view settings for future use*

1.22 Open a different interior elevation view. Pick the View pulldown menu and select Apply View Template. In the Select View Template dialogue, select Interior Elevation – 3/8". Click OK. Repeat for the other two interior elevations (see Figure 6–17). Note that there is an option to apply the template you choose to new views as you create them.

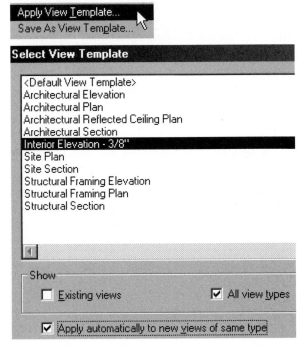

Figure 6–17 *Apply the new View Template to the other interior elevation views*

NOTE In this case you are using the View Template to simply store the scale and detail level of the Interior Elevations, but this tool can be used to manage any number of visibility settings including linestyles, lineweights, and colors. Just about anything you can change regarding view characteristics or visibility can be stored and managed using view templates. It will be well worth your time to explore these settings on this or any other model.

1.23   Save your project file.

EXERCISE 2. TEXT AND NOTES

As you recall, you are hypothetically in Schematic Design, and all of our annotation text so far has used the Comic Sans MS font. You will need to create a note style that matches.

2.1   Open or continue working in the file from the previous exercise. Open the 1ST FLOOR – FLOOR PLAN view. Zoom to include the washrooms and open area to the right.

2.2   From the Basics tab of the Design Bar, pick the Text tool. Select Text: 3/32″ Arial from the Type Selector pulldown, and select the Properties icon. Choose Edit/New. Click Rename and change the name of the Text Type to **3/32″ Note**. Click OK in the Rename dialogue. Change the Leader Arrowhead value to Heavy End 1/8″. Change the Text Font to Comic Sans MS and check the Italic checkbox, as shown in Figure 6–18. Click OK twice.

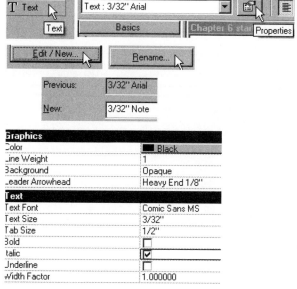

Figure 6–18 *Modifying text properties*

2.3 While still in the text command, pick the Two Segment leader and the Left Justified Text Alignment from the Options Bar, as shown in Figure 6–19.

Figure 6–19 *Leader type and Text Alignment*

2.4 To start a text leader, pick a spot on the bottom washroom wall, then a point down and to the right of that, and finally select a point directly to the right of the second point. Type **8″ MASONRY WALL, FULL HEIGHT. PROVIDE FULL FIRE SEALANT**. Pick a point outside of the text to end the text string, and hit ESC twice to get back to the Modify tool.

2.5 Select the text you just entered to bring up the blue grips and drag the right-hand grip to the left to adjust the width of the text box.

 NOTE There are many ways to move the text you have created with different results. First pick the text and put the cursor over the highlight around the text. Drag the text box to drag the text and leader as a stationary unit. Now select the text and drag the blue four-headed arrow at the top-center of the text box. Note that the text moves but the leader arrowhead remains stationary (see Figure 6–20).

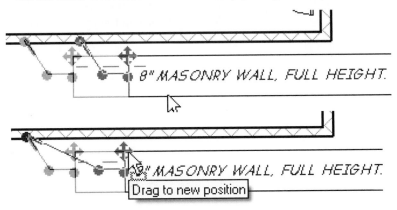

Figure 6–20 *Moving text two ways*

2.6 Continue adding notes as shown in Figure 6–21. Revit Building automatically snaps the second and endpoints of the leader line to align with the adjacent leaders and notes.

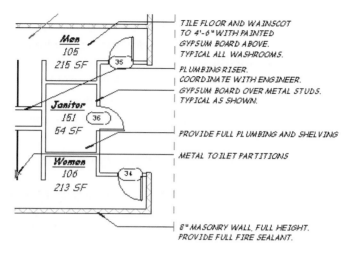

Figure 6–21 *Schematic Design notes and leaders*

DIMENSIONS: LINEAR

 NOTE For the purpose of linear dimensions, you will use an enlarged plan of the washrooms.

2.7 From the View Tab on the Design Bar, select the Callout tool. Pick a point at the upper-left of the washrooms and pick again at the bottom right. Move the callout head to the bottom of the callout lines by selecting the callout and dragging the grip at the intersection of the callout head and the leader to the bottom of the callout, as shown in Figure 6–22. Hit ESC once to get back to the Modify command.

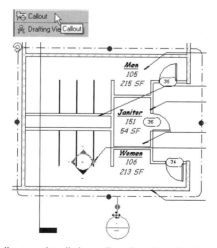

Figure 6–22 *Create a callout and pull the callout head to the bottom of the callout*

2.8 Open the callout by double-clicking the blue callout head. In the callout view, type **VP** to open the View Properties dialogue. Set the View Name to **ENLARGED WASHROOM PLAN**, the View Scale value to ¼"=1' - 0", and clear the Crop Region Visible checkbox, as shown in Figure 6–23.

2.9 Select Edit from the Visibility Parameter (this is the same as typing **VG** from the view). Pick the Annotation Categories tab and clear the Elevations checkbox. Click OK twice to see the results.

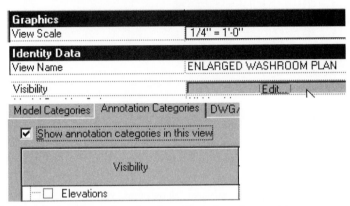

Figure 6–23 *View Properties settings for the enlarged washroom plan*

2.10 From the Menu Bar select View>Save As View Template. Name the template **ENLARGED WASHROOM PLAN**. Choose OK twice to create the template. From the Menu Bar select View>Apply View Template. Select the new ENLARGED WASHROOM PLAN template you just created. Check the option to apply this template to new views of the same type. See Figure 6–24.

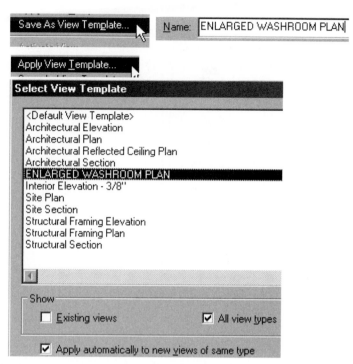

Figure 6–24 *Use the new view settings as a template*

2.11 Add room tags to the Men's, Women's, and Janitor's rooms in this view. The Room Tag tool is on the Basics tab of the Design Bar. See Figure 6–25 for appropriate locations. Notice that the room information is filled in for you. You have previously defined each space as a room and identified it using a tag; Revit Building recognizes it as an object and therefore "reports" the information through the new room tag.

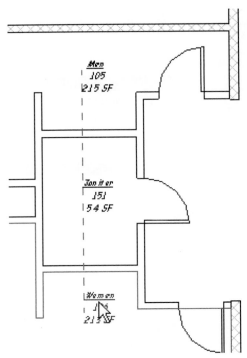

Figure 6–25 *Add room tags*

2.12 From the Basics or Drafting Design Bar, select the Dimension tool. Add dimensions to the plan similar to Figure 6–26. (You will need to switch between Prefer wall centerlines and Prefer wall faces on the Options Bar during the dimension picks.)

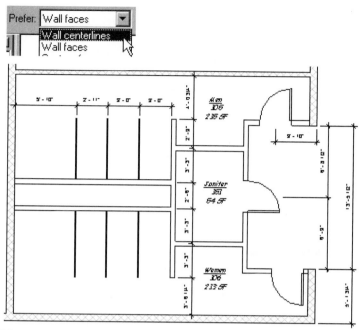

Figure 6–26 *Place dimensions*

2.13 Select Modify to terminate the dimension tool. Select the entire view, use the Filter tool, and clear all of the categories except Dimensions, as shown in Figure 6–27.

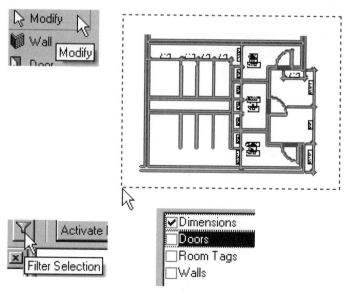

Figure 6–27 *Filter out all categories except the Dimensions*

2.14 From the Options toolbar, select the Properties icon. Select Edit/New, then Rename. In the New Name field, name the dimension style **Linear - SD** and click OK. Set the Parameters per Figure 6–28. Click OK twice when you are finished to see the results as shown in Figure 6–29. (You will notice that there is no Italic checkbox for the text.)

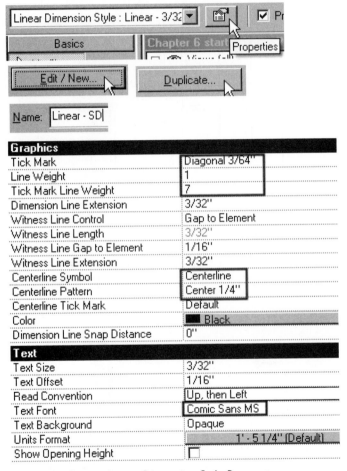

Figure 6–28 *Schematic Design Linear Dimension Style Parameters*

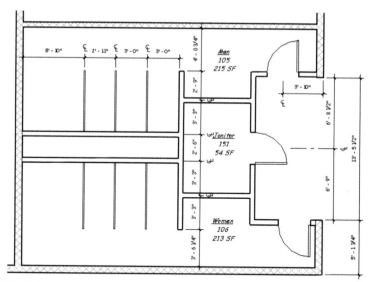

Figure 6–29 *Schematic Design dimensions on the washroom callout plan*

ANGULAR DIMENSIONS

2.15   Go back to the 1ST FLOOR – FLOOR PLAN view and Zoom to the curtain wall near the doors. From the Basics Design Bar, select the Dimension tool and pick the Angular icon, as shown in Figure 6–30.

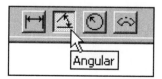

Figure 6–30 *Angular icon*

2.16   Dimension the curtain wall grids as shown in Figure 6–31. You will have to pick sides of the mullions carefully. Pick Modify from the Basics Design Bar and select the dimension string. Modify the dimension type Parameters using the technique outlined in Step 2.14. Name the type **Angular – SD**, and use all of the same Parameters you used for the linear dimensions. Choose OK twice.

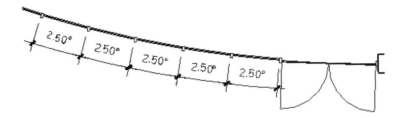

Figure 6–31 *Angular Schematic Design dimensions at curved curtain wall*

2.17 Now it is time to dimension the radius of the curtain wall. From the Basics toolbar, select the Dimension tool and check the Radial icon. Pick the centerline of the curtain wall, it will pre-highlight with a light dashed line and the status bar will read Walls: Curtain Wall: Curtain Wall 1. If it does not automatically pre-highlight, tap the TAB key while the cursor is over the center of the wall. When the wall centerline does highlight, pick the wall and then pick a location for the dimension line location (see Figure 6–32). Hit ESC twice to exit the command and put you back in the Modify tool.

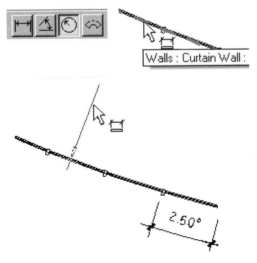

Figure 6–32 *Status bar feedback and wall centerline highlight*

2.18 Select the radial dimension you just created. Choose the Properties icon from the Options Bar. Click Edit/New, and then pick Rename .Rename the Dimension Type **Radial – SD**. Modify its Parameters to match Figure 6–28. Before closing the Type Properties dialogue box, select Units Format, clear the Use project settings checkbox, and change the Rounding value To the nearest 1″, as shown in Figure 6–33. Click OK three times to exit the dimension dialogue boxes.

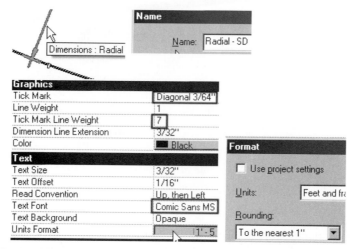

Figure 6–33 *Radial Dimension Parameters and Units/Rounding Format*

You now have Schematic Design dimension styles for all three dimension types, to use anywhere in this project or transfer to other projects.

GRIDS

Just like some of these other annotation families you have been modifying, Revit Building's grids also hold the parametric controls necessary to change their appearance with a few simple modifications. In this next exercise you will create a new grid head from scratch.

2.19 Use the File pulldown and select File>New>Family, as shown in Figure 6–34. The New Family browser will open for you to select a template file.

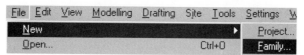

Figure 6–34 *New Family File*

2.20 Navigate to the *Chapter 6/Library* folder and open *Grid Head.rft*. This will open the template file to create any grid head geometry you want. In this case you will create a Schematic Design head to match our other annotations.

2.21 From the Family tab (the only one) on the Design Bar, pick the Label tool. Make sure that the Center and Middle icons are selected in the Text Alignment section of the Options Bar, as shown in Figure 6–35. Place the label centered on the vertical reference plane and above the horizontal reference plane. The label will snap weakly to the reference planes. Revit Building will ask you for the label name; there is only one Parameter to choose from on a grid, Name. Click OK.

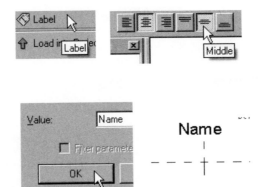

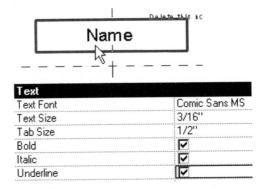

Figure 6–35 *Label Text Alignment*

2.22 Select the Modify tool from the Design Bar and choose the Name label. Pick the Properties icon from the Options toolbar. Pick Edit/New in the Element Properties dialogue, and change the Font to Comic Sans MS. Check the Bold, Underline, and Italic checkboxes, as shown in Figure 6–36. Click OK twice to see the results.

Text	
Text Font	Comic Sans MS
Text Size	3/16"
Tab Size	1/2"
Bold	☑
Italic	☑
Underline	☑

Figure 6–36 *Label text properties*

2.23 The red text and "dummy" line are there for your use in orienting the grid. They are no longer necessary—delete them. Save the file as **Grid Head – SD.rfa** in a location determined by your instructor.

2.24 Select the Load into Projects icon, as shown in Figure 6–37. If you have more than one project file open you will see a dialogue with a list of projects from which to choose. The screen will change to the project file. Return to the family file and close it.

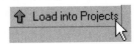

Figure 6–37 *Load grid head into the project*

2.25 From the Basics tab on the Design Bar, pick the Grid tool and draw a gridline by selecting a point at the bottom left of the project and another straight above the first. The grid head will appear as a circle with the number 1 in it. Select the Modify tool.

2.26 Select the grid, pick the Properties icon from the Options toolbar, and choose Edit/New in the Element Properties dialogue. Pick Duplicate and name the new type **Grid – SD**. In the Grid Head Value, select Grid Head – SD, as shown in Figure 6–38. Click OK twice to see the changes.

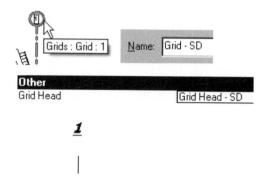

Figure 6–38 *New Grid properties*

2.27 From the Basics Design Bar, select the Grid tool. In the Type Options selector, pick Grid – SD, and continue placing vertical grids from left to right on your project. Specific locations are not critical. Note that the end points will snap to alignment; additionally, when you drag grid endpoints all of the other aligned grid endpoints will maintain their alignment.

2.28 Choose Modify. Select any grid line and pick the Copy tool. Choose an arbitrary point in space, drag the cursor horizontally to the right, type 1, and hit ENTER. You have just copied a grid line 1′ – 0″ to the right of the original. The text at the grid head will be overlapped. Zoom in Region to an area just around the grid heads.

2.29 Pick the new grid and drag the squiggle grip on the grid line to the right. The text will move and there will be an angled extension line on the grid line, as shown in Figure 6–39. Select the Grid label text to edit it.

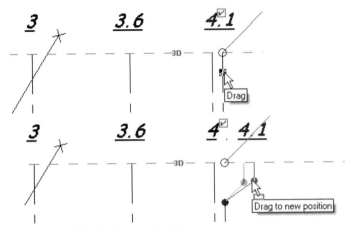

Figure 6–39 *New Schematic Design grids with offset heads*

2.30 Save your file.

EXERCISE 3. TITLEBLOCKS

NOTE One of the most common requirements in customizing standard annotation stems from the fact that every firm has an individualized titleblock or sheet frame to incorporate into plan sets. We are going to show you how to create a Schematic Design (horizontal) titleblock easily using an existing titleblock and its labels.

3.1 From the File menu, select File>Open. Navigate to the *Library/Chapter 6* folder and open the file *E 34 x 44 Horizontal.rfa*. This is a generic Revit Building titleblock that you will use to create our custom Schematic Design titleblock.

3.2 From the File pulldown, select File> Save As and save the file as **34 x 44 Horizontal – SD.rvt**, in a location determined by your instructor.

3.3 Zoom in around the lower-right corner of the sheet. While holding down the CTRL key, pick the text and labels circled in Figure 6–40.

Figure 6–40 *Text and labels to be reused in the Schematic Design titleblock*

3.4 Use the Move command to move the selection to the middle of the sheet. Repeat this technique to move the Autodesk Revit logo and web address from the top right of the titleblock to the middle of the sheet. Your arrangement will look similar to Figure 6–41.

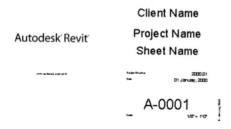

Figure 6–41 *Place all of the necessary labels and text in the middle of the sheet to modify them*

3.5 Delete all of the lines and remaining text from the right side of the sheet; leave the outside perimeter line for size reference. You are now going to change the appearance of the remaining text and labels.

3.6 Select the Autodesk Revit logo and pick Properties. This is a bitmap image with a height and width. Change the Width value to **7″** and click OK. See Figure 6–42.

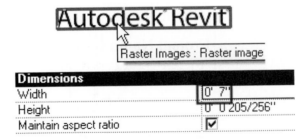

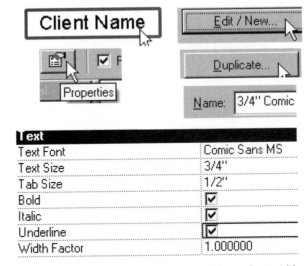

Figure 6–42 *Resize the logo image*

   3.7 Select the Address label and change the value in the Type Selector to Text: ¼".

   3.8 Select the Client Name label and click the Properties icon from the Options toolbar. Pick Edit/New in the Element Properties dialogue. Select Duplicate in the Type Properties dialogue. Name the new type **3/4" Comic** and click OK. Change the Type Parameter Values as shown in Figure 6–43. Click OK twice.

Figure 6–43 *¾" Comic label Parameters for the Client Name, Project Name, and Sheet Name labels*

   3.9 Drag the right-hand grip well to the right to allow for a long client name.

NOTE Remember that the labels represent information that will come from the project and may be longer in some projects than others. If you plan to re-use title-blocks in the future, allow enough space for labels.

   3.10 While holding the CTRL key down, select the Project Name and Sheet Name labels. Change the value in the Type Selector to Label: ³/₈" Comic.

3.11 Edit the Project Number label (not the text) similarly. Create a **3/8″ Comic** Label type, using the Parameters shown in Figure 6–44. Change the type of the Date and Scale labels by selecting them and changing the value in the Type Selector to Label: 3/8″ Comic. Stretch the grips on any label that is not long enough to contain its contents in one line.

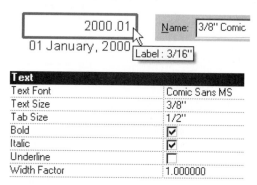

Figure 6–44 *3/8″ Comic label Parameters used for the Project Number, Sheet Date, and Scale labels*

3.12 Repeat the above step on the Sheet Number label. Create a **1″ Comic** Label Type using the Parameters from Figure 6–45. Before closing the Element Properties dialogue, set the Horz. Align value to Right and the Vert. Align value to Bottom. Click OK.

Text	
Text Font	Comic Sans MS
Text Size	1″
Tab Size	1/2″
Bold	☐
Italic	☑
Underline	☐
Width Factor	1.000000

Graphics	
Horz. Align	Right
Keep Readable	☑
Visible	☑

Other	
Sample Text	A-0001
Label	Sheet Number
Format	Edit...
Vert. Align	Bottom

Figure 6–45 *1″ Comic label Parameters used for the sheet number label*

3.13 Edit the vertical Date/Time Stamp label similarly. Create a **1/4″ Comic** Label Type using the Parameters shown in Figure 6–46.

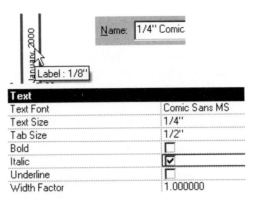

Figure 6–46 *1/4" Comic label Parameters used for the vertical Date/Time Stamp date label*

3.14 Finally, for the plain text, select the Project Number text and pick the Properties icon from the Options toolbar. Select Edit/New, and then pick Rename. Rename the type **3/8" Comic** and change the Parameters as shown in Figure 6–47. Click OK twice. The Date and Scale text will also display the Type changes you just made.

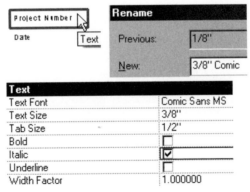

Figure 6–47 *3/8" Comic text Parameters used for the straight text*

3.15 Stretch the three text objects as necessary to make each fit on one line with the text border of an appropriate size. You now should have a mess in the middle of your sheet—maybe even more congested than what's shown in Figure 6–48. Relax—you will straighten that out right away.

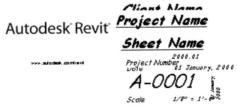

Figure 6–48 *New label and text types at the middle of your sheet—quite a mess!*

3.16   Zoom to Fit (**ZF**) to see your whole sheet.

3.17   From the Options toolbar, select the Offset tool. Set the type to Numerical, the Offset to 1″ (be sure to add the ″), and check the Copy checkbox, as shown in Figure 6–49.

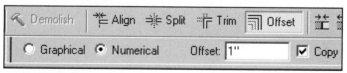

Figure 6–49 *Offset settings*

3.18   Move your cursor over the top of one of the boundary lines. A green dashed line will appear on one side or the other, indicating where Revit Building is going to place your offset line. Before selecting, tap the TAB key once to link all four boundary lines and select the interior side of the line. You will use this line as a guide and delete it later.

3.19   Offset another, single line, 3½″ up from the bottom (change the value of the Offset on the Options Bar). This will also be a guideline to be deleted later.

3.20   Draw a line from the midpoint of the horizontal boundary, vertically, the full height of the sheet.

3.21   Now move the text and labels into position, as shown in Figure 6–50. Location is not too critical. You may notice that Revit Building text and labels do not have snap handles per se—eyeballing the text into location is acceptable here. Revit Building text and labels do, however, align with each other, so once one is in place the others will align.

Figure 6–50 *Text and label locations for our new titleblock; guidelines are still in place*

Many firms have solid or otherwise wide linework on their titleblocks. The most effective way to re-create this look is by using fills in various shapes and colors.

3.22   From the Design Bar, select the Filled Region tool. From the Options toolbar, select the Rectangles option, as shown in Figure 6–51. If you do not want a black border on the edge of your band, make sure you are sketching with <Invisible lines> selected in the Type Selector drop-down list.

Figure 6–51 *Rectangle sketch tool*

3.23 You will be creating a long narrow rectangular blue band from the horizontal guideline, the full width of the printable area of the sheet, 3/8″ high. Pick the end point on the guideline above the Autodesk logo, and pick another point on the right-hand vertical guideline above the horizontal guideline. You will be left with a rectangle that is 3′ – 5-1/2″ wide by an ambiguous height. Hit ESC twice to get back to the Modify tool, select the top line of the rectangle, and change the temporary dimension value between the top and bottom of the rectangle to **3/8″**.

3.24 From the Sketch Design Bar, select the Region Properties tool and choose Edit/New. Change the Color and Cut Fill Parameters to the values shown in Figure 6–52. Click OK twice, and then select Finish Sketch on the Sketch Design Bar.

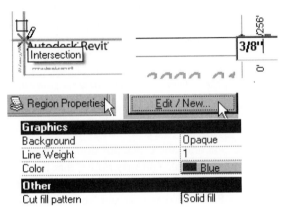

Figure 6–52 *Solid fill Parameters for solid blue band on titleblock*

3.25 Select Label from the Family Design Bar. Set the Type to 3/8″ Comic. Select the Center and Bottom alignment icons, as shown in Figure 6–53.

Figure 6–53 *Text Alignment icons for label alignment*

3.26 Pick a point directly below the Client Name label. Revit Building will show you green dashed alignment lines to assist in the placement. In the Select Parameter dialogue, select the Project Status Parameter as shown in Figure 6–54, and click OK.

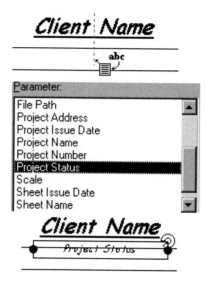

Figure 6–54 *Parameter selection dialogue*

    3.27 Hit ESC to terminate the Label tool. Select the label you just created. Adjust its position if necessary.

    3.28 Zoom to Fit, erase all of the guidelines and save the file.

    3.29 Congratulations, you have now created a custom titleblock that can be used on any project in the Schematic Design phase (see Figure 6–55). Close the file.

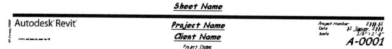

Figure 6–55 *Finished Schematic Design titleblock family*

EXERCISE 4. SHEET LAYOUT

    4.1 Now you'll use that Schematic Design style Titleblock in our project file. Open or return to the *Chapter 6 annotations.rvt* file from the previous exercises. From the File pulldown menu, select File>Load from Library>Load Family. Navigate to the *34 x 44 Horizontal – SD* family you just created, select it, and click Open.

    4.2 Repeat the last step. Navigate to the *Library/Chapter 6* folder and load the generic titleblock family file *34 x 44 Horizontal.rfa* into the project. You will use both when making the transition from Schematic Design to Design Development annotations. See Figure 6–56.

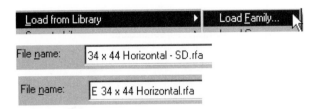

Figure 6–56 *Load two titleblocks*

    4.3 Activate the View tab on the Design Bar and select the Sheet tool. Select the 34 x 44 Horizontal – SD sheet from the list. Revit Building creates and opens a Sheet view named A101 – Unnamed.

    4.4 Click on the 1ST FLOOR – FLOOR PLAN view from the Project Browser. Drag it onto the sheet and locate it in the center towards the bottom. Drag the ENLARGED WASHROOM PLAN view from the browser to the area in the upper-right side of the sheet.

    4.5 Select the Titleblock. Pick the Properties icon from the Options toolbar and change the Parameters as shown in Figure 6–57.

Graphics	
Scale	As indicated
Identity Data	
Sheet Name	Floor Plans - 1st Floor
Sheet Number	A1
Sheet Issue Date	11/22/05
Checked By	LCF
Designed By	You
Approved By	Your Instructor
Other	
Date/Time Stamp	06/22/05
File Path	
Drawn By	JJB

Figure 6–57 *Titleblock Parameters for sheet A1*

 NOTE The Drawn By, Checked By, Designed By, and Approved By labels are not used in the Schematic Design titleblock but will appear in the next phase.

You will notice that the labels for the floor plan views on this sheet do not use our Schematic Design look with the Comic Sans font. We have preloaded a Schematic Design view title for your use.

    4.6 Choose either floor plan view (which will display a red border), and select the Properties icon from the Options toolbar. Pick Edit/New, and change the Title Parameter to View Title – SD as shown in Figure 6–58. Click OK twice to exit the dialogues.

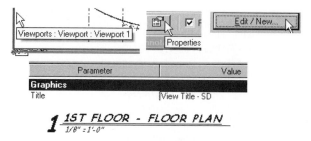

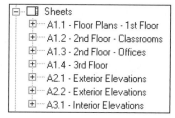

Figure 6–58 *Change the View Title to the SD style*

4.7   From the View Design Bar, use the Sheet tool to create a new sheet using the SD titleblock. Note that Revit Building numbers the new Sheet A2—it picks up the numbering system from the existing sheet. Delete the new sheet. Change the number of the existing sheet to **A1.1**

4.8   From the View Design Bar, use the Sheet tool to create six new sheets. Modify the sheet names and numbers per Figure 6–59.

```
⊟─☐ Sheets
     ⊞── A1.1 - Floor Plans - 1st Floor
     ⊞── A1.2 - 2nd Floor - Classrooms
     ⊞── A1.3 - 2nd Floor - Offices
     ⊞── A1.4 - 3rd Floor
     ⊞── A2.1 - Exterior Elevations
     ⊞── A2.2 - Exterior Elevations
     ⊞── A3.1 - Interior Elevations
```

Figure 6–59 *New sheets names and numbers*

4.9   Once the new sheets have been added, rename the four interior elevation views and create an enlarged floor plan view of the elevator area.

4.10 Drag the appropriate views to the new sheets and arrange them as you see fit. Figure 6–60 shows the sheet names, numbers, and views on each of the sheets in our completed set. Save the file.

TIP Before you move views to sheets, right-click on the name of each floor plan view, right-click, and pick Properties, which opens up the View Properties dialogue. Select the Edit button in the Visibility field, and make sure that Elevations is not checked on the Annotation Categories tab. Do this for each floor plan, and they will take up less space on the sheets. For the Elevation views, clear Topography on the Model Categories tab.

Figure 6–60 *All sheets, names, numbers, and subsequent views*

You now have what represents a mini-set of Schematic Design drawings. Take a few minutes to look at each sheet; print them out if possible. This set could be 10 sheets or it could be 100 sheets, but they all look the same and are coordinated with one another. Now it is time for that phase switch mentioned at the beginning of the chapter. This exercise so far has been focused on the Schematic Design look and feel. It is now time to jump to the Design Development phase of our hypothetical project. With that jump comes a completely different set of graphics, tags, notes, and sheets—everything needs to look "tighter" (more condensed and composed on the page) and "hard-lined" (not hand-drawn or sketchy in appearance).

It is important to note that you will be using two different techniques to make this change. In exercises in the first part of this chapter, when in the Element Properties dialogue you selected Edit/New and then did one of two things. One option was picking Duplicate (to create a new type) and then modifying Parameters for the new type. The other option was just to modify the Parameters of the existing object. The reason this is important now is that you will be using two different, corresponding techniques to change the appearance of annotation objects. For the items you duplicated (created new annotation sets), you will "swap out" the Schematic Design annotation for the Design Development annotation. For the annotation objects you edited individually, you will re-edit them to change their appearance accordingly.

EXERCISE 5. CHANGE ANNOTATION TYPES

SWAP OUT DUPLICATE OR NEW TYPES

5.1 Open or continue working in the file from the previous exercise. Open the view IST FLOOR - FLOOR PLAN. From the Window menu, pick Window>Close Hidden Windows. In the Project Browser, scroll down towards the bottom and pick the + sign next to Families. The family list will expand, showing all of the family categories. Pick the + sign next to Annotation Symbols to expand this category. Expand the Room Tag with the Number - Area – SD category, and right-click on the Room Tag with Area – SD type, as shown below. Pick Select All Instances, as shown in Figure 6–61.

It is important to note that this is a global selection and will select room tags from all views whether they are visible or not. It is a very powerful tool.

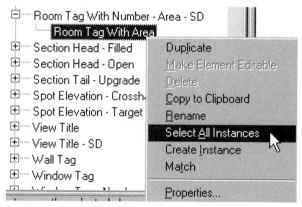

Figure 6–61 *Select All Instances of the Room Tag with Number – Area – SD*

5.2 You will notice that all of the room tags in the adjacent view are now selected and highlighted in red. From the Type Selector pulldown, change the type to Room Tag as shown in Figure 6–62. Notice that the fonts in the room tags are now all Arial, there is a rectangular box around the room number, and the area field is no longer shown.

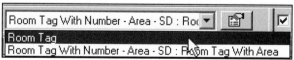

Figure 6–62 *Room Tag Type*

5.3 Repeat the selection step as in Step 5.1 with the Titleblocks, at the top of the list under Families>Annotation Symbols in the Project Browser. Select 36 x 48 Horizontal – SD, right-click to Select All Instances, and use the Type

Selector drop-down list to select 34 x 44 Horizontal, as shown in Figure 6–63. Revit Building will apply this titleblock to all sheets.

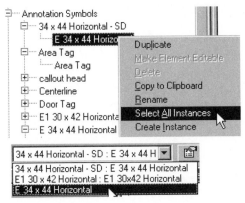

Figure 6–63 Select All Instances of 34 x 44 Horizonatal – SD titleblock and change type

5.4 Open the view ENLARGED WASHROOM PLAN. Select one of the dimensions. Right click and choose Select All Instances. Change the Type to Linear: 3/32″ Arial. See Figure 6–64.

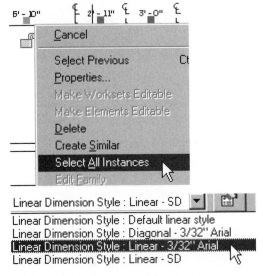

Figure 6–64 *Change all linear dimensions to another style*

5.5 Open the view IST FLOOR – FLOOR PLAN. Repeat the changes in step 5.4 on the angular and radial dimensions.

CHANGING EXISTING PARAMETERS

The second technique you will use to manage the annotation graphics is simply changing the Parameters of the annotation families.

> 5.6 Select one of the grid lines in the Floor Plan view. Pick the Properties icon on the Options toolbar. In the Element Properties dialogue, pick Edit/New. Make the Grid Head value Grid Head – Circle, as shown in Figure 6–65. Click OK twice.

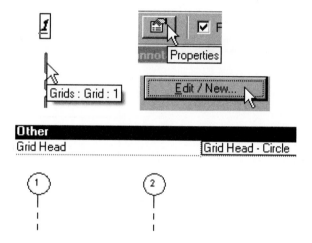

Figure 6–65 *Change the Grid Head to the original Grid Head – Circle type*

> 5.7 Select one of the notes you created at the washrooms. Pick the Properties icon from the Options toolbar. Select Edit/New; change the Text Font to Arial, clear the Italic checkbox, and change the Leader Arrowhead to Arrow Filled 30 Degree, as shown in Figure 6–66. Click OK twice.

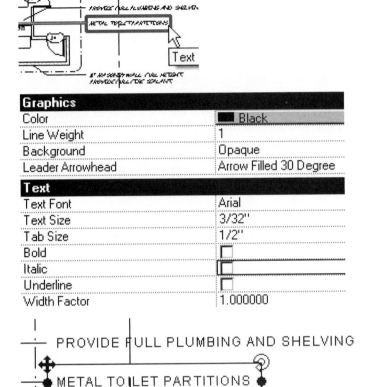

Graphics	
Color	Black
Line Weight	1
Background	Opaque
Leader Arrowhead	Arrow Filled 30 Degree
Text	
Text Font	Arial
Text Size	3/32"
Tab Size	1/2"
Bold	☐
Italic	☐
Underline	☐
Width Factor	1.000000

PROVIDE FULL PLUMBING AND SHELVING

METAL TOILET PARTITIONS

Figure 6–66 *New Parameters for 3/32" Note*

5.8 Repeat the previous step on the interior elevation symbol, changing the Elevation Tag to 1/2" Square, as shown in Figure 6–67. Note that you must select the center of the symbol, as the arrows are separate entities and have properties of their own. Click OK twice to exit the dialogue.

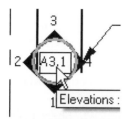

Graphics	
Elevation Tag	1/2" Square
Callout Tag	Callout Tag 1
Reference Label	Sim

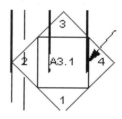

Figure 6–67 *Interior elevation Parameters*

    5.9 Open any sheet view and select any viewport. Repeat the previous steps on the viewport, changing the Title value to View Title, as shown in Figure 6–68. Click OK twice to exit the Type Properties and Element Properties dialogues.

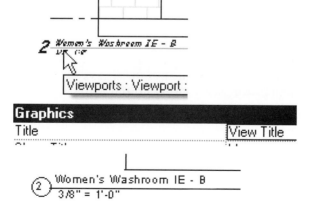

Figure 6–68 *Viewport Parameters*

    5.10 Save the file.

In a working environment, well-designed project templates will carry annotation style or appearance information to eliminate even the little bit of item-by-item editing that you just went through.

The replacement of our SD titleblock that has sheet information along its bottom with the generic version that has its project data fields along the right side will make the views on sheets appear a little unbalanced. There is still a little house cleaning to be done to make these sheets completely ready for client presentation, but you have dramatically changed the appearance of an entire set of drawings in a few quick and easy steps.

Keep in mind, these techniques can be used on any annotation family, not just the few you used today. Imagine that these 7 sheets were a 250-sheet set (not an unusual size for a building of the size of our hypothetical project), and you changed everything about its appearance in 20 minutes. We promised you a few short steps!

SUMMARY

You have just walked through a small portion of Revit Building's annotation capabilities. While there is not enough room in this book to cover all of the possibilities and techniques with each annotation tool, what you have just completed will set a solid foundation for creating, editing, and managing the other annotation types.

REVIEW QUESTIONS – CHAPTER 6

MULTIPLE CHOICE

1. Changing Type Parameters in a Room Tag family lets you control
   a) The appearance of all instances of that Tag
   b) The size of rooms
   c) The room numbers
   d) The appearance of Interior Elevations for that room

2. Adding a Callout to a Floor Plan
   a) Creates a new sheet in the project
   b) Deletes any dimensions that the callout touches
   c) Creates a separate plan view showing the area enclosed by the callout border
   d) Means the end of civilization as we know it

3. Titleblock Text and Labels

   a) can only be erased from a Titleblock family file, not added

   b) can only appear in one Titleblock family at a time in any single project

   c) can only be Left Justified

   d) none of the above

4. Titleblock Labels that are going to read-in values from Project Parameters

   a) have to be created before any text

   b) should be of appropriate size for long names

   c) can be formatted Left, Right, or Center

   d) b and c, but not a

5. View Templates

   a) are only for elevation views, not plans or sections

   b) save and apply common view settings

   c) can't be applied after they are changed

   d) can only be used once per file

TRUE/FALSE

6. You can only adjust a grid line header offset when you first create the grid line.

7. Elevation indicators can be square or circular.

8. Dimension Styles for Linear, Angular, and Radial dimensions are edited separately.

9. Revit Building will align text leaders to model objects, but not to other leaders.

10. You create a room when you place a room tag into a space whose boundary consists of either three or more room-bounding walls or three or more room separation lines.

 Answers will be found on the CD.

CHAPTER 7

Schedules

INTRODUCTION

It has been said that one of the most tedious and unrewarding tasks in an Architectural/Engineering/Construction firm is compiling, counting, and organizing schedules. Whether a schedule lists doors, windows, vents, parking stalls, sheets, or anything else, time spent devising and populating the schedule is much better spent elsewhere.

If you are a working designer, does the thought that "today I am going to count, categorize, and organize all of the windows on this project" sound depressingly familiar? If you are a student, does the prospect of spending your first years at work preparing lists from plan pages feel like the right preparation for your landmark design?

Computers are unambiguously excellent at these tasks. The developers of Revit Building understood this when they set out to develop a software package for the AEC industry. One of the founding principles of Revit Building is to let the computer handle the tedious, trivial tasks, and let the designer design. The Scheduling module is the perfect example of this concept. Because Revit Building is a central database of building information, scheduling is quick, accurate, and simple. The fact that you can create an accurate custom schedule in a matter of minutes means you can spend more time designing and less worrying about the count and sizes of your windows. For that matter, you can create several custom schedules and provide more specific information for better communication, resulting in less confusion in the field. Virtually every object in Revit Building can be scheduled.

The following exercises will take you through a variety of schedule types and illustrate several techniques in creating, editing, and managing your schedules.

OBJECTIVES

- Understanding the Schedule Properties Dialogues
- Create, modify, and manage a single category schedule
- Create, modify, and manage a multiple category schedule

REVIT BUILDING COMMANDS AND SKILLS

Create a Schedule

Add and remove fields

Create new fields

Work with shared and project parameters

Sort Schedule Rows

Create and modify headers and footers

Group Schedule rows and columns

Format Schedule font, alignments, and orientations

Manage Schedule appearance

Create and modify a multi-category Schedule

Scheduling is one of the strongest features of Revit Building. You can add, modify, remove, and change any or all components in the model and Revit Building tracks them, no matter what. Revit Building's scheduling abilities will report location, size, number, or any other parameter associated with virtually any building component.

OVERVIEW OF THE SCHEDULE DIALOGUE BOX

The power of scheduling is controlled by the Schedule Properties dialogue box with its five critical tabs. Because all schedules are controlled by these tabs, it is imperative that you understand what each tab contains and the functions each controls. This chapter will start out by outlining these tabs, and follow with the exercises.

You can examine the Schedule Properties tabs we are about to discuss in any Revit Building project file by picking View>New>Schedule/Quantities from the Menu Bar, or picking Schedule/Quantities from the View tab of the Design Bar. Click OK in the New Schedule dialogue to open the Schedule Properties dialogue, as shown in the following illustrations.

FIELDS TAB

The Fields tab controls the fields that will be in your schedule (see Figure 7–1).

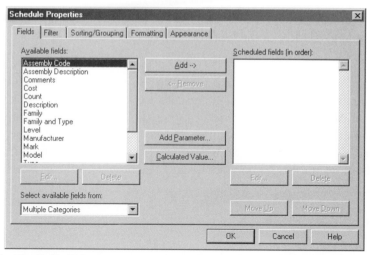

Figure 7–1 *The Fields Tab*

The Available fields area contains a list of the default Family Parameters that can be included in your schedule. This list will be different for each schedule type. The schedule type shown in Figure 7–1 is multi-category. The Schedule Type is selected in the New Schedule dialogue. The Scheduled fields area contains the list of parameters (in order) that will be included in your schedule. The top of the list will be the far-left column, with the next on the list being the next column to the right, and so on.

The Add button adds the highlighted parameter in the Available fields list to the Scheduled fields area, and the Remove button removes the parameter from the Scheduled fields area.

The Add Parameter button allows you to create a custom field, whether it is a project parameter or shared parameter. You will look at these two parameter types later.

The Calculated Value button creates a field whose value is calculated from a formula based on other fields in the schedule. For instance, a Width parameter value could be calculated as 1/2 the Length parameter.

The Edit Field buttons—note that there are two—allow you to edit user-created parameters, and the Delete buttons—also two—allow you to delete a user-created parameter.

The Move Up and Move Down buttons move the highlighted parameter up or down the list and therefore left and right in the schedule.

There is a filter named Select available fields from in the lower left corner of the dialogue. This is a drop-down list of schedule categories related to the one chosen. When a category is selected in this list, the available fields list changes to show fields from the related category, such as Finishes or Occupancy for Rooms, or From Room and To Room for Doors.

THE FILTER TAB

The Filter tab shown in Figure 7–2 allows you to restrict what elements display in simple or multi-category schedules (and also to view lists, drawing lists, and note blocks). You can set up to four filters on this tab, and all the filters must be satisfied for elements to display. You can use displayed or hidden schedule fields as filters. Certain fields—mostly Type parameters—can't be used as filters. An example of a filtered schedule would be a door schedule filtered by floor.

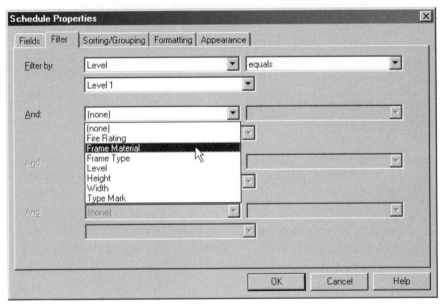

Figure 7–2 *The Filter Tab*

THE SORTING/GROUPING TAB

The Sorting/Grouping tab shown in Figure 7–3 is a very powerful tab that works well with Revit Building's underlying ease of use. There are some concepts that you will have to understand before jumping in. This tab, simply put, controls the order in which the information will be displayed—sorting. An example of this in a door schedule might be sorting by door number. Revit Building will display all of the doors in numerical order by door number (which is most often keyed to the room number).

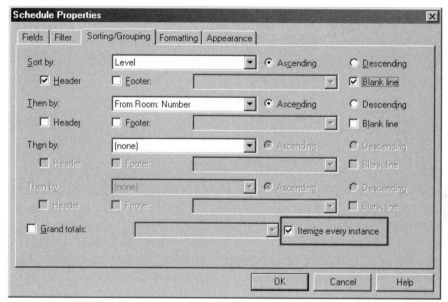

Figure 7–3 *The Sorting/Grouping Tab*

Grouping allows you to group items together. In the door schedule example, you might want to group doors by level or by a particular zone in a building. By virtue of this you could tell Revit Building to group the doors by zone and then sort them by door number. To activate grouping you will need to check the Header or Footer checkbox. The Grand totals checkbox will provide a grand total at the bottom of your schedule, where it applies. The Itemize every instance checkbox allows you to toggle between a list of every item (doors, in our example) and a list that might just show the type and a count of each.

THE FORMATTING TAB

The Formatting tab shown in Figure 7–4 controls the formatting of the individual columns. There is a field to change characteristics of the displayed heading—Heading, Heading orientation, and Alignment. In addition you can format the units shown for numerical fields using the Field Format button, and have Revit Building calculate the totals with a click of the Calculate totals checkbox. The Hidden field checkbox allows you to hide a column. This is useful if you want to group items by field (level, for instance), but don't want to repeat the information in the schedule.

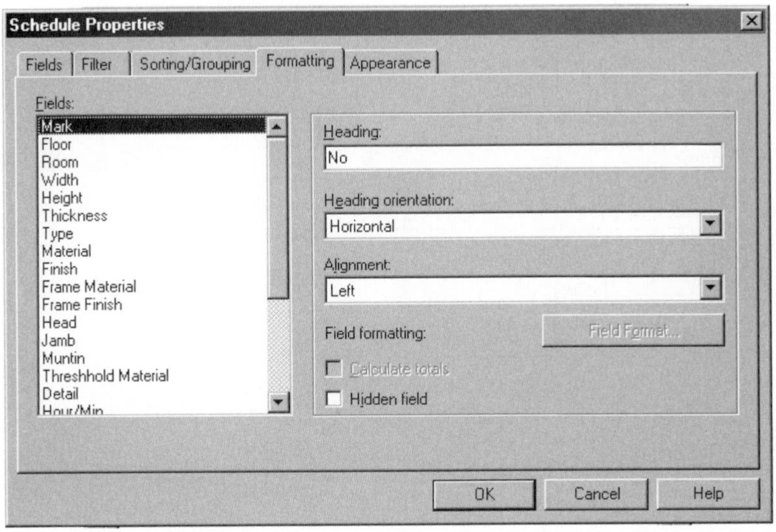

Figure 7–4 *The Formatting Tab*

THE APPEARANCE TAB

The Appearance tab shown in Figure 7–5 controls the overall graphic appearance of the schedule. The Header font controls manage the font type, its size, and whether it is bold, italic, or both. This will affect all headers in the schedule. The Body font controls manage the appearance of the remaining fonts. You have the ability to display grid lines, a title, and column headers or not, by using the checkboxes at the bottom of the dialogue box. You will use a majority of these settings in the exercises to follow.

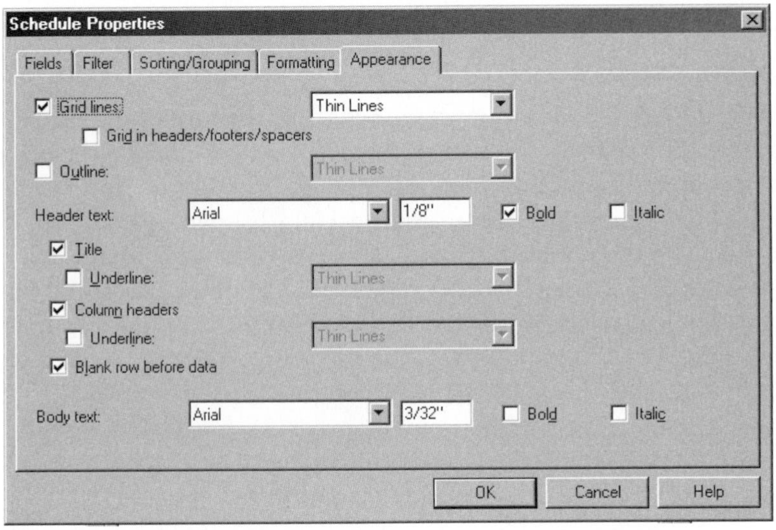

Figure 7–5 *The Appearance Tab*

As stated earlier, Revit Building enables you to schedule virtually any object or combination of objects in your project. For the purposes of this next exercise, we will demonstrate the schedule features using a door schedule. You will begin with the basics and add complexity as you go.

EXERCISE 1. CREATE A SCHEDULE FOR DOORS

1.1 Launch Revit Building and open *Chapter 7 Start.rvt*. This file is the continuation of the exercises in Chapter 6 with a few modifications to door tags and numbers.

1.2 Save the file as **Chapter 7 schedules.rvt** in a location determined by your instructor.

1.3 From the View tab on the Design Bar, select the Schedules/Quantities tool as shown in Figure 7–6.

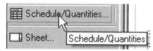

Figure 7–6 *Schedule/Quantities Tool*

1.4 In the New Schedule dialogue, select Doors from the Category list. Leave the Schedule Name as Door Schedule, select the Schedule building components radio button, and set the Phase to Offices, as shown in Figure 7–7. Click OK.

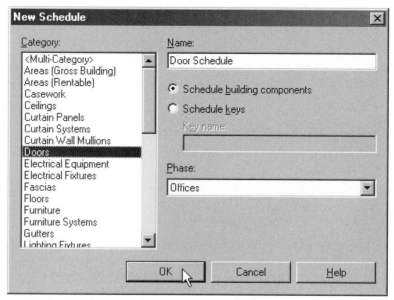

Figure 7–7 *Starting a new Door Schedule for the Offices Phase*

1.5 In the Schedule Properties dialogue, while holding down the CTRL key, select Height, Mark, Thickness, and Width, and click the Add button.

1.6 In the Selected fields list, highlight Mark and click Move Up until Mark is at the top of the list. Using the Move Up and Move Down buttons, arrange the parameters to match what you see in Figure 7–8. Click OK.

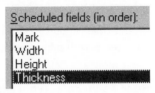

Figure 7–8 *Arranging added fields*

A Schedule view will open. You will now have a list of all of the doors in this building project. Please notice that it is in random order, and that there are some doors labeled using a room number and a sequential letter, and others that are just numbers. This is intentional, to demonstrate the fact that Revit Building schedules recognize phases.

1.7 In the Schedule view, right-click and select View Properties to bring up the Element Properties dialogue. In the Phase Filter pulldown, select Show Previous + New. The Phase parameter was already set to Offices, as shown in Figure 7–9. Click OK to see the results. Your schedule should now include only doors that are numbered by room number, plus a sequential letter designation, but the numbering is still in random order.

151A	3' - 0"	7' - 0"	0' - 2"
212A	3' - 0"	7' - 0"	0' - 2"
38	3' - 0"	7' - 0"	0' - 2"
39	3' - 0"	7' - 0"	0' - 2"
205A	3' - 0"	7' - 0"	0' - 2"
204A	3' - 0"	7' - 0"	0' - 2"
204B	3' - 0"	7' - 0"	0' - 2"
43	3' - 0"	7' - 0"	0' - 2"
44	3' - 0"	7' - 0"	0' - 2"

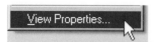

Phasing		
Phase Filter	Show Previous + New	
Phase	Offices	

151A	3' - 0"	7' - 0"	0' - 2"
212A	3' - 0"	7' - 0"	0' - 2"
205A	3' - 0"	7' - 0"	0' - 2"
204A	3' - 0"	7' - 0"	0' - 2"
204B	3' - 0"	7' - 0"	0' - 2"
208A	3' - 0"	7' - 0"	0' - 2"
207A	3' - 0"	7' - 0"	0' - 2"
210A	3' - 0"	7' - 0"	0' - 2"

Figure 7–9 *Schedule Phase Filter Seetings—the Phase was set earlier*

1.8 In the Schedule view, right-click and select View Properties again. In the Element Properties dialogue, select the Edit button for the Sorting/Grouping value. The Schedule Properties dialogue opens to the Sorting/Grouping tab. In the Sort by pulldown, select Mark and click OK twice. You now have a basic door schedule that represents all of the doors in the office phase, in ascending numeric order from top to bottom. See Figure 7–10. Let's now really move.

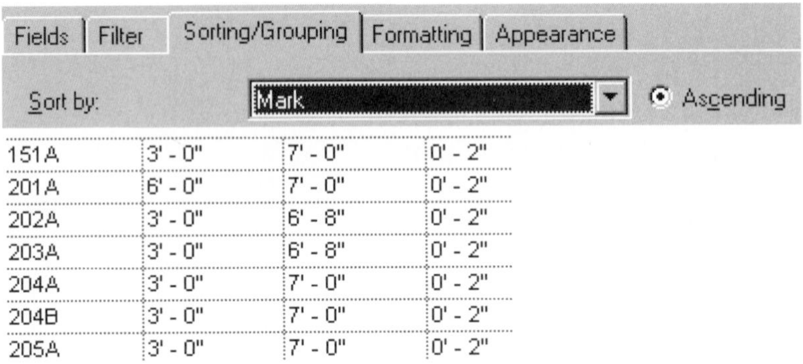

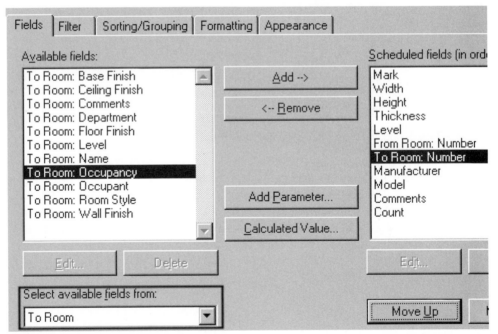

Figure 7–10 *Sort by Mark*

    1.9 Right-click in the view again, select View Properties as before, and set the properties for Fields, Sorting/Grouping, Formatting, and Appearance, as shown in figures 7–11 through 7–14. Click OK twice when the adjustments are complete.

Figure 7–11 *Fields Settings—Add and move the additional fields, and use available Room fields*

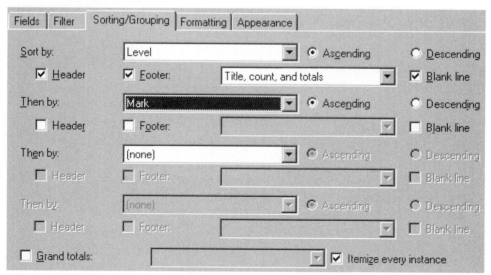

Figure 7–12 *Sorting/Grouping Settings—Sort by Level, check Header and Footer, then Sort by Mark*

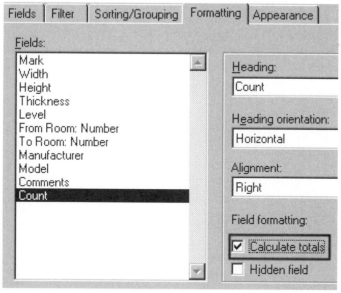

Figure 7–13 *Formatting Settings—set Count to Align at Right, and select Calculate totals*

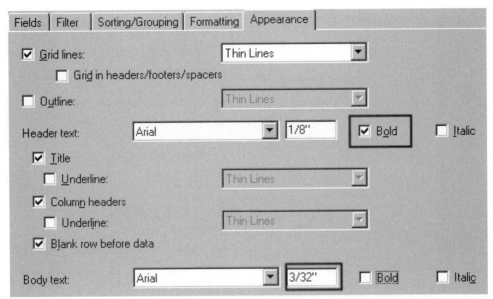

Figure 7–14 *Appearance Settings—set Header font to Bold and Body font to 3/32″*

You will notice that the Appearance tab does not affect the font appearance in the view. Revit Building only displays this when the schedule is placed on a sheet. Additionally, you will see that the level information is redundant, as it is already in the header and in each row.

1.10 Right-click and select View Properties to get back to the Element Properties dialog. Select Edit in the Formatting parameter. In the Formatting tab, select Level and check the Hidden field checkbox; then click OK twice. See Figure 7-15.

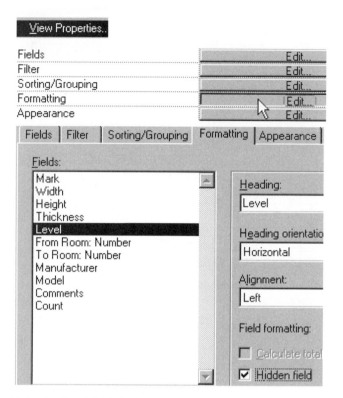

Figure 7–15 *Make the Level field hidden*

 TIP You may also right-click over the top of any of the cells in the level column and select Hide Column(s).

1.11 From the View tab on the Design Bar, select Sheet, select the **36x48 Horizontal – DDCD** titleblock, and click OK. Revit Building will create a new sheet view and open it. Drag the Door Schedule view from the Project Browser onto the new sheet. Notice that the columns are all of equal width. Click the schedule to bring up the display controls. There will be triangles at the top and a tilde (squiggle) line at the mid-point of the far right-hand column, as shown in Figure 7–16.

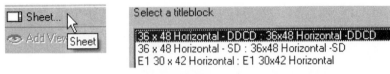

Figure 7–16 *Schedule display control on a sheet—triangles control column width and a sguiggle symbol splits the schedule*

1.12 Drag the triangular grips left and right to set the column widths.

TIP It is best to set widths starting with the columns at the left and move right.

There will be occasions when you will want to display the same schedule but separate the grouped areas in different locations on the sheet. You will separate level 1 from levels 2 and 3.

1.13 Pick the squiggle symbol at the right of the schedule; this will split the schedule into two individual views. Use the Move grip (double-headed arrow) in the middle of the bottom schedule to drag the lower view to the left of the upper portion. Drag the grip at the bottom center of the right-hand view up, until it is set at the gap between levels 1 and 2. The tops of the schedules will snap to alignment. See Figure 7–17.

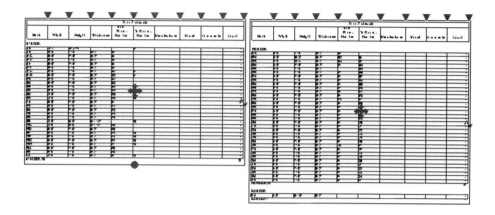

Figure 7–17 *The schedule has been split*

 NOTE These adjustments require you to be mindful of doors as they are added and deleted. As the level 1 schedule grows in length it will begin to populate on the level 2 side, and you will need to stretch the grip to adjust it.

One of the advantages to having a schedule is that information can be viewed and edited in a format conducive to annotative information on a broad brush scale.

1.14 Return to the Door Schedule view. Select the cell under Manufacturer in the door 101B row, and type in **Judson Door and Window Co.** This change will be applied to all doors of this type—a notice will appear. Click OK.

1.15 Repeat the above step for doors 105A – **Fox and Sons**, 125B – **Delmar Door Corp.,** and 204A – **Lucas Woodworks**. Use the same technique to fill out model information. Fill in individual comments as you see fit, as shown in Figure 7–18.

1.16 The column widths may not display all of the information you have input. Move the cursor over the top of one of the vertical column lines to get a double-headed arrow (see Figure 7–18) and drag the column edge to the appropriate width. Note that this does not affect the schedule display on the sheet.

2ND FLOOR								
201A	6' - 0"	7' - 0"	0' - 2"		201	Judson Door and Window Co.	7284-2-ABC	Verify Clearance
202A	3' - 0"	6' - 8"	0' - 2"	202	201	Western Door and Window	1236-KJ-RW	Under cut 1"
203A	3' - 0"	6' - 8"	0' - 2"	203	201	Western Door and Window	1236-KJ-RW	
204A	3' - 0"	7' - 0"	0' - 2"	211	204	Lucas Woodworks	CDW-3672-PSS	Paint Red

Figure 7–18 *Schedule—manufacturer, model, and comments*

1.17 Go back to the sheet view that you placed the schedule on and make sure that the new information is showing correctly—a single line of text per schedule row. Use the triangulated grips to adjust the column widths if necessary.

It is now time to add key card reader information to the schedule, but it is not on the list of parameters for this schedule.

1.18 Right-click on the schedule in the sheet view and select Edit Schedule. The Door Schedule view will open. Right-click and click View Properties. Select the Edit button for Fields and select Add Parameter. See Figure 7–19.

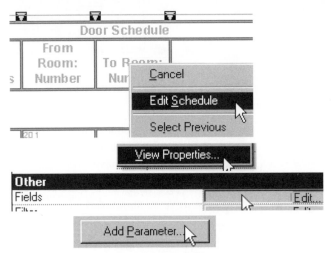

Figure 7–19 *Prepare to customize the schedule fields*

ADD A PARAMETER, CHANGE THE FORMAT

1.19 In the Name field of the Parameter Properties dialogue, type **Key Card Reader**, set the Type to Yes/No, and select the Instance radio button (as shown in Figure 7–20), as doors of the same type will have different Key Card Reader values. Click OK.

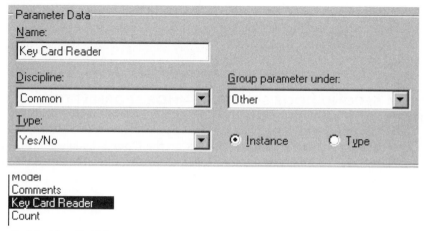

Figure 7–20 *New Field Parameter—Key Card Reader*

1.20   Move Key Card Reader to just above the Count parameter in the order of scheduled fields.

1.21   Go to the Formatting tab and set the Key Card Reader Heading orientation to Vertical and the Alignment to Right. Set the Count Heading orientation to Vertical as well. Click OK twice. Check the boxes in the new Key Card Reader column for various doors.

1.22   Open the sheet with the Door Schedule on it. Adjust the column widths for Key Card and Count to take advantage of the vertical header alignment, as shown in Figure 7–21. Note that the doors for which you checked the Key Card Reader yes/no boxes show Yes in the schedule and others are blank.

Manufacturer	Model	Comments	Key Card Reader	Count
Judson Door and Window	72832-ABC	Verify Clearance	Yes	1
Western Door and Window	1236_KJ_RW	Under cut 1"	Yes	1
Western Door and Window	1236_KJ_RW		Yes	1
Lucas Woodworks	CDW-3672-PSS	Paint Red	Yes	1
Lucas Woodworks	CDW-3672-PSS			1
Lucas Woodworks	CDW-3672-PSS		Yes	1

Figure 7–21 *Vertical heading orientation*

 NOTE Traditionally, architects provided a door schedule once, in one format, and it was up to the contractor and door manufacturers to sort through this information to get the information they needed for their particular use. Because Revit Building is a centralized database, information is always up-to-date; the architect using Revit Building has the ability to group, sort, and provide it in a concise manner. These next few steps fall under the "Revit Building Magic" category, allowing you to manipulate the same data and provide it in a completely different manner for different uses.

CREATE A SECOND DOOR SCHEDULE—MORE FORMATTING CHANGES

1.23 From the Project Browser, right-click on the Door Schedule and select Duplicate. This will make a view called Copy of Door Schedule the active view.

1.24 In the new view, right-click, and select View Properties. Change the settings as shown in figures 7–22 through 7–24.

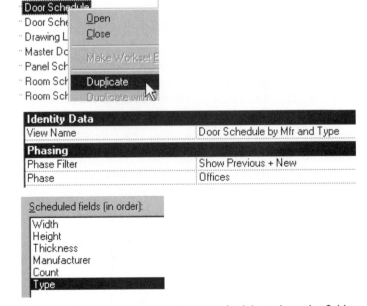

Figure 7–22 *View Properties—name the new schedule and set the fields*

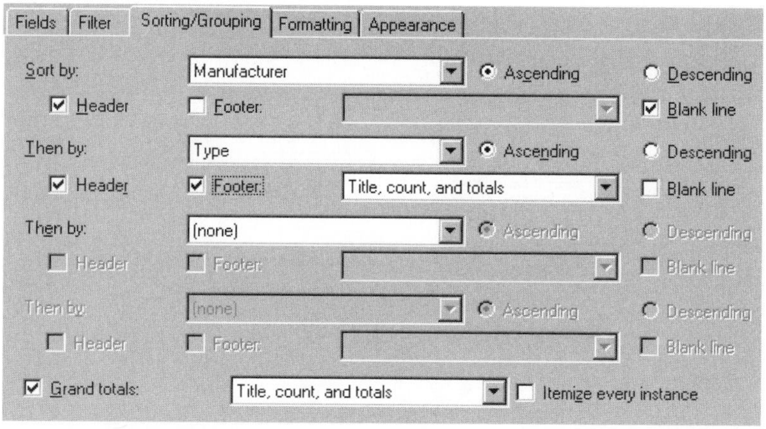

Figure 7–23 *Sorting/Grouping in the new schedule*

Figure 7–24 *Formatting and Appearance changes*

1.25 From the Formatting tab, select Manufacturer and check the Hidden field checkbox. From the Appearance tab clear the Display grid lines checkbox, as shown in Figure 7–24. Click OK twice.

1.26 Open the sheet view with the first door schedule on it and drag the new Door Schedule by Mfr and Type onto the schedule sheet. Note that it appears markedly different from the first schedule. Drag the right hand part of the schedule over the left hand part to combine then into one, as shown in Figure 7–25.

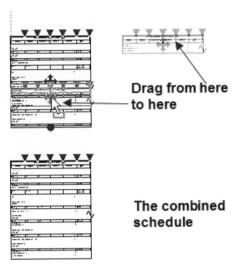

Drag from here to here

The combined schedule

Figure 7–25 *Place and condense the schedule*

1.27 From the Project Browser, open view 2ND FLOOR OFFICES – FLOOR PLAN. Select door number 212A and select properties from the Options bar. Note that all of the type and instance parameter information you input on the schedule is displayed, as well as the Key Card Reader Parameter you defined in the schedule (see Figure 7–26). Clear the Key Card Reader checkbox and click OK.

Parameter	Value
Head	
Jamb	
Muntin	
Detail	
Hour/Min	
Label	No
Glazing	
Louver	
Closer	No
Panic Hardware	No
Hardware Set	43
Threshhold Material	None
Material	Paint
Frame Finish	
Floor	2nd Floor
Key Card Reader	

Figure 7–26 *Door Parameters—note that the Key Card Reader now appears*

On many occasions your schedule will have several columns that are related to each other. Many firms like to call attention to this by putting a header above the related columns. You achieve this by grouping columns.

1.28 From the Project Browser, open the Door Schedule view. Click and drag your cursor over the Width, Height, and Thickness column heads. Select the Group button from the Options bar.

1.29 In the cell that is created, type **Door**. Repeat for To Room and From Room, using **Room** for the group header.

1.30 Highlight the words To Room and type **To** in the Heading box, and do the same for From Room, typing in **From**. See Figure 7–27. Click OK.

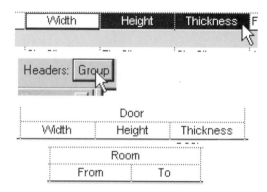

Figure 7–27 *Grouping column headers*

 NOTE This can also be accomplished in the Schedule (View Properties) dialogue, as shown in Figure 7–28.

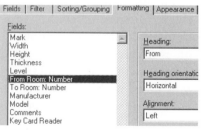

Figure 7–28 *Changing the display name on a parameter*

1.31 Open the schedule sheet to see the results.

1.32 From the Project Browser, open the Door Schedule by Mfr and Type that you just created and modify the width on the 36″ x 84″ doors supplied by

Lucas Woodworks to 3'-6". All of the doors of that type change in both door schedules.

I.33 From the Project Browser, open 2ND FLOOR – FLOOR PLAN, and note that the adjusted door widths of the doors for rooms 212–234 are now represented graphically in the plan as well. See Figure 7–29.

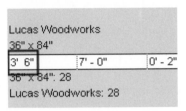

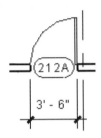

Figure 7–29 *Change one field to edit all doors of that type*

I.34 Save the file.

MULTI-CATEGORY SCHEDULES

Most schedules you produce will be similar to the schedules you have just completed: door, window, wall, curtain panel, furniture, and so on. These types of schedules report information on a single category type. On occasion you will need to schedule components from different categories that might have the same parameter. In addition to single category schedules, Revit Building allows you to schedule components from different categories, known as Multi-Category Schedules. One example that comes to mind is the electrical requirements for particular components or building parts. Any number of building components have electrical requirements and you may be asked or, better yet, *paid*, to assimilate this information. The following exercise is designed to expose you to the process of creating a multi-category schedule.

To enable the multi-category schedule functionality, the categories of the components must have a common parameter. Revit Building calls these parameters "shared parameters." Shared parameters are parameters that you can share with multiple Autodesk Revit Building families and projects. You define them once and save them to a shared parameter file. You can use shared parameters in multi- and single-category schedules, and to tag elements.

There are a few key concepts to understand before using multi-category schedules. Multi-category schedules consider *all* components within the model and return the desired field information. Of course, a schedule on all of the components in a model is of little use. When you set up a multi-category schedule, you will make use of the filter tab in Schedule Properties. This tab allows you to filter out all of the components that do not contain the filter parameter(s), so that you schedule only the components that do.

EXERCISE 2. CREATING A SHARED PARAMETER FILE

2.1 Open or continue working with the file from the previous exercise. From the File pulldown menu, select Shared Parameters.

2.2 In the Edit Shared Parameters dialogue, select Create and type **Chapter 7 – Shared Parameters** in the Name field. Save it in a location as indicated by your instructor.

2.3 From the Groups area, select New and type **Electrical** in the Name field, as shown in Figure 7–30.

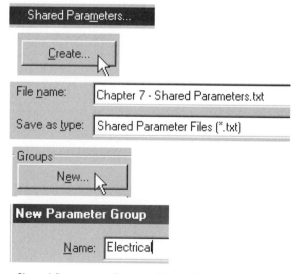

Figure 7–30 *New Shared Parameter Group—Electrical*

2.4 From the Parameters area of the dialogue, select New and type **Amps** in the Name field. Use the pulldown list to tell Revit Building that this parameter will be a Number, and click OK (see Figure 7–31).

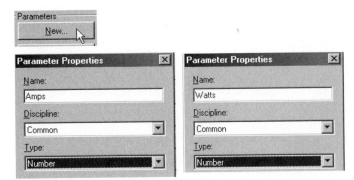

Figure 7–31 *Parameter Properties—it's important to specify the type here, as it is not editable afterwards*

2.5 Repeat the last step adding **Watts** (also a Number) to the Electrical group. Click OK twice.

 NOTE You have now created two new parameters that will be available within this project, as well as other projects you create. This is a great way to create standard custom parameters for a company that you know will be maintained throughout your projects. You will now need to indicate which categories have the parameters.

2.6 From the Settings menu, select Project Parameters, click Add, select the Shared Parameter radio button, click Select, highlight Amps, and click OK.

2.7 Back in the Parameter Properties dialogue, select the Type radio button next to Value stored by. Select Electrical in the Group Parameter Under dropdown list, and check the following categories: Furniture Systems, Lighting Fixtures, Mechanical Equipment, and Specialty Equipment (see Figure 7–32). Click OK.

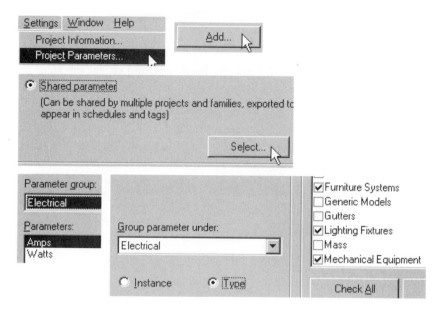

Figure 7–32 *Adding shared paramters to the project.*

You have now added a parameter to the categories checked above that will be included in all components in those categories.

 NOTE Selecting Instance or Type values is only available when you add the parameter to the project.

2.8 Repeat the above steps and add Watts to the project parameters. Remember to select the Type radio button. Click OK twice to exit the dialogues.

2.9 From the Project Browser, open the 2ND FLOOR OFFICES– FLOOR PLAN. Pick one of the Workstation Cubicles and click the Properties icon from the Options bar. Select Edit/New, and enter a numerical value for Amps and Watts. The actual numbers don't matter in this case. See Figure 7–33. Click OK twice.

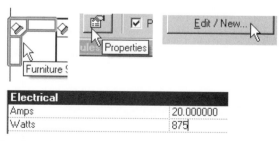

Figure 7–33 *Specify values for the new parameters*

2.10 Repeat the above step for the Computer Monitor in the Workstation and the Copier out in the Work Room area.

2.11 From the Project Browser, open the Roof view, and repeat the steps above on the Air Conditioner and Boiler.

2.12 From the View Design Bar, select View>New>Schedule/Quantities. From the New Schedule dialogue box, choose <Multi-Category> and click OK, as shown in Figure 7–34. There is no option to set the Phase for the schedule at this point.

Figure 7–34 *Create a new Multi-Category Schedule*

2.13 Use the Fields tab to add and order fields for the new schedule, as shown in Figure 7–35: Family, Type, Room, Count, Amps, and Watts.

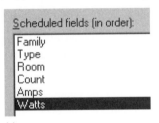

Figure 7–35 *Multi-Category Fields*

2.14 On the Filter tab, select Amps from the drop-down list, as shown in Figure 7–36. Once a parameter is selected, the right-hand drop-down list becomes active, so that you can further filter against the parameter value: equals/does not equal, greater than/less than, and so forth. Don't make any changes to that option.

Figure 7–36 *The Filter Tab*

2.15 On the Sorting/Grouping tab, choose Sort by Family, as shown in Figure 7–37. Clear Itemize every instance.

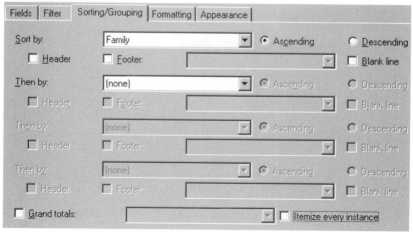

Figure 7–37 *Sorting by Family*

2.16 On the Format tab, set the Count, Amps, and Watts with Right Alignment, and check the Calculate totals checkbox (see Figure 7–38). Click OK.

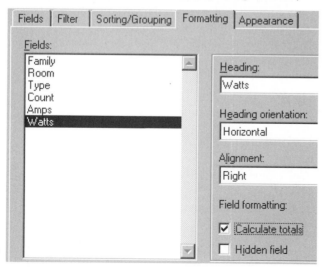

Figure 7–38 *Formatting*

2.17 There you have it—a schedule showing all of the Amperage and Watts for specified equipment in the entire building. If you do not see the Copier, Monitor, and Workstation in your schedule, right-click and select View Properties, then set the Phase for the schedule to Offices (see Figure 7–39). The view that was active (on the screen) controls the initial Phase setting for any Schedule (which is a View) that you create.

Phasing	
Phase Filter	Show All
Phase	Offices

Multi-Category Schedule

Family	Room	Type	Count	Amps	Watts
Air Conditioner-Outside		37" x 28" x 2	24	50	56400
Boiler		21" x 32" x 3	9	65	22500
Copier-Floor		Copier-Floor	5	30	6250
Monitor		17" Monitor	25	20	16250
Work Station Cubicle		96" x 96"	44	20	38500

Figure 7–39 *The new multi-category schedule tracks electrical requirements for different types of equipment*

NOTE Keep in mind that this technique can be applied to just about any categories. You can customize the Filter by creating a new Shared Parameter just for this use. While you have covered considerable ground regarding schedules, you have barely scratched the surface of Revit Building's potential. Continue to explore new uses and combinations; you will never fail to amaze yourself with the power of schedules.

SUMMARY

Congratulations, you now have the base knowledge to create virtually any schedule type imaginable. These exercises have exposed you to the tools you need to create any schedule type and modify it to suit any customized requirements.

Keep in mind, a schedule is just another view into your project data and there is no need to create schedules with "blinders on"—in other words, like the non-linked, non-smart schedules CAD software creates. You have a new tool, so use it effectively. Remember that different people need the same information in different ways. You might have an overall master door schedule and a separate schedule in which you group and total all the doors by hardware set for the hardware installer. The same data can be shown in different ways for different uses (see *Sheet A9.91* in your exercise file).

Additionally, you can display graphical information in coordination with your schedule. For instance, you could create three different room schedules (by Department, Floor Finish, and Occupancy Type) and put them on the same sheet with color-fill diagrams for analysis (see *Sheet A9.92* in your exercise file).

We have included several different schedules and their uses for your review. Open the A9.9 series of sheets and check out some of the schedules that Revit Building provides.

REVIEW QUESTIONS – CHAPTER 7

MULTIPLE CHOICE

1. The Schedule Properties dialogue controls
   a) Fields and Filters
   b) Sorting, Grouping, Formatting, and Appearance
   c) Both a and b
   d) b only

2. To Hide a schedule column
   a) Use the Hidden Field checkbox on the Schedule Properties Formatting tab
   b) Right-click in a field cell in the Schedule View and click Hide Column
   c) Delete the Schedule and start over

d) a or b

3. Which property of the view that is on-screen when you create a schedule affects the schedule output?

    a) zoom

    b) detail level

    c) phase

    d) crop region

4. Multi-Category Schedules make use of

    a) Shared Parameters

    b) Schedule Filters

    c) Cost Estimates

    d) a and b, but not c

5. Project or Shared Parameters can include fields of the following types:

    a) Text, Number, Length

    b) Area, Volume, URL

    c) Material, Angle, Yes/No

    d) all of the above

TRUE/FALSE

6. A Schedule Filter can be set to equal or not equal a certain value

7. Project Parameters can appear in Schedules but not in Tags

8. Shared Parameters can be shared by multiple projects and appear in Schedules and Tags

9. Calculated Parameters are text-only values

10. If you Group Schedule Column Headers, you have to delete the Schedule from any Sheets it appears on and then replace it for the change to appear

 Answers will be found on the CD.

CHAPTER 8

Beyond the Design— Area and Room Plans

INTRODUCTION

Revit Building's intelligent building modeler provides tools that fill a couple of basic needs of the people who pay designers to do their work. Design clients, usually the owners of the building being designed, almost always need to know the relative areas reserved for different spaces and functions within the design. Sometimes this information will be used for taxation purposes, most often for rental or other occupancy classifications. Throughout the lifespan of a commercial building in particular, accurate occupancy information can be critical. Revit Building models can supply this information—as anticipated uses or building configurations change during and after construction—in ways that static 2D drawings or previous 3D models never could. This may have long-range implications for design firms as suppliers of "life-cycle" management (or facilities management) services for buildings long after architectural functions are complete.

One of the easiest ways to collect and display basic area information about a Revit Building model is the color fill diagram, which can be either an area analysis plan or more sophisticated room plan with associated schedules. These plans are displayed in Views of their own. Schedules are also Views, as you have seen in Chapter 7; room and area plans, like schedules, can be located on sheets.

OBJECTIVES

- Create and edit a room schedule
- Work with room tags
- Define room styles
- Apply room styles to rooms
- Create and edit a color fill room diagram
- Create multiple color fill diagrams from one set of room tags
- Place color fills and schedules on a sheet
- Create a gross building area plan

- Create a rentable area plan
- Create a custom area plan
- Create and edit a color fill area diagram
- Export Revit Building information via ODBC
- Export a schedule
- Export to other formats

REVIT BUILDING COMMANDS AND SKILLS

Room Schedules

Schedule Keys

Room Styles

Color Fills

Area Settings

Area Plans

Export ODBC

EXERCISE I. CREATE A ROOM SCHEDULE

1.1 Open the file *Chapter 8 start.rvt*. This model file has been created without Worksets to make the exercises in this chapter simpler than would otherwise be the case.

1.2 Floor Plan: IST FLOOR – FLOOR PLAN should be the active view. Save the file as ***Chapter 8 area plans.rvt*** in a location specified by your instructor.

1.3 Make Floor Plans: 2ND FLOOR OFFICES the active view. Make Drafting the active tab on the Design Bar.

1.4 From the View menu, choose View>New>Schedule/Quantities. In the New Schedule dialogue, select Rooms under Category. Accept the defaults, as shown in Figure 8–1. Click OK.

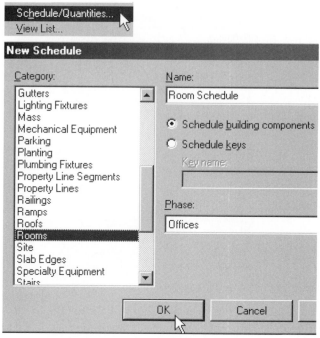

Figure 8–1 *Make the new schedule category Rooms*

1.5 On the Fields tab of the Schedule Properties dialogue, select Area from the Available fields: list. Click Add-> to move it to the Scheduled fields (in order): list.

1.6 Select Department, Name, and Number, and add them to the Scheduled fields, as shown in Figure 8–2.

Figure 8–2 *Move fields from Available to Scheduled*

1.7 Select Name in the Scheduled fields. Click Move Up twice to put Name at the top of the list. Place Number in the second position (see Figure 8–3).

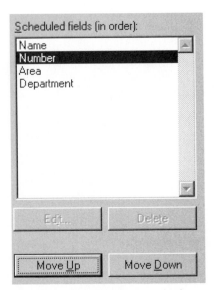

Figure 8–3 *Change the order of the scheduled fields in the list*

1.8 On the Appearance tab, check Bold and Italic for the Header font, as shown in Figure 8–4. This change will appear when the schedule view is placed on a Sheet. Click OK.

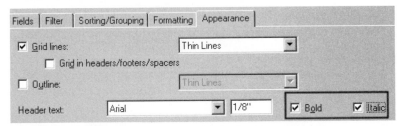

Figure 8–4 *You can adjust the appearance of the schedule on this tab*

Revit Building creates and opens a view named Room Schedule under Schedules/Quantities in the Project Browser and makes it the active view. Eighty rooms are listed.

1.9 Select New on the Option Bar in the Schedule view. Revit Building will place a room with Name **Room**, Number **81**, and Area Value **Not Tagged** in the last row of the schedule. This blank Room definition has not yet had an area tag placed inside a boundary or room.

1.10 Select New again to create a second un-tagged room in the schedule. See Figure 8–5.

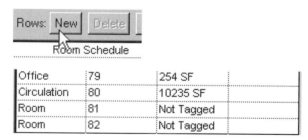

Figure 8–5 adjacent content:

Rows: New Delete		
Room Schedule		
Office	79	254 SF
Circulation	80	10235 SF
Room	81	Not Tagged
Room	82	Not Tagged

Figure 8–5 *Insert two room definitions in the schedule*

1.11 Open the view 2ND FLOOR OFFICES. On the Drafting tab of the Design Bar, select Room Separation. On the Option Bar, check Draw (the pencil icon) and Chain (see Figure 8–6).

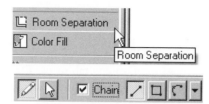

Figure 8–6 *The Room Separation tool and drawing options*

1.12 Sketch two lines to the left of the corridor door as shown in Figure 8–7. The exact dimensions are not critical.

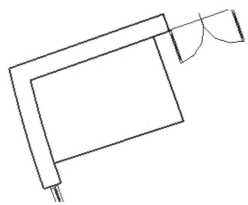

Figure 8–7 *Sketch a room where there are no walls*

1.13 Sketch two more lines as shown in Figure 8–8. The exact dimensions are not critical. Select Modify. Move the Circulation area tag to the right of the new lines.

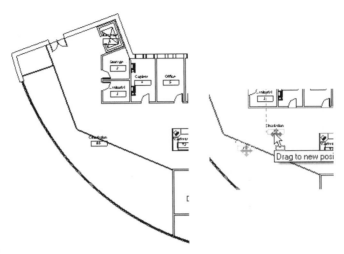

Figure 8–8 *Sketch two more lines to define another room separation without walls*

1.14 Select Room Tag from the Drafting tab of the Design Bar. On the Option Bar in the Room: field, click the drop-down arrow to expose the list. The two new room definitions are available to use. Select 81 Room, as shown in Figure 8–9. Place the tag in the small rectangular area. Select 82 Room from the Room: list and place the tag in the larger boundary area.

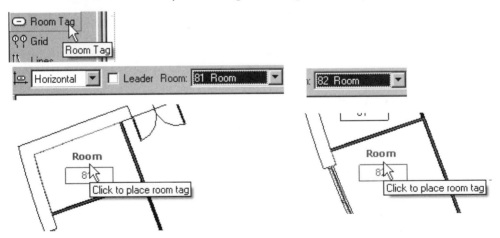

Figure 8–9 *Pick the room tag values from the Option Bar list*

1.15 Select Modify. Select the Room Tag for room 81. Pick the room name field a second time to make it editable. Change the Name Value to **Reception**.

1.16 Select Room Tag 82. Change the Name Value to **Waiting Area**. Click away from the Tag to exit edit mode.

1.17 Select the following Room Tags and change their Department Values according to the following table:

Room Tag Number	Department Values
4–10	Human Resources
11–22	Finances
23, 24	Student Affairs
29–40	Student Affairs
41–55	Finances
56–63	Human Resources
64–72	Fundraising
73–79	Dean's Office

 TIP Use Crossing or Inside selection windows with Filters as necessary. Once a new Value is typed into a field, Revit Building remembers the value and makes it available when that field is selected for editing later in another tag.

1.18 Open the Room Schedule view to see the new Department Values. Parameter values that are not driven by the model can be edited in the Schedule View. Select the Name cell for room number 82 and change it to **Lounge**, as shown in Figure 8–10. Your area value for room 82 may be different than shown.

Office	78	461 SF	Dean's Office
Office	79	254 SF	Dean's Office
Circulation	80	8980 SF	
Reception	81	94 SF	
Lounge	82	1162 SF	

Figure 8–10 *Edit a Name value in the Schedule view*

EXERCISE 2. CREATE A SCHEDULE KEY

2.1 Make the View tab of the Design Bar active. Select Schedule/Quantities.

2.2 In the New Schedule dialogue, select Rooms in the Category: field. Select the Schedule keys radio button. Accept the Schedule name and Key name default values as shown in Figure 8–11. Click OK.

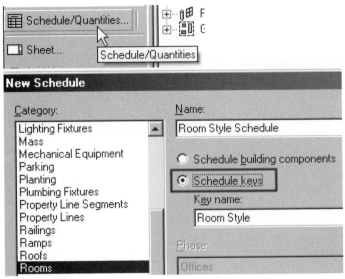

Figure 8–11 *Pick Schedule Keys to make the new schedule a Room Style Schedule*

2.3 In the Schedule Properties dialogue, select Available fields Base Finish, Floor Finish, and Wall Finish, and add them in turn to the Scheduled fields, as in Figure 8–12. Click OK.

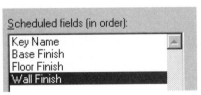

Figure 8–12 *Fields for the Room Style Schedule*

2.4 Revit Building will create a Room Style Schedule view and make it active. Select New three times from the Option Bar to create three rows within the schedule.

2.5 Fill in the row cells as shown in Figure 8–13.

Figure 8–13 *Values for the Room Style Schedule*

2.6 Open the view 2ND FLOOR OFFICES. Select the Room Tags in large Offices 77 and 78, the three Conference Rooms, and the Library. Select the Properties icon from the Options Bar. There will be a Room Style Parameter now available. Select VP Office from the drop-down list, as shown in Figure 8–14. Click OK. Other values will fill in automatically.

Occupancy	
Department	Dean's Office
Base Finish	Vinyl
Ceiling Finish	
Wall Finish	Paint/Panel
Floor Finish	Carpet 1
Other	
Room Style	VP Office
Occupant	

Figure 8–14 *Assiogning a Room Style to multiple rooms using Properties*

2.7 Select Office 79 and all the Offices along the upper wall. Select the Properties icon. Make the Room Style Parameter Std. Office. Click OK.

2.8 Select the room tags for the following rooms: Vending 25, Toilets 26 and 28, Services 27, Storage 2, Janitor 3, and Copiers 4. Pick the Properties icon. Make the Room Style Parameter value Services. Click OK.

2.9 Zoom to Fit. Make the Drafting tab of the Design Bar active. Select Color Fill. When you move the cursor over the Drawing Window, a Color Fill legend will be ghosted under the arrow.

2.10 Place the Color Fill legend to the left of the model. Revit Building will fill in room boundaries with a color fill pattern. Click OK in the notice box that appears about visibility of Floors. See Figure 8–15.

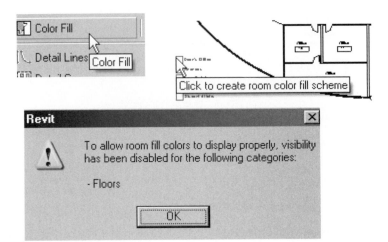

Figure 8–15 *Create a Color Fill*

2.11 Select the Color Fill legend. Pick the Properties icon on the Options Bar. Choose Edit/New. Select Rename. Change the name from Scheme 1 to **Color by Department**. See Figure 8–16. Click OK three times.

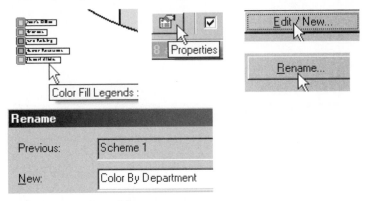

Figure 8–16 *Rename the Color Fill*

2.12 Choose Edit Color Scheme from the Options Bar. In the Edit Color Scheme dialogue, select the Color field for Finances. Select the Red swatch under Custom colors, as shown in Figure 8–17. Click OK.

Figure 8–17 *Editing colors in the fill*

2.13 Select the Color field for Human Resources. Pick the Cyan swatch under Custom colors. Click OK. See Figure 8–18.

Value	Color
Dean's Office	PANTONE 610 C
Finances	Red
Fund Raising	RGB 139-166-178
Human Resources	Cyan
Student Affairs	PANTONE 621 C

Figure 8–18 *The Color shows in the Edit Color Scheme panel*

2.14 Click OK to return to the View window. Hit ESC to terminate the edits (see Figure 8–19). Save the file.

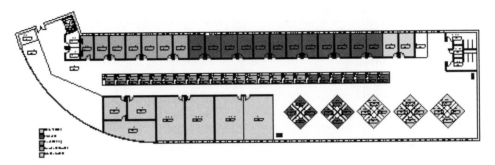

Figure 8–19 *The finished Departmental Color Fill diagram*

EXERCISE 3. WORK WITH COLOR FILLS

CREATE A NEW SCHEDULE FIELD

    3.1 Select the Room Schedule view in the Project Browser. Right-click and select Properties, as shown in Figure 8–20.

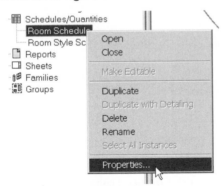

Figure 8–20 *Edit the Room Schedule Properties*

    3.2 In the Element Properties dialogue, select Edit in the Fields value. Choose Add Parameter (see Figure 8–21).

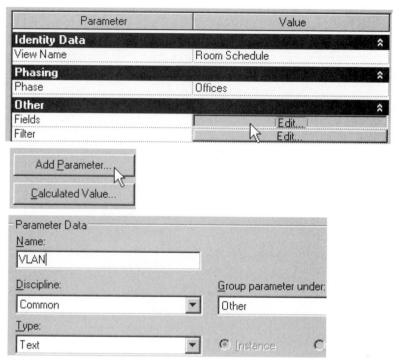

Figure 8–21 *Add a new parameter to this schedule*

3.3 In the Parameter Properties dialogue, type **VLAN** in the Name field. Accept the defaults. Click OK.

VLAN stands for Virtual Local Area Network. The new field will identify ethernet computer hookup nodes for offices and workstations.

3.4 Click OK three times to exit the dialogues. Open the Room Schedule view.

3.5 Give the rooms the following VLAN values:

Room Tag Number	VLAN Value
4–15	1
16–24	2
29–49	3
50–63	4
64–73	1
74–77	4
78–82	1

Figure 8–22 shows a selection of the Schedule view with the VLAN values being applied.

Room Schedule				
Name	Number	Area	Department	VLAN
Janitorial	3	80 SF		
Copiers	4	145 SF	Human Resources	1
Office	5	165 SF	Human Resources	1
Office	6	169 SF	Human Resources	1
Office	7	169 SF	Human Resources	1
Office	8	169 SF	Human Resources	1
Office	9	169 SF	Human Resources	1
Office	10	169 SF	Human Resources	1
Office	11	169 SF	Finances	1
Office	12	169 SF	Finances	1
Office	13	169 SF	Finances	1
Office	14	169 SF	Finances	1
Office	15	169 SF	Finances	1
Office	16	169 SF	Finances	2
Office	17	169 SF	Finances	2
Office	18	169 SF	Finances	2

Figure 8–22 *Assign VLAN values to rooms*

CREATE NEW COLOR FILLS FROM THE SAME INFORMATION

3.6 Pick the 2ND FLOOR OFFICE view in the Project Browser. Right-click and select Duplicate, as shown in Figure 8–23.

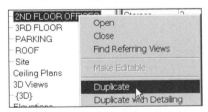

Figure 8–23 *Duplicate the view*

NOTE When you duplicate a view, the annotations do not duplicate by default. Use Duplicate with Detailing to keep notes, etc. visible in the new view. In this case you don't need to see the Room Tags.

3.7 Revit Building will open the new view, named Copy of 2ND FLOOR OFFICES. Choose the new view name in the Project Browser, right-click, and select Properties, as shown in Figure 8–20. Change the View Name to

2ND FLOOR OFFICES BY DEPARTMENT. Change the View Scale value to **1″ = 20′-0″** (see Figure 8–24). Click OK.

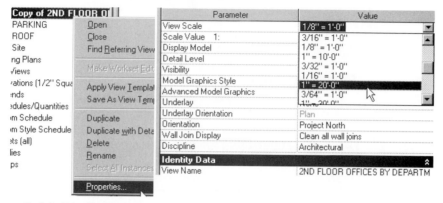

Figure 8–24 *Edit the View properties*

3.8   Select the 2ND FLOOR OFFICES BY DEPARTMENT view in the Project Browser. Right-click and pick Duplicate.

3.9   Revit Building will open the new view. Select the new view name in the Project Browser, right-click, and choose Rename. Change the view name to **2ND FLOOR OFFICES BY STYLE**. See Figure 8–25. Click OK.

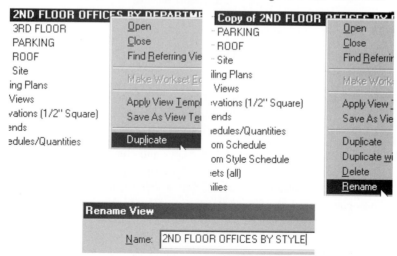

Figure 8–25 *Duplicate a view and rename the new view*

3.10   Select the Color Fill Legend in the new view. Right-click and pick Properties. Choose Edit/New, and then select Duplicate. Make the new Name value by **Color by Style**. Click OK. Select Edit in the Color Scheme Value field. In the Edit Color Scheme dialogue, change the Color by: value to **Room Style**.

Change the (none) Color to Basic Color White, as shown in Figure 8–26. Click OK three times to exit the dialogues.

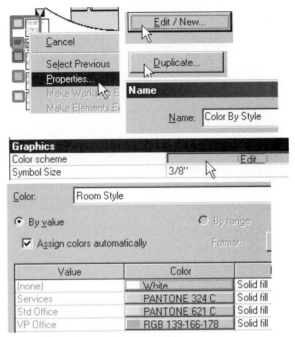

Figure 8–26 *The colors for the Room Style color fill*

3.11 Pick the 2ND FLOOR OFFICE BY STYLE view in the Project Browser. Right-click and select Duplicate.

3.12 Revit Building will open the new view. Pick the new view name in the Project Browser, right-click, and pick Rename. Change the view Name to **2ND FLOOR VLAN**.

3.13 Select the Color Fill Legend. Right-click and choose Properties. Pick Edit/ New, and then choose Duplicate. Make the new Name value **VLAN**. Click OK. Choose Edit in the Color Scheme Value field. In the Edit Color Scheme dialogue, change the Color by value to **VLAN,** as shown in Figure 8–27. Click OK three times to exit the dialogues.

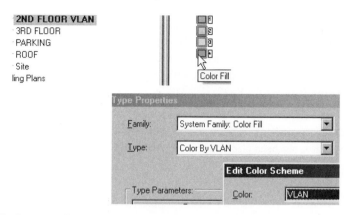

Figure 8–27 *Set up a duplicated view to show color by VLAN*

PUT COLOR FILLS AND SCHEDULES ON A SHEET

3.14 Expand Sheets in the Project Browser. Open view A801.

3.15 In the Project Browser, select view 2ND FLOOR OFFICES BY DEPARTMENT. Drag the view onto the sheet and place it in the upper-left corner.

3.16 Repeat for the views 2ND FLOOR OFFICES BY STYLE and 2ND Floor VLAN (see Figure 8–28).

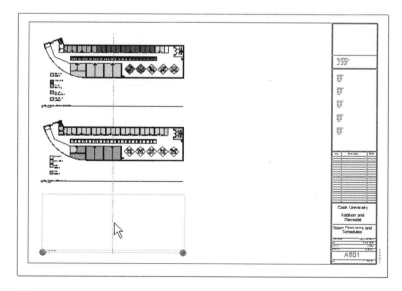

Figure 8–28 *Drag the views onto the page—Revit Building will supply alignment planes*

3.17 Pick and Drag the Room Schedule view onto the right half of the sheet. The schedule is a little too long to fit. Zoom in Region to the top rows of the schedule. Click the blue arrow-shaped grip that controls the width of the Department field and drag it to the right so that the row text fits on one line per row (see Figure 8–29).

 NOTE The appearance edits you made to the Room Schedule header earlier appear now that it has been placed on a sheet.

Figure 8–29 Adjust the column width for better fit

TIP Revit Building provides a control point on schedule views to allow you to split a long schedule to fit on a sheet, as shown in Figure 8–30.

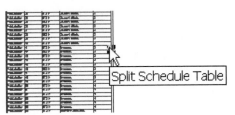

Figure 8–30 You can split a long schedule

3.18 Drag the Room Style Schedule onto the sheet. Place it below the Room Schedule, as shown in Figure 8–31.

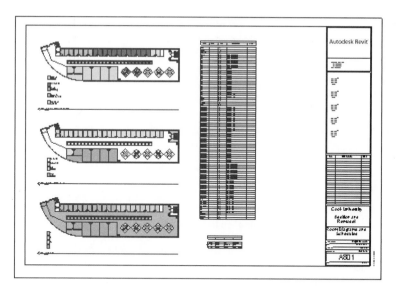

Figure 8–31 *The new page with color fills and associated schedules*

3.19 Save the file.

You have just quickly created and displayed five views of the same data. Using the floor plan, room tags, and color fills, you have provided graphic information that will be valuable to firms working on the building—interior finish contractors and systems electricians, for example. This same information will also be useful to the facilities manager during the life of the building. The room and room style schedules provide other matrix views of the same information shown in the color fills.

AREA PLANS AND COLOR FILLS

Revit Building's Area Plans are views that will appear in the Project Browser once created. Area plan views show spatial relationships of floor-plan levels based on Area Schemes. Area Schemes are definable. Revit Building creates two types of Area Schemes by default: Gross Building, which cannot be deleted or edited, and Rentable, which can be copied and edited to create additional Area Schemes. Area Boundaries and Area Tags are the building blocks of area plans.

EXERCISE 4. CREATE A GROSS BUILDING AREA PLAN

4.1 Open or continue working with *Chapter 8 area plans.rvt*. Make the 1ST FLOOR view active.

4.2 Place the cursor over the Design Bar and right-click to bring up the list of available tabs. Select Area Analysis to make that tab available and active, as shown in Figure 8–32.

Figure 8–32 *Open the Area Analysys tab*

4.3 From the Area Analysis tab of the Design Bar, select Area Settings. In the Area Calculations Setting dialogue, choose New.

4.4 Click in the Name Field for the new Area Scheme, and change the Value to **Classrooms and Labs**. Change the Description Value to **Measurements – Temporary Conditions** (see Figure 8–33). You will use this new Area Scheme definition later. Click OK.

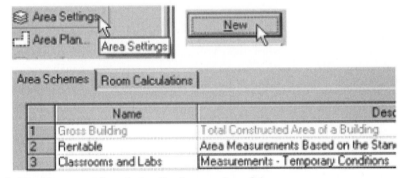

Figure 8–33 *Name and Description for the new Area Scheme*

4.5 Select Area Plan from the Area Analysis tab of the Design Bar. In the New Area Plan dialogue, select Gross Building for the Type and 1ST FLOOR for the level, as shown in Figure 8–34. Click OK. Click Yes to create area boundary lines from all external walls.

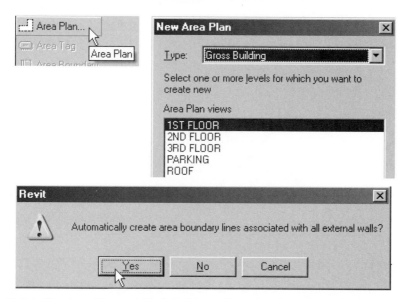

Figure 8–34 *The Area Plan tool. Revit Building will read and use the external walls for you.*

4.6 Revit Building will create a new view named Area Plan (Gross Building): 1ST FLOOR, as shown in Figure 8–35, and make it active. Select the Toposurface and Type **VH** to adjust its Visibility settings of the new view. Open the View Properties dialogue and set the Underlay value to None. This will clean up the screen. Zoom to Fit.

Figure 8–35 *The new area plan appears in the Project Browser*

4.7 From the Area Analysis tab of the Design Bar, pick Area Tag. A tag will appear under the cursor. Zoom in Region to the left side of the model so you can read the tag text.

4.8 Hold the cursor outside the model. The second line of the tag will read Not Enclosed—it is not within an area boundary. Move the cursor inside the exterior walls of the model. The tag will read an area value. This value will be the same whether the cursor is within a room or not, as shown in Figure 8–36.

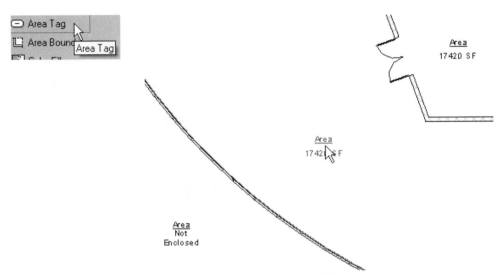

Figure 8–36 *Inside the model, an area tag displays the area of whatever area boundary it is within*

4.9 Place the cursor inside the model near the curved curtain wall and left-click to place it. Pick Modify from the Area Analysis tab of the Design Bar. Select the new tag. Select the Name text to open the field for editing. Change the Name Value to **Gross Area 1st Floor**, as shown in Figure 8–37. Click OK. Zoom to Fit.

Figure 8–37 *The edited tag name*

CREATE A RENTABLE AREA PLAN

The Gross Building Area scheme contains two simple Area Types: Gross Building Area and Exterior Area. The Rentable Area scheme contains the Exterior area type, plus five others, defined according to the following table:

Office Area	Contains tenant personnel, furniture, or both
Store Area	Suitable for retail occupancy and use
Floor Area	Areas available primarily for use of tenants of the given floor: washrooms, janitorial spaces, utility or mechanical rooms, elevator lobbies, public corridors
Building Common Area	Lobbies, conference rooms, atriums, lounges, vending areas, food service facilities, security desks, concierge areas, mail rooms, health/daycare/locker/shower facilities
Major Vertical Penetration	Vertical ducts, pipe or heating shafts, elevators, stairs, and their enclosing walls

4.10 Select Area Plan from the Area Analysis tab of the Design Bar. In the New Area Plan dialogue, choose Rentable for the Type and 1ST FLOOR for the level, as shown in Figure 8–38. Click OK. Answer Yes in the question box that appears to create area boundary lines from all external walls.

Figure 8–38 *Settings for a Rentable area plan*

4.11 Revit Building will create a new view named Area Plan (Rentable): 1ST FLOOR (as shown in Figure 8–39) and make it active. Type **VP** to adjust the Visibility settings of the new view. Change the Underlay to None. Click Edit in the Visibility field. On the Model Categories tab of the Visibility/Graphic

Overrides dialogue, clear Topography. On the Annotations Categories tab, clear Elevations. Click OK twice. Zoom to Fit.

Figure 8–39 *The new area plan appears in the Project Browser*

4.12 From the Area Analysis tab of the Design Bar, select Area Tag. A tag will appear under the cursor. Place it in the same location as on the previous plan. See Figure 8–40.

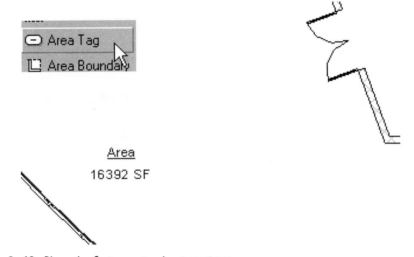

Figure 8–40 *Place the first area tag in open space*

4.13 Select Area Boundary from the Area Analysis tab of the Design Bar. On the Options Bar, leave the Pick option selected and Apply Area Rules checked. Zoom in Region to the left side of the model.

4.14 Select the two walls of the elevator enclosure facing the interior of the building, as shown in Figure 8–41. Note that once you choose the second wall, the area value of the existing area tag changes.

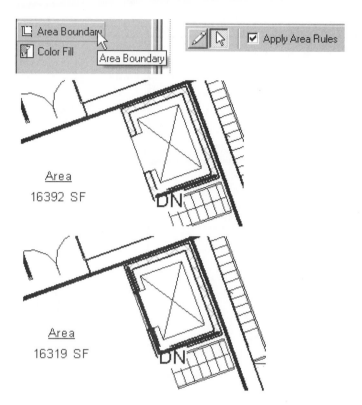

Figure 8–41 *Select only two walls to define the elevator area*

4.15 Continue selecting all the room-defining interior walls (see Figure 8–42). Do not pick walls inside the washroom areas.

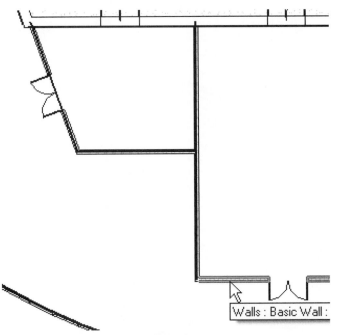

Figure 8–42 *Select a wall and Revit Building will create the corresponding boundary line*

4.16 Select the Draw option on the Option Bar. Draw two lines, as shown in Figure 8–43 below, to create an area boundary not defined by walls.

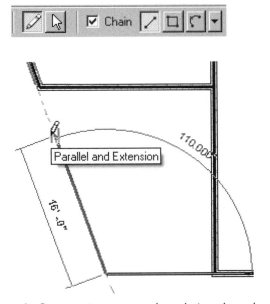

Figure 8–43 *Switch to the Draw option to create boundaries where there are no walls*

4.17 Draw two sides of a boundary, as shown in the right side of Figure 8–44. Draw a line across the entrance to the washroom area.

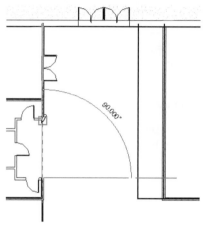

Figure 8–44 *Boundaries in a wide hallway*

4.18 Draw two sides of a boundary, as shown in Figure 8–45. The actual dimensions are not important.

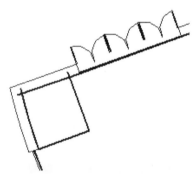

Figure 8–45 *Another boundary near the double doors*

4.19 Draw two sides of a boundary at the exterior walkway, as shown in Figure 8–46.

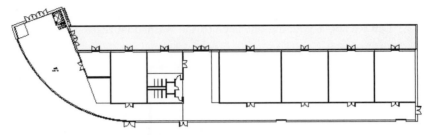

Figure 8–46 *Boundary lines include the walkway outside the walls*

    4.20 Select Area Tag from the Area Analysis tab of the Design Bar. Place area tags inside each room and boundary area you have created. Include the outside walkway (see Figure 8–47).

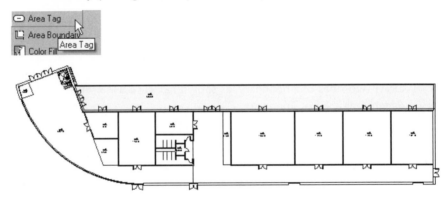

Figure 8–47 *Area tags in place — 14 in all*

    4.21 Edit each area tag Name Value and Area Type in turn, based on the information contained in figures 8–48 and 8–49 below and the table shown in Step 4.23.

Identity Data	
Name	Elevator
Comments	
Other	
Area Type	Major Vertical Penetration

Figure 8–48 *Using Properties to edit the area tag in the elevator*

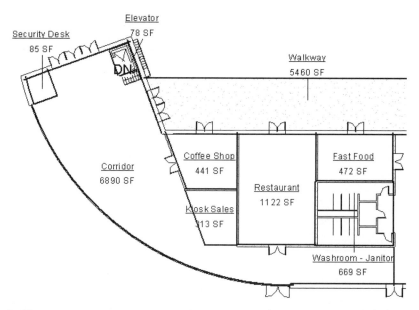

Figure 8–49 *Renamed area tags—they have have been enlarged in this illustration*

4.22 For the area tag in the Vending-Seating area, select Vertical orientation from the Option Bar (see Figure 8–50). For the tag inside the walkway, pick the Leader option from the Option Bar, then drag the tag outside the walkway boundary (as shown in Figure 8–49), so that it will be sure to show when the color fill is applied in the next steps.

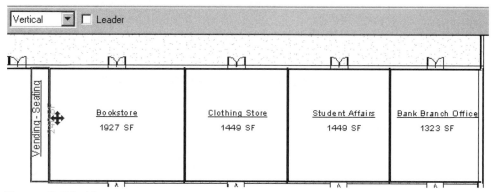

Figure 8–50 *More area tags—note the vertical orientation at Vending*

4.23 Edit the Area Type values for the area tags according to the following table:

Security Desk	Building Common Area
Elevator	Major Vertical Penetration
Corridor	Floor Area
Walkway	Exterior Area
Kiosk Sales	Store Area
Coffee Shop	Store Area
Restaurant	Store Area
Fast Food	Store Area
Washroom—Janitor	Floor Area
Vending—Seating	Building Common Area
Bookstore	Store Area
Clothing Store	Store Area
Student Affairs	Office Area
Bank Branch Office	Store Area

4.24 Zoom to Fit. Choose Color Fill from the Area Analysis tab of the Design Bar. A legend will appear under the cursor. Left-click to locate it to the left of the model, as shown in Figure 8–51. Click OK in the notification box about floor visibility.

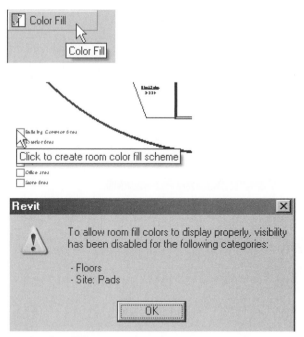

Figure 8–51 *Locate the Color Fill scheme*

4.25 Revit Building will supply color fill to the tagged areas, as shown in Figure 8–52. The Color scheme is editable. You have edited a color scheme in a previous exercise.

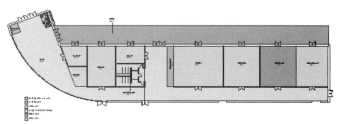

Figure 8–52 *The Color Fill in place*

4.26 Save the file.

EXERCISE 5. CREATE A CUSTOM AREA PLAN

5.1 Open or continue working in the file from the previous exercise. Select Area Plan from the Area Analysis tab of the Design Bar. In the New Area Plan dialogue, choose Classrooms and Labs as the Type, and 2ND FLOOR as the Area Plan view's level (see Figure 8–53). Click OK. Choose Yes in the query box about creating boundary lines from the external walls.

438

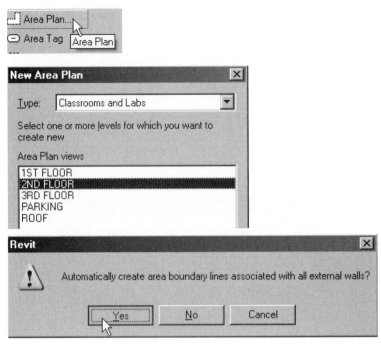

Figure 8–53 *Settings for the custom Area Plan*

5.2 Revit Building will create a new view named Area Plans (Classrooms and Labs): 2ND FLOOR, as shown in Figure 8–54, and make it active.

Area Plans (Classrooms and Labs)
  2ND FLOOR
Area Plans (Gross Building)
  1ST FLOOR
Area Plans (Rentable)
  1ST FLOOR
Schedules/Quantities

Figure 8–54 *The new Area Plan appears in the Project Browser*

5.3 Type **VP** to open the View Properties dialogue. Set the Phase value to Temporary Classrooms. Set the Underlay to None. See Figure 8–55. Click OK.

Underlay	None
Underlay Orientation	Plan
Phasing	
Phase Filter	Show All
Phase	Temporary Classrooms

Figure 8–55 *Set the View Phase and Underlay*

5.4 On the Area Analysis tab of the Design Bar, select Area Boundary. Pick walls as shown in figures 8–56 and 8–57. Do not select the walls of the washroom in the upper-right corner of the model or the elevator shaft walls. Do not select partition walls between similarly sized spaces.

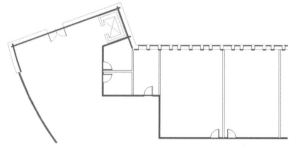

Figure 8–56 *The first wall picks do not include the elevator shaft walls or walls inside comparable spaces*

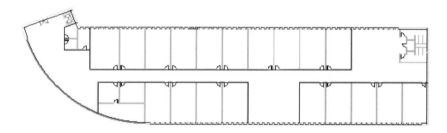

Figure 8–57 *The interior walls selected for boundaries*

5.5 On the Area Analysis tab of the Design Bar, select Area Tag. Place a tag inside each major area whose walls you previously picked (four in total). See Figure 8–58. Select Modify to terminate tab placement.

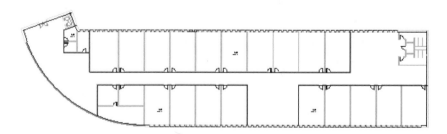

Figure 8–58 *Place four tags in the areas you enclosed*

5.6 Zoom in Region to the left side of the model. Choose the area tag shown in Figure 8–59. Pick the Properties icon. Make the Name Value **Permanent Utility Area**. Leave the Area Type Value as Building Common Area. Click OK.

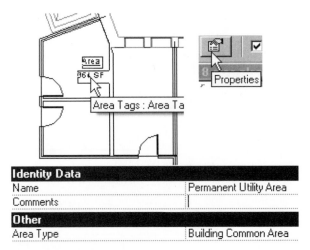

Figure 8–59 *Name and Area Type values for the area tag*

5.7 Select the area tag shown in the Figure 8–60 and edit its properties. Make the Name Value **Temporary Classroom**. Make the Area Type Value **Floor Area**. Click OK.

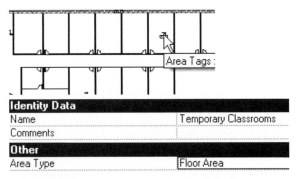

Figure 8–60 *Edit the tag in the room areas on the north wall*

5.8 Select the area tag shown in Figure 8–61 and edit its properties. Make the Name Value **Temporary Labs**. Make the Area Type Value **Store Area**; these spaces will be considered rentable during their existence (see Figure 8–61). Click OK.

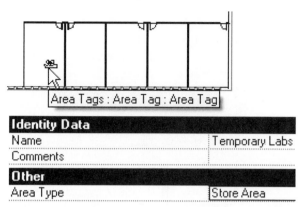

Figure 8–61 *Edit this tag in the rooms at the southeast corner*

5.9 Select the remaining area tag and edit its properties. Make the Name Value **Permanent Offices**. Make the Area Type Value **Office Area**. See Figure 8–62. Click OK.

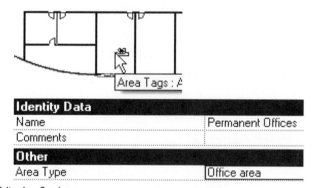

Figure 8–62 *Edit the final tag*

5.10 Zoom to Fit. Choose Color Fill from the Area Analysis tab of the Design Bar. Locate the Color Fill Legend to the left of the model. Click OK in the visibility notice.

5.11 Select the Color Fill Legend. Choose Edit Color Scheme from the Options Bar. In the Edit Color Scheme Dialogue, change the Color by: Value to **Name**, as shown in Figure 8–63. Click OK.

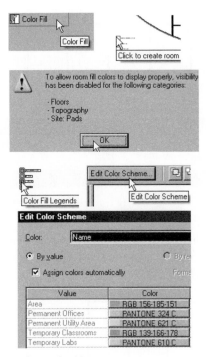

Figure 8–63 *Color the new scheme by Name*

    5.12 The Color Fill Legend will change to show the names you just assigned to the area types in this area plan, as shown in Figure 8–64.

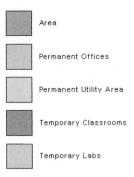

Figure 8–64 *The legend now shows the assigned Name values*

    5.13 Zoom to Fit (see Figure 8–65). Save the file.

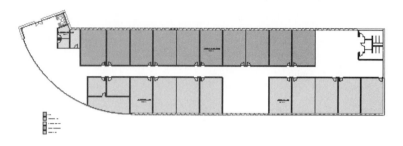

Figure 8–65 *The custom Area Plan with color fill*

EXERCISE 6. EXPORTING SCHEDULE INFORMATION

Schedule information is valuable inside Revit Building, but even more useful to many firms when it's taken out to other applications. Revit Building will export data to any ODBC-compatible (Open Database Connectivity—the Microsoft standard for accessing data from a variety of sources) database management application. This requires setting up the ODBC interface by specifying drivers to use. For this exercise you will use the driver supplied by Revit Building to export to Microsoft Access. The process would be similar for any ODBC-compliant application.

EXPORTING MODEL INFORMATION VIA ODBC

6.1   Open or continue working in the file from the previous exercise. From the File menu, pick File>Export>ODBC Database, as shown in Figure 8–66.

Figure 8–66 *Export Database information to an external application*

6.2   The Select Data Source dialogue opens. It contains two tabs and a file browser window.

ODBC requires Data Sources with Data Source Names (DSN) for connection. Data Sources can be located on the current workstation (Machine Data) or anywhere on the network (File Data). Machine Data Sources cannot be shared, and can be limited to individual users on any given machine. File Data Sources can be shared. Revit Building supplies a DSN file that appears on the File Data Source tab of the dialogue (see Figure 8–67).

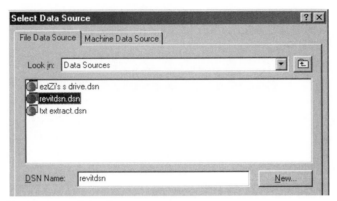

Figure 8–67 *Create a new Data Source*

6.3 Select the file *revitdsn.dsn* that appears in the dialogue window. This will place the file name in the DSN Name field. Click OK.

6.4 The ODBC Microsoft Access Setup dialogue will appear, as shown in Figure 8–68. Leave the System Database option on None. Select Create.

Figure 8–68 *Next, create a database file to load*

6.5 In the New Database dialogue, give the new database the name ***chapter 8.mbd***, in a location as directed by your instructor. Accept the other defaults in the dialogue. Click OK.

The Locale list provides a list of available languages. The Format selection refers to the Jet engine version inside Access (to allow you to convert to older formats if necessary). System Database and Encryption options allow you to define the database as a System file, which then requires system Administrator status to manage, or to password-protect the file.

6.6 Revit Building will display a confirmation box, as shown in Figure 8–69. Click OK twice to exit the setup.

Figure 8–69 *Click OK at the notification*

6.7 If Microsoft Access is installed on your computer, launch that application and open the *chapter 8.mdb* file you just created (or use your File Manager/ Computer Browser to find the new filename and double-click on it to open Access).

6.8 Access will open to a window listing the tables created by Revit Building's export, as shown in Figure 8–70.

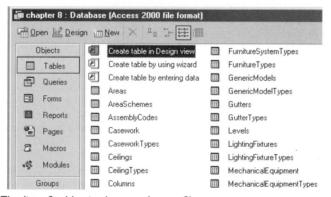

Figure 8–70 *The list of tables in the new Access file*

6.9 Select **Rooms** in the Table list and right-click. Choose Open. Revit Building's table assigns an ID to each room and shows the contents of each cell value as you saw in the schedule, except for Level and Room Style, which are identified by a numerical ID rather than a name (see Figure 8–71).

WallFinish	FloorFinish	Area	Number	Name
Paint	Carpet 2	13.4637620215	4	Copiers
Paint	Carpet 2	15.3209802926	5	Office
Paint	Carpet 2	15.6870637058	6	Office
Paint	Carpet 2	15.6870637058	7	Office
Paint	Carpet 2	15.6870637058	8	Office
Paint	Carpet 2	15.6870637058	9	Office
Paint	Carpet 2	15.6870637058	10	Office
Paint	Carpet 2	15.6870637058	11	Office

Figure 8–71 *Revit Building's room schedule information in the Access database*

6.10 Revit Building exports the Room Style Schedule you created as a separate table that uses the Room Style key ID shown in the Rooms table (see Figure 8–72).

RoomStyleSchedule : Table

	Id	KeyName	BaseFinish	FloorFinish	WallFinish
▶ ⊞	77045	Services	Wood	Tile	Vinyl WC
⊞	77046	Std Office	Vinyl	Carpet 2	Paint
⊞	77047	VP Office	Vinyl	Carpet 1	Paint/Panel

Figure 8–72 *TheRoom Style Schedule is translated as well*

6.11 Exit Access.

6.12 Revit Building does not maintain a live link to this external file, so design changes must be re-exported. To update a previously created database with new design information, select File>Export>ODBC Database from the File menu, as before.

6.13 In the Select Data Source dialogue, select the *revitdsn.dsn* file as before. Click OK.

6.14 In the ODBC Microsoft Access Setup dialogue, choose Select. In the Select Database dialogue, pick the name of the file you previously created (see Figure 8–73). Click OK twice to update that file. At this point you could also select another database file to load the Revit Building information into, or create a new one.

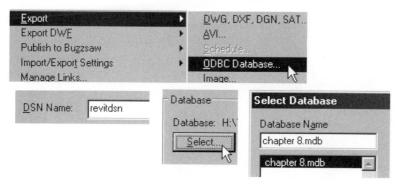

Figure 8–73 *Click OK to update the listed file, pick another, or create a new one*

EXPORTING A SINGLE SCHEDULE

6.15 Open the Room Schedule view. From the File menu, select File>Export>Schedule. This option is only available in a Schedule view. The Schedule export creates an ASCII text file, a very simple format that is read by many applications.

6.16 Locate the output file in a location you can find. See Figure 8–74.

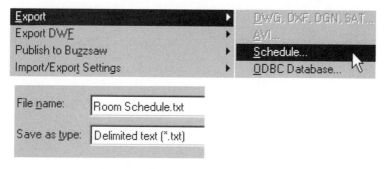

Figure 8–74 *Export a txt file from the schedule*

6.17 In the Export dialogue, accept the defaults. See Figure 8–75. There are options for what to include and how to format the text file indicators for fields and their contents. These options will affect how applications such as Excel import the file. Choose OK.

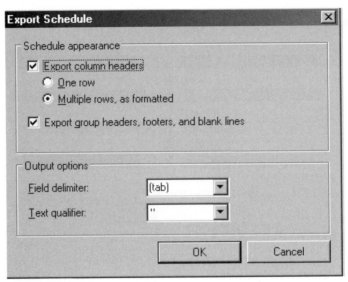

Figure 8–75 *Output options*

6.18 Revit will create the file. If you have Microsoft Excel on your system, you can use that application to open the txt file and use the schedule data. This is a quick and efficient way to export information to spreadsheets. See Figure 8–76.

A	B	C	D	E
Room Schedule				
Name	Number	Area	Department	VLAN
Elevator	1	51 SF		
Storage	2	101 SF		
Janitorial	3	80 SF		
Copiers	4	145 SF	Human Resources	1
Office	5	165 SF	Human Resources	1
Office	6	169 SF	Human Resources	1
Office	7	169 SF	Human Resources	1
Office	8	169 SF	Human Resources	1
Office	9	169 SF	Human Resources	1
Office	10	169 SF	Human Resources	1
Office	11	169 SF	Finances	1
Office	12	169 SF	Finances	1
Office	13	169 SF	Finances	1

Figure 8–76 *The text file opened in Excel.*

6.19 Like the ODBC, this is export is not a live link. The schedule must be re-exported after changes so that the spreadsheet file can be updated.

EXPORTING TO OTHER FORMATS

6.20 As preparation for the next steps, select Settings>Project Information from the Menu Bar. Choose the Edit button for Project Address. Provide this project file with a hypothetical address, complete with Zip code, as shown in Figure 8–77. Post Code information is used in the export you will perform next.

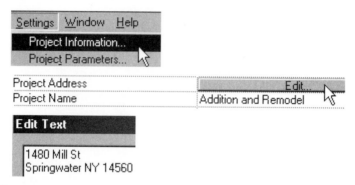

Figure 8–77 *Modify the Project Address*

6.21 From the File menu, choose File>Export>gbXML. See Figure 8–78. This creates a report in XML (extended markup language) format. There are no options; this is an automatic process. Export the file to a convenient location.

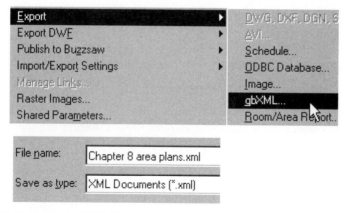

Figure 8–78 *Generate an XML file*

6.22 Use your file browser to navigate to the folder where you created the XML file. Double click on the file name, and it will open in Microsoft Internet Explorer, showing a list of field data, as shown in Figure 8–79.

```
<?xml version="1.0" ?>
- <gbXML temperatureUnit="F" lengthUnit="Feet" areaUnit="SquareFeet"
    xmlns="http://www.gbxml.org/schema">
  - <Campus id="cmps-1">
    - <Location>
        <Name />
        <ZipcodeOrPostalCode>14560</ZipcodeOrPostalCode>
      </Location>
    - <Building id="bldg-1" buildingType="Office">
        <Area>24832.231691</Area>
      - <Space id="sp-1-Elevator">
          <Name>1</Name>
          <Description>Elevator</Description>
          <Area>61.221964</Area>
          <Volume>1571.363734</Volume>
          <CADObjectId>73502</CADObjectId>
```

Figure 8–79 *XML export viewed in a browser*

XML is a developing format for sharing data across platforms. It is recognized by a growing number of applications. Figure 8–80 shows the same file when opened in Excel 2003.

A	B	C	D	E
/gbXML				
/@areaUnit	/@lengthUnit	/@temperatureUnit	/@useSIUnitsForResults	/@volumeUnit
SquareFeet	Feet	F	FALSE	CubicFeet
SquareFeet	Feet	F	FALSE	CubicFeet
SquareFeet	Feet	F	FALSE	CubicFeet
SquareFeet	Feet	F	FALSE	CubicFeet
SquareFeet	Feet	F	FALSE	CubicFeet
SquareFeet	Feet	F	FALSE	CubicFeet
SquareFeet	Feet	F	FALSE	CubicFeet

Figure 8–80 *XML export viewed in Excel*

6.23 Open view Area Plans (Classrooms and Labs) 2ND FLOOR. From the File menu, select File>Export>Room/Area Report. Accept the supplied name and locate the file in a location you can find later. Select Current View. See Figure 8–81.

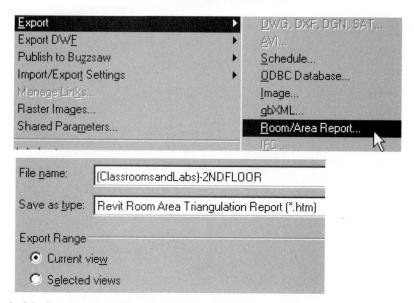

Figure 8–81 *Export a Room/Area Report in htm format*

6.24 Select Settings to view format possibilities for the report, as shown in Figure 8–82. Choose Cancel.

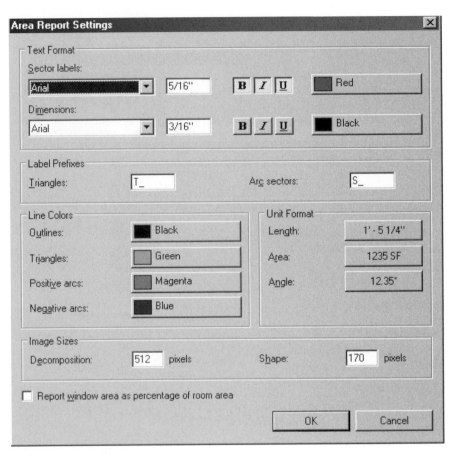

Figure 8–82 *Setting for the export*

6.25 Choose Save in the Export dialogue. Revit Building will work for a while. Click Delete Element if an error box opens, as shown in Figure 8–83.

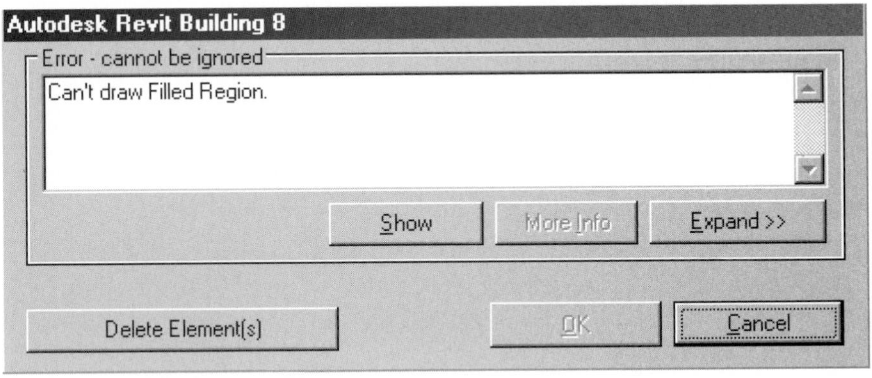

Figure 8–83 *There may be an error*

6.26 Revit will not display the export. Use your File browser to navigate to the folder where you saved the file, and double click on the file name to open it in Microsoft Internet Explorer. See Figure 8–84.

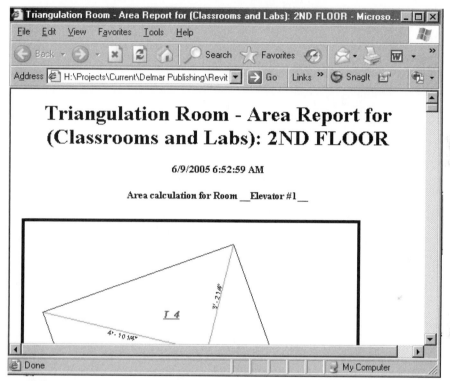

Figure 8–84 *The area report opened in a browser*

6.27 Microsoft Excel can also open *htm* files, as shown in Figure 8–85.

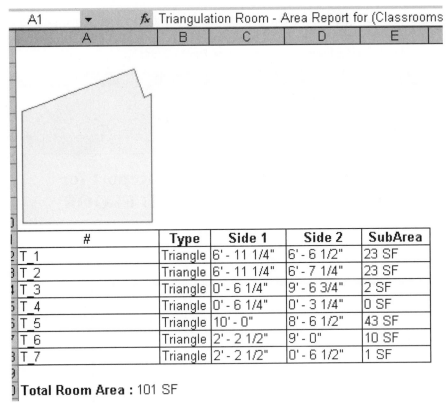

| A1 | ▼ | f_x | Triangulation Room - Area Report for (Classrooms |

#	Type	Side 1	Side 2	SubArea
T_1	Triangle	6' - 11 1/4"	6' - 6 1/2"	23 SF
T_2	Triangle	6' - 11 1/4"	6' - 7 1/4"	23 SF
T_3	Triangle	0' - 6 1/4"	9' - 6 3/4"	2 SF
T_4	Triangle	0' - 6 1/4"	0' - 3 1/4"	0 SF
T_5	Triangle	10' - 0"	8' - 6 1/2"	43 SF
T_6	Triangle	2' - 2 1/2"	9' - 0"	10 SF
T_7	Triangle	2' - 2 1/2"	0' - 6 1/2"	1 SF

Total Room Area : 101 SF

Figure 8–85 *The area report opened in Excel*

6.28 Room Area reports are primarily for the European market.

6.29 Revit Building will also export an Industry Foundation Classes (IFC) file, as shown in Figure 8–86.

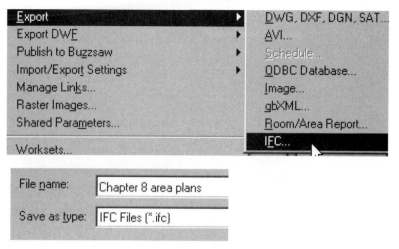

Figure 8–86 *The menu pick to export an IFG file*

This file type is under development by the International Alliance for Interoperability for cross-platform Building Information Modeling sharing. Revit Building objects are translated into IFC container classes—a Revit Building window and its parameters would be exported to an IFC container called IFCWindow.

The export to IFC classes can be configured to match certain standards.

6.30 Select File>Import/Export Settings>IFC Options from the Menu Bar. This opens the IFC Export Classes Dialogue, as shown in Figure 8–87.

Figure 8–87 *Open the IFC Options dialogue*

6.31 The IFC Export Class dialogue opens. See Figure 8–88. Certain Revit categories are not exported to IFC. Most model objects are, but some Revit object classes require mapping to IFC containers, as with Layer Standards mapping, which is discussed in other chapters. Select Standard. This opens a box where you can select from a list of pre-defined export settings to use as starting points for your own setup. Choose Cancel.

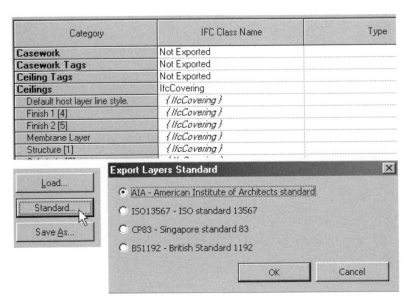

Figure 8–88 *IFC Export Settings*

6.32 At press time there are no standard office or design applications that open, import or display the contents of IFC files, so we cannot demonstrate the output format. IFC export seems to be a format that will be of use in the near future. Select Cancel.

6.33 Save the Revit Building file, or close without saving as directed by your instructor.

SUMMARY

In this chapter you used Revit Building's tools for cataloguing areas and rooms. Area Plans and Room Schedules provide quick graphic representations in color-fill diagrams, which are customizable and update automatically (always in Revit Building) as the model changes. Revit Building will export component and schedule information to database applications using a variety of formats.

REVIEW QUESTIONS – CHAPTER 8

MULTIPLE CHOICE

1. Room definitions can be created

    a) before any Room Schedules are created, not after

    b) after all Room Schedules are created, not before

    c) only while a Room Schedule is the active view

    d) any time

2. Schedules and Color Fills are created from Plan views, therefore you need to be aware of

    a) the Underlay of the original view

    b) the Detail level of the original view

    c) the Phase of the original view

    d) the Area of the original view

3. Area Plans are based on Area Schemes. Revit Building supplies these Area Schemes by default:

    a) Gross Building (not editable)
        and Rentable (editable)

    b) Custom (editable)
        and Standard (not editable)

    c) Exterior (not editable)
        and Interior (editable)

    d) all of the above

4. The Rentable Area Scheme contains the following Area Types:

    a) Exterior, Office

    b) Store, Floor

    c) Building Common, Major Vertical Penetration

    d) all of the above

5. Color Fill color schemes can be keyed by

    a) default schedule field parameters only

    b) text schedule field parameters only

    c) numeric schedule field parameters only

    d) any schedule field parameter

TRUE/FALSE

6. You can select walls to make Area Boundary lines, but not Room Separation lines.

7. Area Plans, unlike Room Plans, can't be placed on Sheets.

8. Color Fills are not editable—once you place a fill, you have to delete it to change a color.

9. Revit Building will create Room definitions in Schedules without having Room Tags placed in Floor Plan views.

10. Revit Building can export schedule information into external database files in all recognized formats except Microsoft Access.

 Answers will be found on the CD.

CHAPTER 9

Illustrating the Design–Graphic Output from Revit Building

INTRODUCTION

By now you have worked with a number of building models using Revit Building and have built pages for construction document sets. Revit Building's ability to produce informative views of building models is hardly restricted to the plans, elevations, and sections used for standard construction documents. Since all Revit Building models are built in three dimensions, with information about materials and surfaces as integral parts of the model objects, 3D-perspective views can be quickly and easily obtained. Revit Building contains a rendering module so that colored, lit (day or night, interior or exterior) pictures can be captured showing any part of the model at any point during design development. Revit Building incorporates a walkthrough tool that allows the user to create an animation based on a camera moving along a path. For users familiar with standalone rendering software applications, Revit Building will export files—images, models, or animations—that these other products can receive and use.

OBJECTIVES

- Create and change perspective views, interior and exterior
- Apply and edit material textures to building objects
- Work with render settings—environment, lighting, location
- Create an exterior daylight rendering
- Capture a rendering image
- Export Revit Building views to external files
- Use the View Section Box to restrict the scope of a view and rendering
- Work with the radiosity tool for interior lighting
- Create an interior nighttime rendering
- Create and edit a walkthrough path/camera combination
- Export a walkthrough animation file
- Plot Revit Building views using the Revit Building PDF writer

- Export 2D and 3D DWF
- Prepare Revit Building views for plotting using standard printers
- Export the Revit Building model for use in Autodesk VIZ

REVIT BUILDING COMMANDS AND SKILLS

Camera properties

New camera

View grip editing

View Properties Eye/Target editing

Export Image

Dynamic View Walkthrough controls

Region Raytrace

Raytrace exterior settings: environment, sun, lighting

Site components

Materials: create and assign

Section Box

Entourage components

Raytrace interior settings: limit model geometry

Light Groups, Daylights

Radiosity

Walkthrough: key frames, camera edit, path edit

Export walkthrough

Print/Plot

Export PDF

Export DWG for rendering

VIEWING THE MODEL

EXERCISE 1: THE CAMERA TOOL

1.1 Launch Revit Building. Open the file *Chapter 9 start.rvt*. The file will open to show the 3D View: {3D} at the default southeast orientation. Pick File> Save As from the Menu Bar. Save the file as **Chapter 9 render.rvt** in a location determined by your instructor.

1.2 Open Floor Plan: Level 1. Make a right-to-left selection of the Toposurface and a tree, as shown in Figure 9–1. Type **VH** at the keyboard to turn off the visibility of Planting and Topography categories in the view. This will simplify the screen.

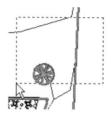

Figure 9–1 *Select topography and a tree to turn off in this view for clarity*

1.3 3D Views have already been created in this file. Choose 3D View 1 in the Project Browser tree. Right-click and pick Show Camera, as shown in Figure 9–2.

Figure 9–2 *Show the view camera*

1.4 Place the cursor over the camera/ field of view symbol (see Figure 9–3). Right-click and pick Properties. Study the properties of this view tool for a moment (see Figure 9–4). Click Cancel to exit the dialogue without making any changes.

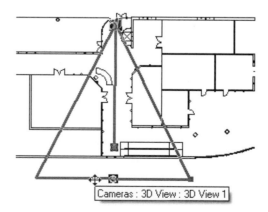

Figure 9–3 *The camera displayed and selected*

Parameter	Value
Extents	
Crop Region Visible	☑
Far Clip Active	☑
Far Clip Offset	85' 0 57/64"
Crop Region	☑
Section Box	☐
Camera	
Perspective	☑
Eye Elevation	5' 6"
Target Elevation	5' 6"
Camera Position	Explicit
Render Scene	None
Render Image Size	Edit...

Figure 9–4 *Properties of the camera that controls 3D View 1*

1.5 3D View 1 in the Project Browser tree is pre-selected. Right-click and pick Open. The view will open with its border highlighted red and blue control dots (grips) displayed, as shown in Figure 9–5. Clicking on any of the grips will allow you to change the field of view.

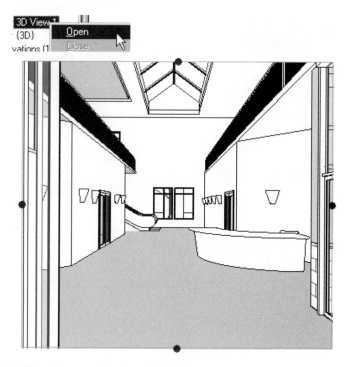

Figure 9–5 *3D View 1—note the grips for adjusting the field of view*

1.6 Open the Level 1 Floor Plan. From the View menu, choose View>New>Camera, as shown in Figure 9–6. Left-click to place the camera eye position at the lower-right side of the model. Left-click inside the model near the curved desk in the main entry.

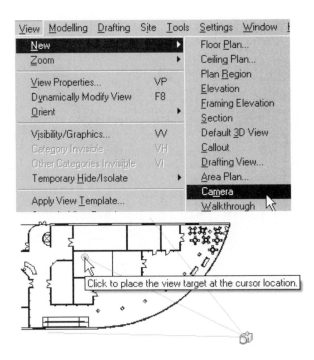

Figure 9–6 *Create a new camera; the first click places the camera at the eye of the viewer.*

1.7 Revit Building will create and open a new view named 3D View 2, which will need some adjustments (see Figure 9–7).

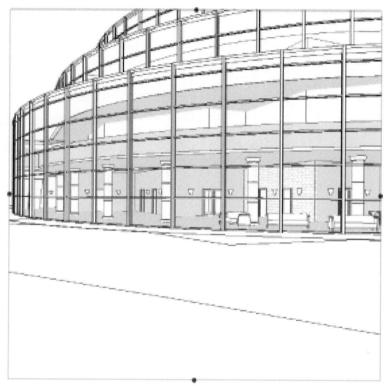

Figure 9–7 *The new 3D view in its rough state*

1.8 Use the grips on the view to display the model completely, as shown in figures 9–8 and 9–9. As you change the shape of the view, you may find it useful to Zoom Out (2X) to give yourself room to work, then Zoom to Fit when the adjustments are complete. The appearance of your view may be slightly different from the illustrations shown here, since your camera will not be in exactly the same place.

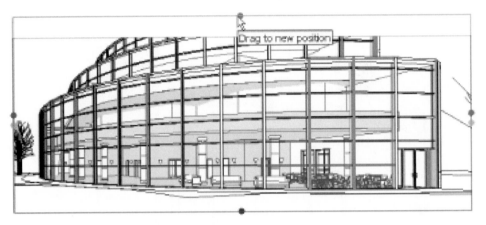

Figure 9–8 *Drag the grips to alter the field of view*

Figure 9–9 *The view now shows the entire model from this angle*

1.9 With the View grips still active, open the Level 1 Floor Plan. Note the changed appearance of the camera field of view indicator, as shown in Figure 9–10.

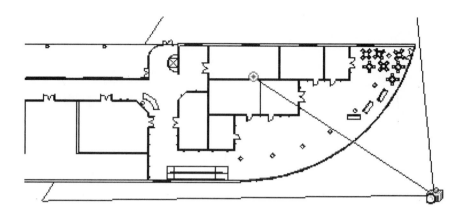

Figure 9–10 *Note how the field of view has altered*

> 1.10   Return to 3D View 2. From the View menu, select View>Shading with Edges. Revit Building will take a few seconds to process the display, shown in Figure 9–11.

TIP The two-stroke typed shortcut for the Shading with Edges display is **SD**. **HL** sets the display to Hidden Line.

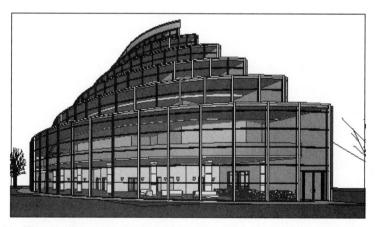

Figure 9–11 *The perspective view, shaded*

> 1.11   From the File menu, select File>Export>Image, as shown in Figure 9–12.

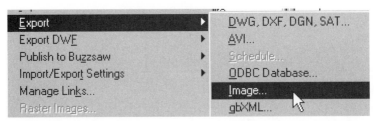

Figure 9–12 *The Export Image menu pick*

1.12 In the Export Image dialogue, give the file the name ***Chapter 9 east exterior*** and place it in a folder that you can find again easily. Select Visible portion of current window in the Export Range section. Leave the other settings. Note the output choices for file format, shown in Figure 9–13. Click OK to create the file.

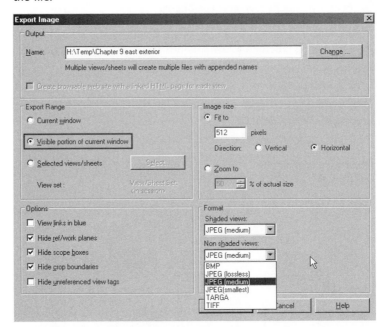

Figure 9–13 *The Export Image dialogue selections*

1.13 Type **HL** at the keyboard or use the View Control bar to return to Hidden Line mode in the 3D View. Right-click in the view window and select View Properties. In the Element Properties dialogue, change the Target Elevation value to **25′ 6″**, as shown in Figure 9–14. Click OK. The angle of the view will change.

Figure 9–14 *Change the camera target elevation*

1.14 Open the View Properties dialogue again. Change the Target Elevation value back to **5' 6"** and change the Eye Elevation value to **55' 6"** (see Figure 9–15). Click OK. The angle of the view will change.

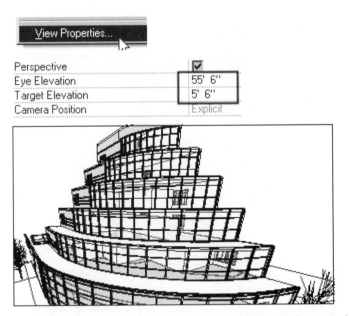

Figure 9–15 *Now change the height of the camera while looking at the original target*

1.15 Select the Undo tool on the Toolbar to undo the last view edit and lower the camera to eye level. Pick the Dynamically Modify View tool on the Toolbar, as shown in Figure 9–16, to open the Dynamic View controls.

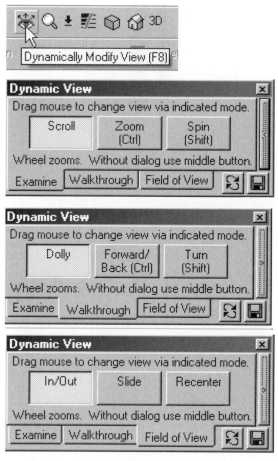

Figure 9–16 *The Dynamic View tool—note the different controls for each tab*

1.16 The Dynamic View control dialogue contains three tabs, with three different types of camera motion controls.

1.17 The view also contains a control to undo a view change, a control to save the current state of the view, and an expandable panel with controls to orient the view using directions, views or planes, as shown in Figure 9–17. The control to orient the view to a plane opens a dialogue you have used before when sketching profiles in elevation views.

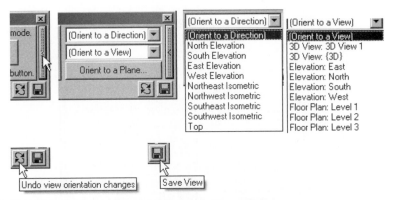

Figure 9–17 *Additional controls on the Dynamic Viewe dialogue*

1.18 On the Dynamic View dialogue, select the Walkthrough tab. Make the Dolly control active, as shown in Figure 9–16. Note the instructions for the controls.

1.19 Hold down the left mouse button and Dolly (move laterally) the camera up/down and left/right to adjust the view. Moving the cursor up has the effect of moving the camera position down, and vice versa (see Figures 9–18 and 9–19).

Figure 9–18 *Preparing to Dolly the camera*

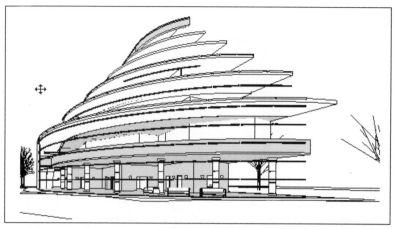

Figure 9–19 *Revit Building simplifies the display during motion*

1.20 Hold down the CTRL key to make the Forward/Back control active and repeat—alter the camera position in and out (see Figures 9–20 and 9–21).

Figure 9–20 *The Forward/Back cursor has a different appearance from the Dolly cursor*

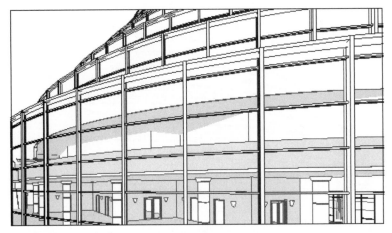

Figure 9–21 *Results of moving the camera Forward*

1.21 Make the Turn control (SHIFT) active and repeat the previous motions (see Figures 9–22 and 9–23). Note that the camera holds position and turns right as the cursor moves left.

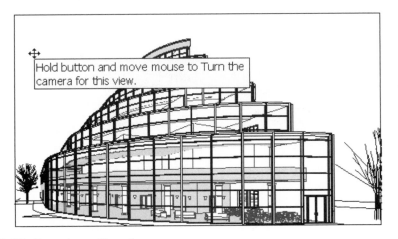

Figure 9–22 *Preparing to Turn the camera*

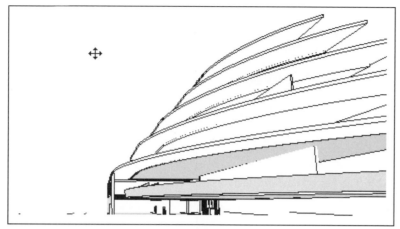

Figure 9–23 *Turning the camera left*

    1.22 Hold the cursor over the blue title bar of the Dynamic View control. Pick the X controls (or right-click and select **Close**) to close the dialogue. See Figure 9–24. If necessary, use View Properties to return the Target and Eye Elevation values to **5′ 6″**, and readjust the field of view using the grips to view the entire model. Your camera may not be in its original position because of your left/right and forward/back adjustments.

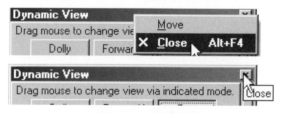

Figure 9–24 *Ways to close the Dynamic View dialogue*

RENDERING

Now that you have experimented with camera placement and have changed a perspective view to show the model from the angle or angles that suit your purposes, you can work on making the model come alive with color rendering.

EXERCISE 2. EXTERIOR SETTINGS

    2.1 Make the Rendering tab of the Design Bar active. Pick the Region Raytrace tool, as shown in Figure 9–25.

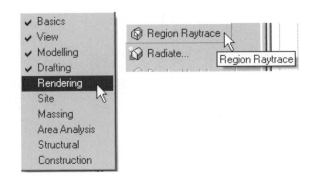

Figure 9–25 *The Region Raytrace tool*

    2.2 Pick two points around the double door, as shown in Figure 9–26.

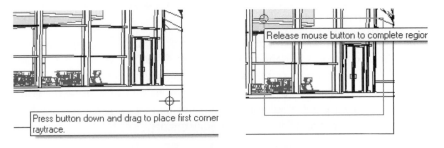

Figure 9–26 *The picks to define a region raytrace*

    2.3 Revit Building will display the Scene Selection dialogue. Select Exterior, as shown in Figure 9–27. Click OK.

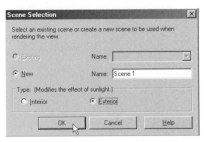

Figure 9–27 *Select Exterior Scene, then click OK*

    2.4 In the query box that appears, click Yes to turn off lights inside the building for the render process, as shown in Figure 9–28. The model already contains light fixtures—since this is an exterior scene, you have the choice to turn the lights off to save rendering time or leave them on for realism (as in a night scene).

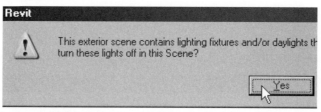

Figure 9–28 *Turning off lights inside the building*

Revit Building will produce a rendered region bounded by your previous picks (see Figure 9–29).

Figure 9–29 *The first rendered test area*

2.5 Choose Settings from the Rendering tab of the Design Bar, as shown in Figure 9–30. In the Render Scene Settings dialogue, select Rename. In the Rename dialogue, enter **Exterior – Gallery Side**. Click OK.

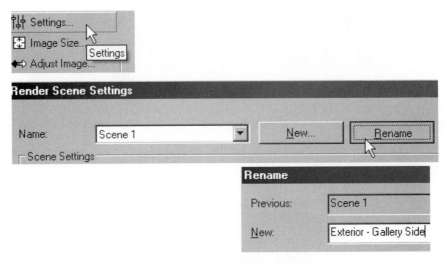

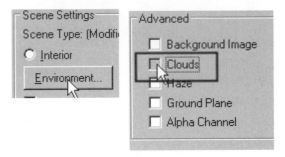

Figure 9–30 *Open Settings and rename the Scene*

2.6 Select Environment. In the Environment dialogue, accept the default Automatic Sky as the Background Color value. Pick Clouds in the Advanced section, as shown in Figure 9–31.

Figure 9–31 *Turn on Clouds*

2.7 Revit Building activates the Clouds tab, as shown in Figure 9–32. Click OK to accept the defaults.

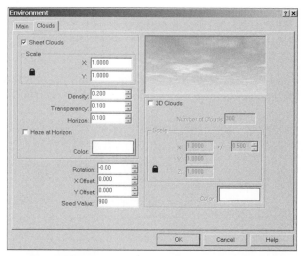

Figure 9–32 *The controls for Clouds*

2.8 In the Scene Settings area of the Environment dialogue, clear Use Sun and Shadow Settings from View. Click Sun. On the Date and Time tab, change the Specify Solar Angles value to By Date, Time and Place. On the Place tab, use the Cities drop-down list to find **Nashville, TN, USA** and select it, as shown in Figure 9–33.

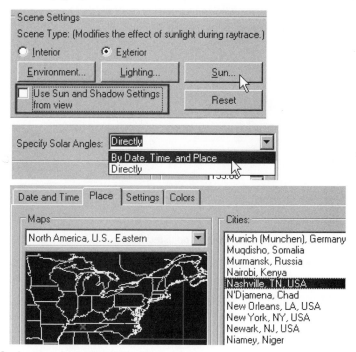

Figure 9–33 *Assign a location on the Place tab*

2.9 Select the Settings tab and Colors tab in turn to view the controls, as shown in Figures 9–34 and 9–35. Accept the defaults. Click OK.

 NOTE The Settings tab allows you to save settings that govern cloudiness, sun and sky intensity, and the direction of solar north, or load in previously saved setting files.

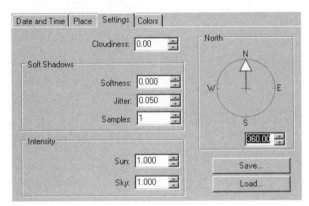

Figure 9–34 *The Settings tab*

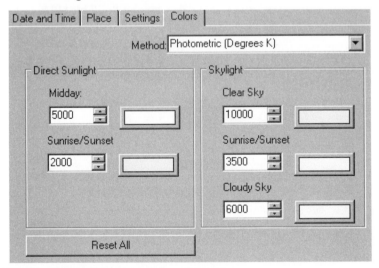

Figure 9–35 *The Colors of the sky can be adjusted here*

2.10 Change the Plant Season value on the main Render Scene Settings dialogue page to **Summer**, to match the sun settings (see Figure 9–36). Click OK to accept the remaining settings.

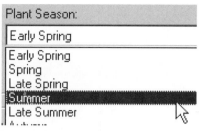

Figure 9–36 *Match the foliage to the sun angle*

2.11 Select Image Size on the Rendering tab of the Design Bar. Enter **300** as the
Resolution dpi value, shown in Figure 9–37. Hit TAB to view the resulting
image size. Click Cancel.

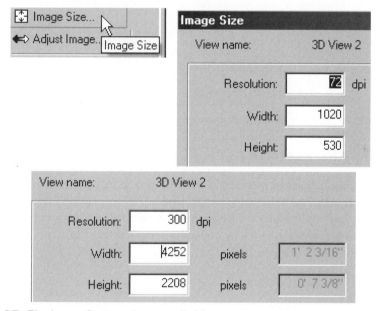

Figure 9–37 *The Image Size can be controlled here*

2.12 Choose Adjust Image on the Rendering tab of the Design Bar. Note the
available Image Controls: Brightness, Contrast, Indirect (ambient light
intensity not coming from the sun or other point light sources), and a toggle
for dynamic range (see Figure 9–38). Click Cancel.

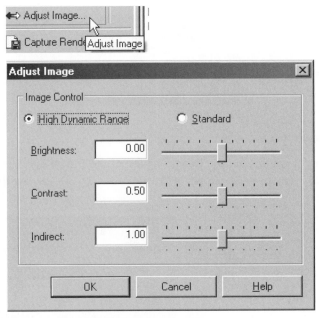

Figure 9–38 *You can tweak overall picture settings here*

MATERIALS AND PLANTINGS

2.13 From the Settings menu on the Menu Bar, choose Settings>Materials, as shown in Figure 9–39. In the Materials dialogue, use the Name drop-down list to find **Finishes – Exterior – Curtain Wall Mullions**. There is no Texture assigned in the AccuRender area. Choose the select arrow to the right of the Texture field.

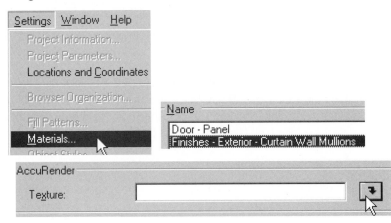

Figure 9–39 *The Materials dialogue*

2.14 Once you're inside the Material Library dialogue, navigate to the _ACCURENDER\Metals\Aluminum, Anodized_ folder in the tree pane. Select **Bronze, Medium** in the Name pane, as shown in Figure 9–40. Click OK.

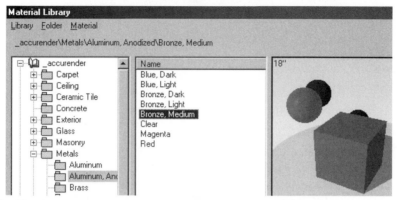

Figure 9–40 *Navigate among the many selections for materials*

2.15 Use the Name drop-down list to find **Glazing – Curtain Wall Glazing**. Pick the select arrow next to the Texture field to change the default value.

2.16 Back inside the extensive Material Library dialogue, navigate to the _ACCURENDER\Glass\Tinted_ folder in the tree pane. Select **Bronze, Medium, Smooth** in the Name pane, as shown in Figure 9–41. Click OK. Click OK again to exit the Settings dialogue.

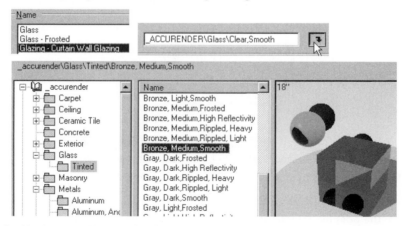

Figure 9–41 *Assign a Material for Curtain Wall glazing*

2.17 On the Rendering tab of the Design Bar, select Display Model to remove the partial rendering. Select one of the curtain wall mullions, as shown in Figure 9–42. (You may have to use the TAB key to cycle through the possible picks—the Type Selector should read **Rectangular Mullion: 2.5″ x 5″ rectangular**.) Choose the Properties icon.

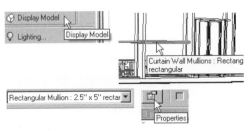

Figure 9–42 *Pick a curtain wall mullion*

    2.18 Click Edit/New. Change the Material parameter Value to **Finishes – Exterior- Curtain Wall Mullions,** as shown in Figure 9–43. Click OK three times to exit the dialogues.

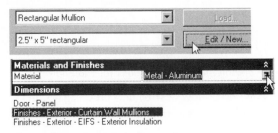

Figure 9–43 *Assign the mullion material via Properties*

    2.19 Carefully choose one of the vertical 5″ x 10″ mullions. Repeat the Materials assignment to **Finishes – Exterior – Curtain Wall Mullions**. (See Figure 9–44.) Click OK three times to exit the Element Properties dialogue.

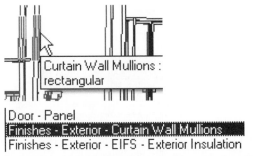

Figure 9–44 *Assign the Curtain Wall Mullion finish to the vertical mullions*

    2.20 Select the double door component. You may have to use TAB to cycle among the possible selections. Choose the Properties icon. Click Edit/New. Change the Materials parameter Value to **Finishes – Exterior – Curtain Wall Mullions**, as shown in Figure 9–45. Click OK three times to exit the Element Properties dialogue.

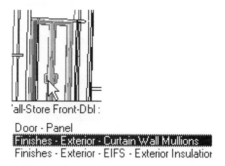

Figure 9–45 *Assign the same material to the Double Door frame*

2.21 Carefully select a curtain wall panel above the door. You may have to TAB through the possible selections. The Type Selector will read System Panel when your selection is correct. Click the Properties icon. Choose Edit/New. Change the Materials parameter Value to **Glazing – Curtain Wall Glazing**, as shown in Figure 9–46. Click OK three times to exit the Element Properties dialogue.

Figure 9–46 *Assign the tinted glazing material to the curtain wall panels*

2.22 Choose Region Raytrace from the Rendering tab. Pick two points around the double door as before. Note the difference in tints now that bronze finishes and glazing have been selected, as shown in Figure 9–47.

Figure 9–47 *The Region Raytrace shows different colors after materials are changed*

2.23 Choose Display Model to remove the rendered image from the view.

2.24 Open the Floor Plan Level 1 view. Type **VG** to open the Visibility/Graphics Overrides dialogue. Check Plantings and Topography. Click OK.

2.25 Make the Site tab active in the Design Bar. Choose Site Component, as shown in Figure 9–48. Pick **Shrub: Rhodo 24″** from the Type Selector on the Options Bar. Pick the Properties icon.

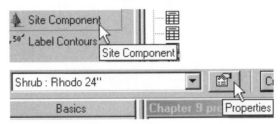

Figure 9–48 *The Site Component tool on the Site tab*

2.26 Pick Edit/New. Click Rename. Change the Type name to **Rhodo 16″**. Change the Plant Height Value to **16″** (Revit Building will convert it to 1′ 4″, as shown in Figure 9–49). Click OK. Click OK to exit the Element Properties dialogue.

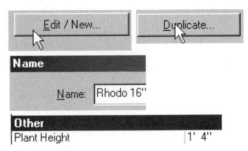

Figure 9–49 *Values for the new Type*

2.27 Place a number of plants beside the concrete walkway, as shown in Figure 9–50. The exact number and placement are not important. Click Modify to terminate the component placement.

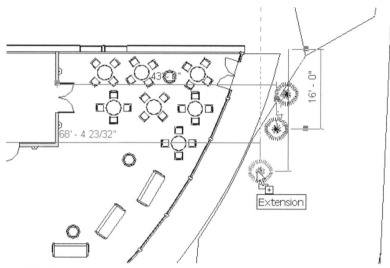

Figure 9–50 *Place shrub instances near the exterior walkway on the east side*

2.28 Open 3D View 2. Make the Rendering tab active in the Design Bar. Select Raytrace on the Rendering tab, then pick GO on the Options Bar.

Revit Building will generate a rendering using the sunlight, environment, and materials settings you have previously defined. This may take a few minutes, so be patient. See Figure 9–51 for an example.

Figure 9–51 *The rendered image*

2.29 When the rendering is complete, select Capture Rendering on the Rendering tab of the Design Bar. Revit Building will add a view named 3D View 2 under Renderings in the Project Browser. See Figure 9–52.

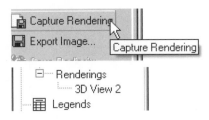

Figure 9–52 *The Capture Image tool*

2.30 Select Export Image on the Rendering tab of the Design Bar. Place a copy of the image in a location you will be able to find later. Note that you can specify the name and file type of the image using this export routine (see Figure 9–53).

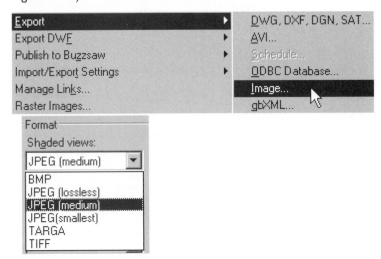

Figure 9–53 *Export the new image to an external file*

2.31 Save the file.

EXERCISE 3. INTERIOR SETTINGS—SECTION BOX CROP

3.1 Open or continue working with the file from the previous exercise. Open the 3D View {3D}. Right-click and select View Properties, as shown in Figure 9–54. In the View Properties dialogue, check Section Box. Click OK. The Section Box will appear around the model.

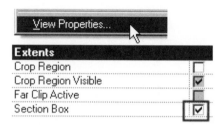

Figure 9–54 *Activate the Section Box for this view*

3.2 Choose the Section Box. It will highlight red, and blue control dots (grips) will appear. Drag the grips and adjust the boundaries of the Section Box to crop the model. Isolate the west (left) half of the top floor of the tower, as shown in Figures 9–55 and 9–56.

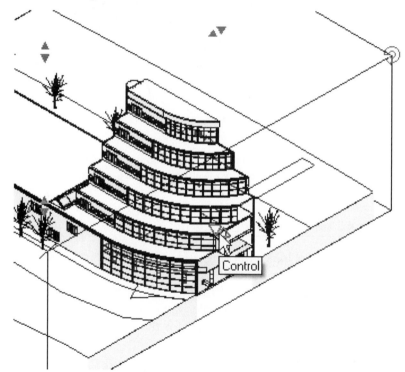

Figure 9–55 *Pick the grip to drag it*

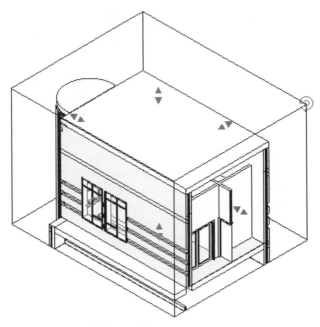

Figure 9–56 *Isolate a portion of the top floor*

 CAUTION This is an important step. If you do not correctly set and use the section box, the Rendering and Radiosity calculations in future steps will run extremely slowly, or not at all.

COMPONENTS – IMAGES OF PEOPLE

Architectural models can look sweeping and grand, or intimate and cozy, when viewed from the proper angle, but even the best spaces tend to appear sterile without human beings or their usual "stuff" in them. Many third-party developers have created quickly-rendered models of people, vehicles, and accessory items to place in rendered scenes to make them come alive for the viewer, and to provide a sense of scale.

3.3 Open the Floor Plan View Level 7. Zoom In Region around the left side of the model, a small lobby or waiting room. From the Rendering tab of the Design Bar, select Component. Choose Load from Library on the Options Bar.

3.4 Navigate to the *Imperial Library\Entourage* folder. Select **RPC Female.rfa**. Hold down the CTRL key and select **RPC Male.rfa**. Click Open. This will load both families into the current project file. See Figure 9–57.

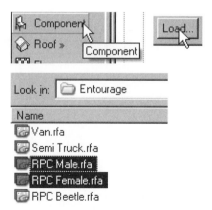

Figure 9–57 *Load in components to represent people*

3.5 Use the Type Selector on the Options Bar to make **RPC Male: LaRon** the current component. Check Rotate after placement on the Options Bar. Place the instance in the lower-left corner of the room, in front of the elevator doors. Rotate it 135° from the default position, which will point the front of the model to the upper-right of the plan.

RPC person components display in the plan as circles with a radius line representing the front. You can rotate the component after placement if necessary.

3.6 Use the Type Selector to place an instance of **RPC Female: Cathy** to the right of the desk as shown. Rotate the component −45° so it faces the other RPC person (see Figure 9–58).

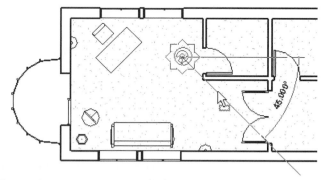

Figure 9–58 *Placing the new components, facing each other*

3.7 From the View menu, select View>New>Camera. Place the camera to the right of the room, in front of the double doors. Place the target behind the desk, as shown in Figure 9–59. Relocate the Cathy component if necessary, so it will be in the camera view.

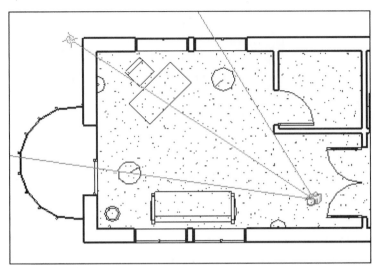

Figure 9–59 *Place and point a camera in the room*

3.8 Revit Building will open the new camera view. There is a size control on the Options Bar when the view is selected. Click the button, which will show the current image size (6″ x 6″). Change the width value to **9″** and the height value to **5″**. Leave the Field of view radio button selected, as shown in Figure 9–60. Click Apply and note the results. Click OK. Use the field of view grips to adjust the view further if necessary, so that it does not display unnecessary amounts of floor or ceiling. See Figure 9–61 for an example.

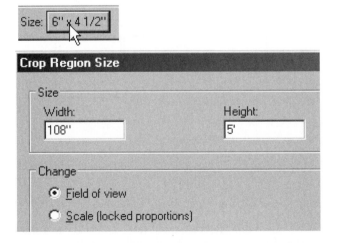

Figure 9–60 *The view size control on the Options Bar and the Crop Region Size dialogue*

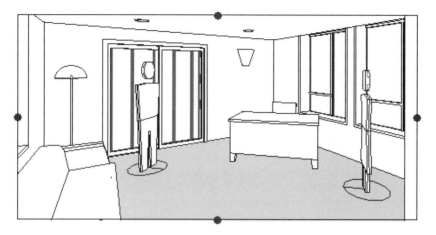

Figure 9–61 *The field of view, adjusted—note the simplified appearance of the person components*

3.9 Choose Settings from the Render tab of the Design Bar. In the Scene Selection dialogue, select New and Interior Scene. Enter the name **7th floor lobby**, as shown in Figure 9–62. Click OK.

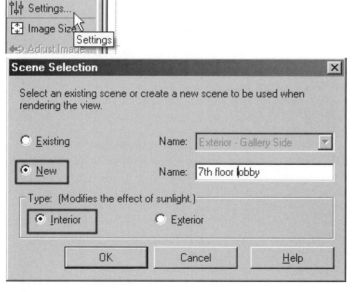

Figure 9–62 *Name the new Scene*

3.10 The Render Scene Settings dialogue will appear. Click OK to close it for now.

LIGHTING, DAYLIGHTS, RADIOSITY

3.11 Select Lighting from the Rendering tab of the Design Bar, as shown in Figure 9–63. Note that a Light Group has been defined in this file—sconce lights on the first floor are all grayed out. Clear the On box for light group **1st floor sconces**. Click OK.

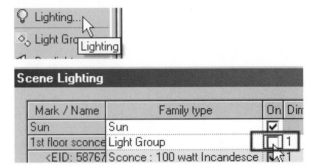

Figure 9–63 *The Lighting tool on the Rendering tab*

Light Groups function as component groups do in Revit Building, and are useful for managing groups of lights when you want to study or display alternate lighting conditions. You will not explore Light Groups in any detail in this exercise.

3.12 Choose Daylights from the Rendering tab of the Design Bar, as shown in Figure 9–64. There is no dialogue associated with this tool. Pick the two windows at the right side of the view. Pick the double door to the elevator tower. Click Modify to terminate the selections.

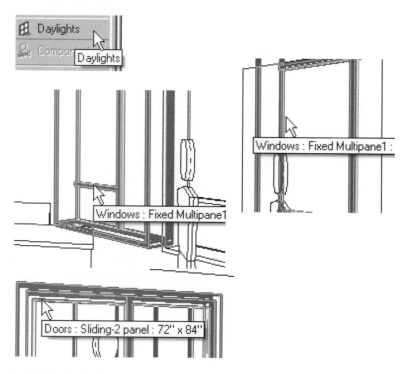

Figure 9–64 *The Daylights tool*

Daylights are sources of exterior light. Glass daylights both admit light and reflect according to the properties of their glass (clear, frosted, or tinted).

3.13 Choose Settings from the Rendering tab of the Design Bar. Select Lighting in the Render Scene Settings dialogue. Scroll to the end of the light list. Note that the new Daylights are now included. See Figure 9–65. Click Cancel to exit the Lighting dialogue.

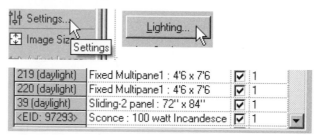

Figure 9–65 *The daylights now show in the lighting list, where they can be turned on and off*

3.14 Choose Environment. Pick Clouds in the Advanced section on the Main tab of the Environment dialogue. See Figure 9–66. Click OK.

Figure 9–66 *Turn on clouds in the sky outside the windows*

3.15 Clear the option to Use Sun and Shadow Settings from view. Select Sun. Set Specify Solar Angles to Date, Time and Place. Set the Place to **Nashville, TN, USA**, as before. Use the time slider to set the time to **8:15 pm** (just after sunset) on **June 30**. Accept all the other defaults. See Figure 9–67. Click OK.

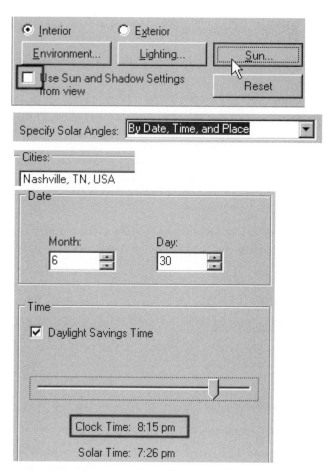

Figure 9–67 *Set the day and time*

3.16 On the main page of the Render Scene Settings dialogue, set the Use View's Section Box drop-down list to display **{3D}**. This applies the Section Box crop you performed earlier in the exterior view to this render setup. Check Back Face Culling and View Culling to save some time in the rendering process (see Figure 9–68). Click OK.

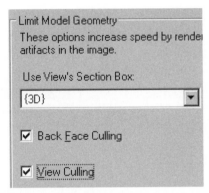

Figure 9–68 *Limit the render geometrically*

3.17 Select Radiate from the Rendering tab of the Design Bar, as shown in Figure 9–69. Revit Building will display a Radiosity Information pane. Click OK.

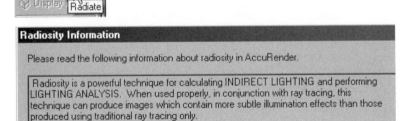

Figure 9–69 *The Radiate tool—useful, but not all-purpose*

Revit Building will do a Radiosity calculation on the space and change the view gradually to a completely lighted rendering. This will take some time, so be patient. Note that the RPC persons will not display in the Radiosity (see Figure 9–70). They will display in the Raytrace.

Figure 9–70 *The radiosity solution displayed graphically*

3.18 When the Radiosity is complete, pick Save Radiosity, as shown in Figure 9–71. Save the file in a convenient location. This creates a file with a .rad extension to hold the settings for this calculated Radiosity solution. This solution can be loaded later, to save creating another Radiosity solution for this view.

You can click the Continue button on the Options Bar to cycle the Radiosity through more steps (25 is the default cycle). This will refine the calculations, but will not change the appearance of the radiated view.

Figure 9–71 *The Save Radiosity tool and Continue option*

3.19 Pick Raytrace from the Rendering tab of the Design Bar. If a warning appears, click OK. This warning appears because you have saved the Radiosity, so Revit Building considers that it may now be out of date.

3.20 Set the Resolution field on the Options Bar to Medium (150 dpi). Note that the Image Size (grayed out to indicate that you cannot change it independently) changes as you change the Resolution value. Click GO! from the Options Bar (see Figure 9–72).

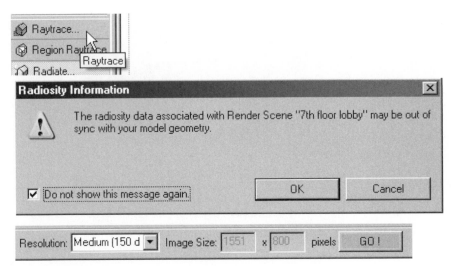

Figure 9–72 *Options Bar controls for the Raytrace rendering*

    3.21 Revit Building will take a few minutes to produce an image of the space (see Figure 9–73). Note that the window glass is both reflective and transparent— a few clouds are visible outside.

Figure 9–73 *The raytrace image with RPC people*

    3.22 Capture or export the image as your instructor directs. Save the project file.

EXERCISE 4. CREATE A WALKTHROUGH

Still images of the model are not the only graphic output possible. The rendering module in Revit Building contains a tool to create a walkthrough, or series of images recorded by a camera that travels along a path. You can view any of these images and export a file that collects the images so they can be played in media players. These files can be sent to clients or located on web sites for playing/ download.

Media-savvy designers can, with very little effort and no additional software, provide clients with completely editable, 3D motion pictures of designs at any stage of progress. This capability enables far better decision making by people with strong design sense but limited ability to visualize from standard orthographic plans, elevations and sections. The walkthrough tool will soon elevate client expectations of the quality of the visual information they can expect from building designers. Besides, this is fun.

4.1   Open or continue working with the file *Chapter 9 render.rvt*. Open the Floor Plan Level 1. Select the Walkthrough tool on the View Design Bar tab.

4.2   The cursor will change to a pencil shape and the tooltip will display the message Click to place Walkthrough key frame. The Options Bar will display Walkthrough controls, as shown in Figure 9–74.

Figure 9–74 *Walkthrough controls on the Options Bar*

4.3   Zoom in Region to the right side of the building. Place the first key frame inside the double door that opens north. Place key frames down the corridor past the curved desk, turn right down the hall past the escalator, and continue placing key frames between the columns and interior walls, moving towards the cafeteria seating (see Figures 9–75 and 9–76).

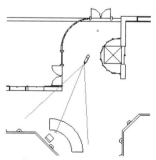

Figure 9–75 *Start in the lobby*

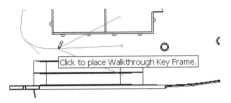

Figure 9–76 *Move around the corner to the right*

4.4 Place 10–12 key frames in all. When the field of view indicator reaches the seating area, shown in Figure 9–77, click Finish from the Options Bar.

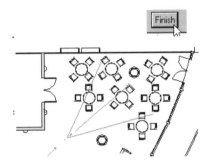

Figure 9–77 *Finish in the seating area*

4.5 Select Edit Walkthrough from the Options Bar, as shown in Figure 9–78. A second set of controls appears on the Options Bar. Note that the Controls drop-down list lets you choose the type of walkthrough component you can edit.

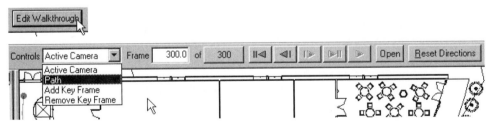

Figure 9–78 *Edit the walkthrough*

4.6 Set the control drop-down list to Path. Select the Frame settings button (it shows the number 300 as its caption). The Walkthrough Frames dialogue opens, as shown in Figure 9–79. This allows you to set the number of frames and the distance between key frames if you do not want a uniform travel speed. Change the Total Frames value to **100**. Click OK.

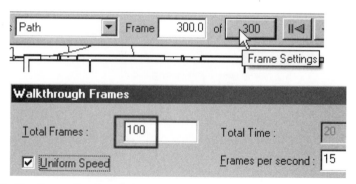

Figure 9–79 *The Frame Settings dialogue*

4.7 The camera is located at the last key frame. The Frame step controls on the Options Bar change the position of the camera along the path. You can step per Frame or key frame. Select the Previous key frame control one time. The Camera will move back one dot (key frame point) on the walkthrough path line. Set control to Active Camera. Pick the camera Target Point control and pull the camera target to the right, pointed towards the couches, as shown in Figure 9–80.

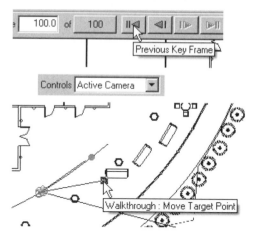

Figure 9–80 *Change the camera direction by moving the target control point*

4.8 Step back two more key frames and point the camera target to the left, as shown in Figure 9–81.

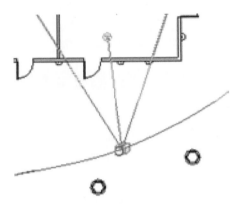

Figure 9–81 *At this key frame, point the camera left*

  4.9 Step back to a key frame in the middle of the entry corridor and point the camera target to the left of the path line, as shown in Figure 9–82.

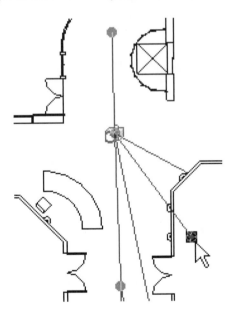

Figure 9–82 *Point the camera left at this key frame*

  4.10 Change the Controls value to **Path** and shift the grip on the 5th key frame from the beginning to the left of the original path line, slightly. Don't make too big a distortion of the path, or your animation will wobble. See Figure 9–83.

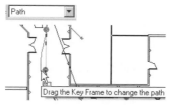

Figure 9–83 *Change the path slightly*

If you accidentally pick a point off of the walkthrough path, Revit Building will ask if you are finished editing the Walkthrough.

 NOTE The Reset Directions button at the right of the Options Bar will undo all the camera target edits, so that the same path can be set to different view directions and thus create different walkthroughs quickly.

4.11   With the Walkthrough highlighted, choose Open on the Options Bar, as shown in Figure 9–84. The current camera view will display. Its border will be highlighted and the grips available. You can edit the field of view—this setting will then hold for the entire walkthrough, not just the frame or key frame.

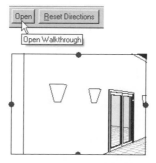

Figure 9–84 *Open the Walkthrough to check the camera view*

4.12   Use the key frame step controls to check the view at each key frame in your walkthrough. You can type in the number of the frame you want to adjust (0 is the first). You can also play the walkthrough using the key frame controls. See Figure 9–85.

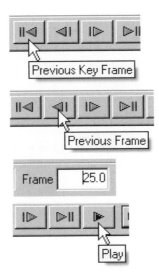

Figure 9–85 *Specific frame controls*

If you need to change the camera location or direction for clarity, open the Level 1 Floor Plan and select Edit Walkthrough.

If the Walkthrough path is not visible in the Level 1 Floor Plan, expand the Walkthroughs section of the Project Browser, select Walkthrough 1, right-click, and pick Show Camera. See Figure 9–86.

Figure 9–86 *You can show or open the Walkthrough from the Browser*

4.13 When you have finished editing the camera location and direction, choose Walkthrough 1 in the Project Browser, right-click, and pick Open (see Figure 9–87). You must do this to make the next step possible.

EXPORT THE WALKTHROUGH TO AN AVI FILE

4.14 From the File menu, select File>Export>AVI. Locate the new file in a suitable folder. Change the Frames per second value to **5**. This will make your walkthrough run for 20 seconds (see Figure 9–87).

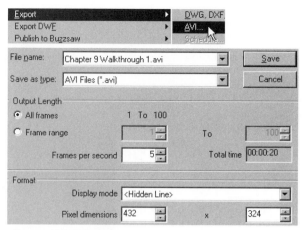

Figure 9–87 *The Export AVI dialogue*

Accept the Hidden Line display mode, but note that Wireframe, Shading, Shading with Edges, and AccuRender displays are available.

A rendered walkthrough in a complex model can take a very long time to create. Try experimenting with some settings and render your walkthrough overnight!

4.15 Pick Save in the Export AVI dialogue to create the file. In the Video Compression dialogue, choose a compression method, as shown in Figure 9–88.

Some compression routines are configurable. Some are not supported by certain media players. For our purposes, any one will do (see Figure 9–88). Using Full Frames (Uncompressed) creates a very big file. Click OK. Revit Building will create the walkthrough frame by frame.

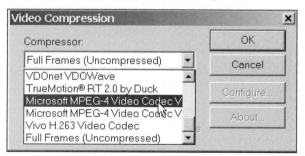

Figure 9–88 *Select a compression method to keep file size down*

If the Export AVI dialogue appears a second time, click Close.

4.16 When the walkthrough has been created, find the new avi file in your file manager and open it in your media player software application (see Figure

9–89). Play it through a few times; use the player's controls to move from view to view.

Figure 9–89 *The AVI plays in a media player*

4.17 Save the Revit Building file.

EXERCISE 5. OUTPUT FORMATS

EXPORT GRAPHICS VIA PLOTTING

Exciting graphics aside, Revit Building provides standard plotting and printing controls for putting images or sheets on paper. Revit Building's installation CD currently includes a PDF writer option so you can generate images or pages in that format, a standard for exchanging electronic documents. Revit Building exports DWF in 2D and 3D. DWF is Autodesk's format for exchanging engineering documents, and allows more control over display of the complex contents of design files than does PDF.

5.1 Open or continue working in the file *Chapter 9 render.rvt*. Open view 3D View 2 (the exterior perspective of the curtain wall). Zoom to Fit. From the Window menu, select Window>Close Hidden Windows, as shown in Figure 9–90.

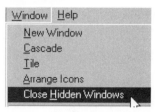

Figure 9–90 *Close other windows*

5.2 Select the Print icon on the Toolbar. Revit Building will open the Print dialogue. Select an available network printer, which will be different from what is shown in Figure 9–91. Check the Visible Portion of Current Window radio button under Print Range. Choose Setup under Settings.

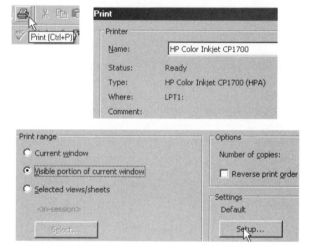

Figure 9–91 *The Print dialogue*

5.3 The Print Setup dialogue opens. Note the available print settings, as shown in Figure 9–92. You can create different named setups for each printer and save them. Check the Fit to page radio button. Select Landscape, as this view is wider than it is tall. Click OK.

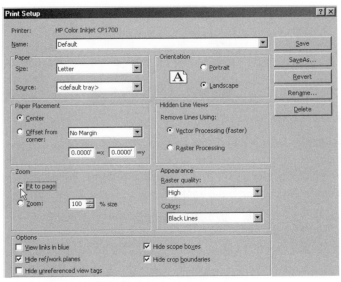

Figure 9–92 *The Print Setup dialogue*

5.4 Select Preview in the Print dialogue. Revit Building will provide a print preview page of the active view based on the print settings (see Figure 9–93). Click Close to exit the print dialogue without generating a printed copy.

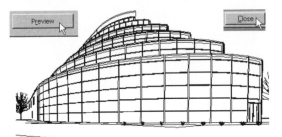

Figure 9–93 *The Print Preview of the View 3D 2*

EXPORT VIA PDF

5.5 Previous versions of Revit shipped with a module to create Adobe PDF output. Revit Building 8.0 does not include this option, but it does not uninstall the PDF Writer if it was previously installed. Click the Print icon on the Toolbar. Choose **Revit Building PDF Writer 4.2** from the Name list, if it is available. If the PDF Writer is not in the selection list, move to step 5.10.

5.6 Select Properties. Note the two tabs available to establish settings in the PDF Writer 4.2 Document Properties dialogue, as shown in Figure 9–94. Choose Cancel.

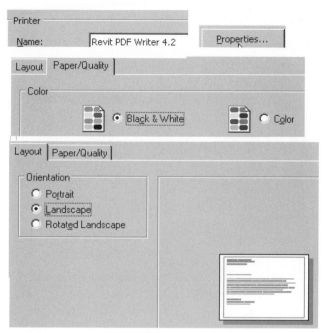

Figure 9–94 *The PDF Writer Properties dialogue, with two tabs*

5.7 In the Print dialogue, select Setup. Check that the plot will be landscape orientation on a letter size page, centered and fitted to the page, as shown in Figure 9–95. Choose OK. If a question box asks about saving settings, Choose No.

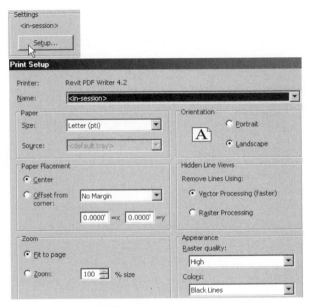

Figure 9–95 *Print settings for the PDF writer*

    5.8 Set the Print Range in the Print dialogue to Current Window. Choose Preview. The image should fill the page. Select Print to return to the Print dialogue. Give the output file a name and location you can find again. Click Save. Click OK to generate the PDF (see Figure 9–96). Revit Building will not display the new PDF.

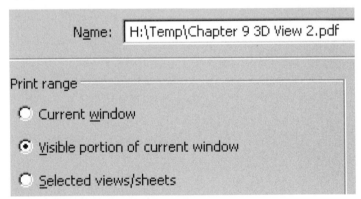

Figure 9–96 *Prepare to print the PDF file*

    5.9 Revit Building will generate a PDF file of the existing view, as you can see in Figure 9–97. Find the new file in your file manager and open it. If you are unable to open the file, Adobe Acrobat reader is a free download.

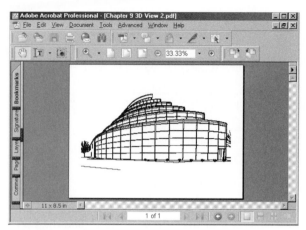

Figure 9–97 *The PDF file in Acrobat*

EXPORT VIA DWF

Autodesk is promoting the use of the DWF (Design Web Format) file output from its design products as a file format made for electronic sharing. DWF files are smaller than native (dwg, rvt) files, and can be sent as email attachments or embedded in Web pages for viewing with Autodesk's DWF Viewer. The DWF Viewer is a free download from Autodesk that, once installed, acts both as a stand-alone DWF viewer and a plug-in for browsers. Autodesk markets DWF Composer, a markup tool for the DWF format

Revit Building will export 3D DWF. The install CD contains an optional DWF Writer to create 2D DWF.

5.10 Type **SD** to turn on the Shaded with Edges display in the current 3D view.

5.11 Select File>Export DWF>3D DWF from the Menu bar. There are no options to configure in this process.

5.12 Give the output file a name and location you can find again. Click Save in the Export dialogue (see Figure 9–98). Click OK to generate the DWF. It may take a couple of moments.

514

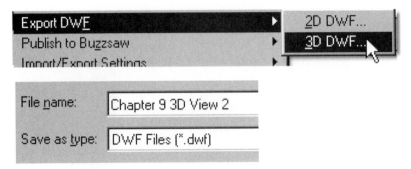

Figure 9–98 *Specify a location for the output file*

   5.13 Revit Building will not display the DWF. Launch the Autodesk DWF Viewer, if it is installed on your machine, navigate to find the DWF file you just created, and open it (or navigate to the file location in your file browser and double-click on the file name). The Viewer will allow you to move through standard views, zoom, pan and orbit in the viewer pane, turn object categories on and off, and print (see Figure 9–99).

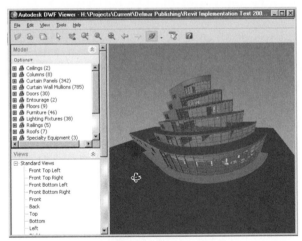

Figure 9–99 *The DWF shows color from the Shaded View*

   5.14 Select File>Export DWF>2D DWF from the Menu bar if it is available. If the option is grayed out, move to step 5.20. A Print dialogue opens.

   5.15 In the Print Range section, choose Selected Views/Sheets, then Select. In the list of available views and sheets, select Elevations East, North, South and West. Pick OK. See Figure 9–100.

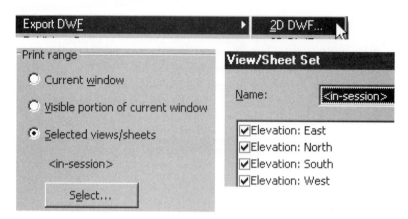

Figure 9–100 *Select views to print to DWF*

5.16 Choose Setup. In the Print Setup dialogue, make the output fit landscape orientation on a letter size sheet, centered and fit to the page, as shown in Figure 9–101. Click OK. Choose No in the question about saving settings.

Paper

Size: ANSI A: 8.5 x 11 in

Source: <default tray>

Orientation

A

○ Portrait

◉ Landscape

Paper Placement

◉ Center

○ Offset from corner: Printer limit

0.0000' =x   0.0000' =y

Hidden Line Views

Remove Lines Using:

◉ Vector Processing (faster)

○ Raster Processing

Zoom

◉ Fit to page

○ Zoom: 100 % size

Appearance

Raster quality:

High

Colors:

Figure 9–101 *Print Setup for the DWF*

5.17 Select Combine multiple views/sheets into a single file and choose Browse to locate the file in a place you can remember. See Figure 9–102.

◉ Combine multiple selected views/sheets into a single file

○ Create separate files. View/sheet names will be appended to the specified name

File name: Chapter 9 elevations

Save as type: DWF Files (*.dwf)

Figure 9–102 *Create a multi-sheet file*

5.18 Choose Save. Choose OK to create the DWF.

5.19 Revit Building will not display the DWF. Launch the Autodesk DWF Viewer, if it is installed on your machine, navigate to find the DWF file you just created, and open it (or navigate to the file location in your file browser and double-click on the file name). The Viewer will allow you to step through the four views, pan and zoom in the viewer pane, and print (see Figure 9–103).

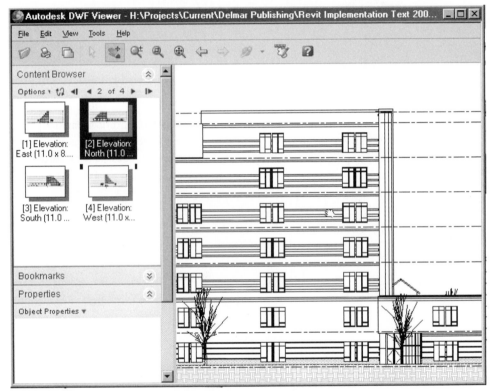

Figure 9–103 *2D DWF output in the viewer*

IMPORT DWF MARKUPS AND EDIT THE MODEL

Revit will import 2D DWF markup files have been processed by Autodesk DWF Composer, but not DWF files that contain no markups. The imports link to the views that were used to create the original 2D DWF file or files, so long as the views have been placed on sheets.

The links can be loaded, unloaded, removed, turned on/off and saved back to the original DWF, This provides round-trip markup capability for dispersed design teams and their reviewers, who do not then need to have Revit Building installed.

5.20   Choose File>Import/Link/Link DWF Markup Set from the Menu Bar. Navigate to the folder that holds *Chapter 9 elevations.dwf*. This file was exported from the file *Chapter 9 start.rvt* and marked up using Autodesk DWF Composer. See Figure 9–104.

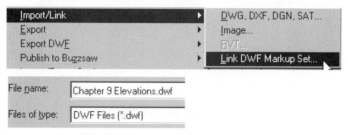

Figure 9–104 *Import a DWF markup*

5.21   A dialogue will open listing views for linking. See Figure 9–105. The two sheets in this file will be referenced. Click OK.

DWF View	Revit View
Drawing Sheet: A101 - Elevations 1	Drawing Sheet: A101 - Elevations 1
Drawing Sheet: A102 - Elevations 2	Drawing Sheet: A102 - Elevations 2

Figure 9–105 *The DWF link dialogue*

5.22   Open the two sheet views in this file in turn. They will now display red markup symbols from the referenced DWF. See Figure 9–106.

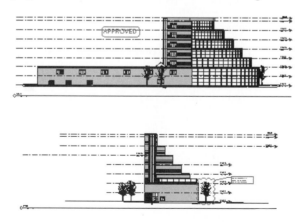

Figure 9–106 *The DWF markups appear in the sheet views*

5.23   Select one of the red DWF symbols. Type **VH** at the keyboard. The DWF objects will disappear. Type VG to open the Visibility Graphics dialogue. On the DWG/DWF/DGN Categories tab the DWF objects are listed according to the sheets, as shown in Figure 9–107. Check the objects that have been cleared and choose OK. The DWF symbols will be restored to the sheet view. You can turn off DWF objects by class (outlines, text, dimensions) in each view.

Model Categories	Annotation Categories	DWG/DXF/DGN Categories	
☑ Show imported categories in this view			

Visibility	Line
	Projection
⊟ ☑ Drawing Sheet: A101 - Elevations 1	By Category
☑ WT_Solid	By Category
☑ WT_Solid	By Category
☑ WT_Solid	By Category
⊟ ☑ Drawing Sheet: A102 - Elevations 2	Override...
☑ WT_Solid	By Category
☑ WT_Solid	By Category
☑ WT_Solid	By Category
☑ Imports in Families	By Category

Figure 9–107 *Turn the DWF on in the sheet view*

5.24 Open sheet view A101. Zoom in on the markup at the tree. Right click and select Activate View so that you can edit the model from the sheet view. See Figure 9–108.

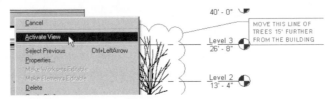

Figure 9–108 *Activate the view*

5.25 Select the trees (there are two, one behind the other). Choose Move from the Toolbar. Move the trees 15′ to the right, as shown in Figure 9–109. Right click and deactivate the view.

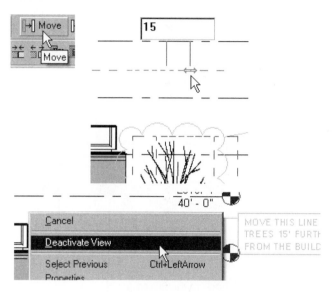

Figure 9–109 *Move the trees according to the markup instruction*

5.26 Select the markup note. Select the Properties icon. Change the Status to Done. See Figure 9–110.

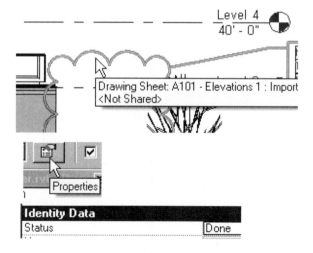

Figure 9–110 *Change the status of the markup once its instruction has been completed*

5.27 Open sheet view A102. Select the corresponding markup at the tree and change its Status property to Done, as you did in step 5.26.

5.28 From the Menu Bar, select File>Manage Links. On the DWF Markups tab, select each of the two linked markups (there are two instances, one for each view that the multi-page markup appears in) and choose Save Markups. This will update the status of the markup objects in the original file. See Figure 9–110.

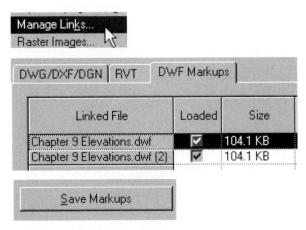

Figure 9–111 *Save the markups back to the linked file*

Figure 9–112 shows the markup opened in DWF Composer after this step was completed. The highlighted markup object now shows Status Done. This is how a reviewer without Revit Building installed can check status of work done in advance of receiving an updated set of DWF graphics.

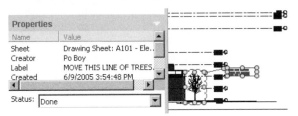

Figure 9–112 *The DWF updates Status*

5.29 Saving the Revit Building file also updates DWF links. Graphics in the DWF will not update until a new DWF is printed from the affected sheets. If completed DWF notes are turned off and a new DWF printed using the same file name, designers and reviews can have an efficient electronic conversation about project progress.

EXPORT THE MODEL TO AN EXTERNAL RENDERER

For users who prefer to work with an external rendering application such as Autodesk VIZ 2006 or Architectural Desktop's VIZ Render, export to DWG will create files that can be imported into VIZ with a significant amount of object and material information retained. A 2D Revit Building view (floor or ceiling plan, section elevation) will create a 2D DWG export. Make a 3D view current in Revit Building to generate a 3D DWG file.

When exporting to DWG files for import into VIZ, all the model entities should be designated by either layer or color so that materials can be assigned. This preparatory work is done in Revit Building's Export Layer Settings. We have looked at layer export settings before in another context.

5.30 From the File menu, select File>Import/Export Settings>Export Layers DWG/DXF.

5.31 The Export Layers dialogue will open. It uses a txt file as its source. Revit Building has a number of default files to use as a basis for your own custom settings. The file shown in Figure 9–113 is *exportlayers-dwg-default.txt*. Note that no color IDs have been assigned. Click Load.

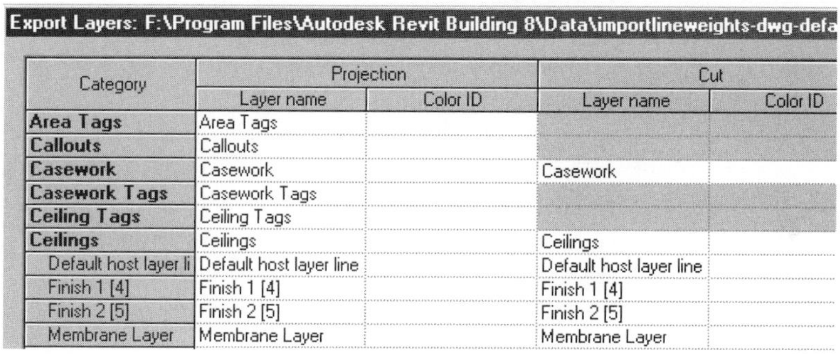

Category	Projection		Cut	
	Layer name	Color ID	Layer name	Color ID
Area Tags	Area Tags			
Callouts	Callouts			
Casework	Casework		Casework	
Casework Tags	Casework Tags			
Ceiling Tags	Ceiling Tags			
Ceilings	Ceilings		Ceilings	
Default host layer li	Default host layer line		Default host layer line	
Finish 1 [4]	Finish 1 [4]		Finish 1 [4]	
Finish 2 [5]	Finish 2 [5]		Finish 2 [5]	
Membrane Layer	Membrane Layer		Membrane Layer	

Export Layers: F:\Program Files\Autodesk Revit Building 8\Data\importlineweights-dwg-defa

Figure 9–113 *The exportlayers-dwg-default.txt export settings*

5.32 Navigate to the *Program Files/Revit Building 8.0/Data* folder. Choose the file *exportlayers-dwg-AIA.txt*, as shown in Figure 9–114. Click OK. Layer names and Color ID values in the Export Layers will change.

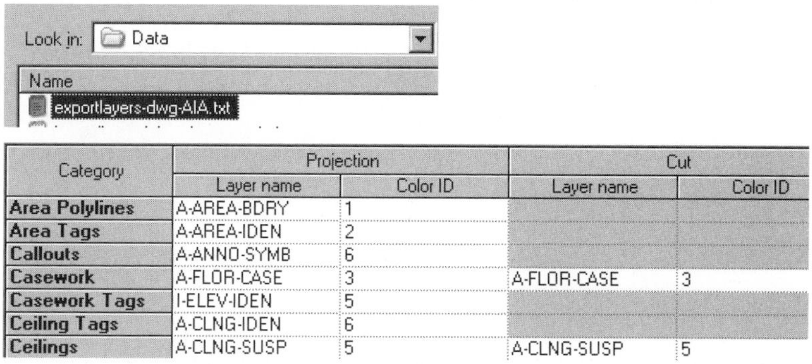

Look in: Data

Name
exportlayers-dwg-AIA.txt

Category	Projection		Cut	
	Layer name	Color ID	Layer name	Color ID
Area Polylines	A-AREA-BDRY	1		
Area Tags	A-AREA-IDEN	2		
Callouts	A-ANNO-SYMB	6		
Casework	A-FLOR-CASE	3	A-FLOR-CASE	3
Casework Tags	I-ELEV-IDEN	5		
Ceiling Tags	A-CLNG-IDEN	6		
Ceilings	A-CLNG-SUSP	5	A-CLNG-SUSP	5

Figure 9–114 *The exportlayers-DWG-AIA.txt export file*

5.33 Note that this file contains layer names based on the AIA CAD standards and color IDs. For export into VIZ, layers and Color IDs should be unique for each object that will receive a different material.

5.34 Enter new Door component Layer name and Color ID properties under Projection and Cut, based on the information shown in Figure 9–115.

Doors	A-DOOR	1	A-DOOR	1
Elevation Swing	{A-DOOR}	1	{A-DOOR}	1
Frame/Mullion	A-DOOR-FRAM	12	A-DOOR-FRAM	12
Glass	A-DOOR-GLAZ	13	A-DOOR-GLAZ	13
Opening	A-DOOR-OTLN	6	A-DOOR-OTLN	6
Panel	A-DOOR-PANL	14	**A-DOOR-PANL**	14
Plan Swing	{A-DOOR}	1	{A-DOOR}	1

Figure 9–115 *Change object layers and colors to provide unique identifiers*

5.35 Click Save As to save your newly edited file for repeated use later. Give it a name to distinguish it from the AIA file you modified, as shown in Figure 9–116. The new file and any others you create can be accessed via the Load button in the import-export settings dialogue.

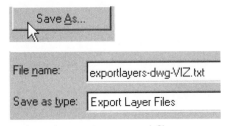

Figure 9–116 *Save the settings changes in a named file*

5.36 To export a file as DWG, select File>Export>DWG, DXF, DGN, SAT.

5.37 The Export dialogue allows you to select a location and name for the new file (see Figure 9–117).

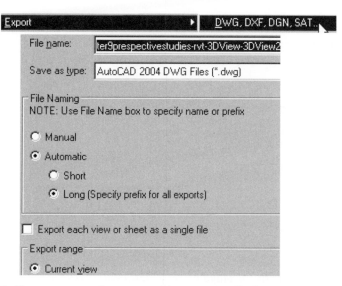

Figure 9–117 *Choose a name, location, and layer setting file for your exported DWG*

Note that Revit Building adds the name of the current view to the end of the file name. This is important to remember for file exchange purposes. As a Revit Building model develops you may export it numerous times. Be sure to keep the many files you can create from one model organized!

5.38   Select Options. You can specify the layer export file to apply to the file you are going to generate, and there are other options for rendering and area analysis. See Figure 9–118.

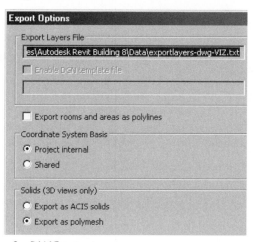

Figure 9–118 *Options for DWG export*

5.39   If you have AutoCAD, Architectural Desktop, or VIZ on your system, Save the export to a DWG file and open it in one of those applications.

Otherwise, click Cancel to exit the dialogue. An information box opens when you export a 3D view as shown in Figure 9–119. Choose OK. Figures 9–120 and 9–121 show exported DWG information as received in VIZ and AutoCAD.

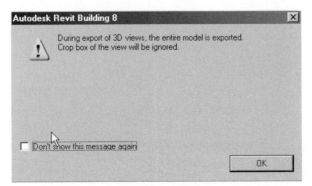

Figure 9–119 *Revit exports the entire model in 3D*

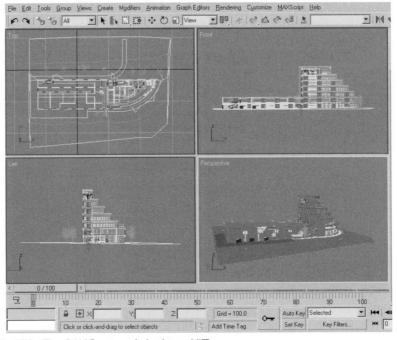

Figure 9–120 *The DWG export linked into VIZ*

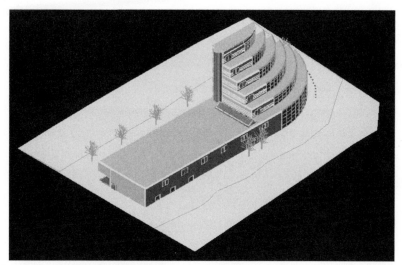

Figure 9–121 *The export DWG opened in AutoCAD*

5.40 Close the Revit Building file.

SUMMARY

The lessons in this chapter have shown you how to make use of Revit Building's many capabilities for viewing your design within Revit Building itself, and for making your standard or enhanced views available as images, printed pages, or raw material for external graphics applications.

As with all the exercises in this book, you have seen only a few of the many possible alternative settings and combinations available. Particularly with rendered images, time spent working on the details and nuances of small portions of a building model can make a big difference in how people will see and understand the design intent behind the model. Practice with the render and walkthrough routines in Revit Building—you may surprise yourself and delight your clients!

REVIEW QUESTIONS – CHAPTER 9

MULTIPLE CHOICE

1. The Edit Walkthrough control allows you to

   a) Change the Camera orientation at any frame point on the walkthrough path

   b) Change the path location at any frame

c) Add or remove key frames at any point on the path

d) all of the above

2. To make building components appear as desired in a rendering, you

a) Apply a Texture definition to a Material

b) Apply a Material definition to the building components

c) a and b in that order only, or the definitions won't apply correctly

d) a and b in any order

3. A Section Box is applied to a view to

a) define custom spaces

b) provide a restricted volume for rendering calculations

c) make a link between floor plans and section views

d) all of the above

4. You can change properties of a Camera

a) using its Properties

b) using View Properties of the Camera View

c) using the Dynamic View control in the Camera View

d) all of the above

5. To prepare a Revit Building file for export to rendering software, use the following menu selection:

a) Settings>Materials

b) View>New>Camera

c) File>Import/Export Settings>Export Layers DWG/DXF (or Export Layers DGN)

d) Render>Raytrace

TRUE/FALSE

6. You must adjust the size of a camera view before you place any RPC people components in the model.

7. The Print Setup dialogue lets you make adjustments to printer settings, but not PDF or DWF output.

8. You use Date, Time, and Place settings for an exterior rendering to adjust the angle of the sunlight.

9. DWF output from Revit Building is can be 2D or 3D.

10. Revit Building can export a Walkthrough to an animation (avi) file in Hidden Line, Shaded, or Rendered view mode.

 Answers will be found on the CD.

CHAPTER 10

Augmenting the Design — Design Options and Logical Formulas

INTRODUCTION

Building construction happens over time, as a result of many directed and realized choices. Revit Building provides design tools that take care of both space and time (phasing), and also allows explicitly for the process of choice. You can create, order, and display alternatives using Design Options. At any stage of the process you can resolve an option into the model and discard alternatives.

Revit Building uses parameters to drive any relationship the user wants to define. Parameters can hold formulas, which now include logical operators such as yes/no and if conditions. Families can now hold features or arrays that can be called or suppressed by family type or instance—cabinets with or without end panels, for example, or brackets that appear under a length of counter.

OBJECTIVES

- Create, manage, and resolve Design Options
- Create two different stair styles with custom rails
- Create a custom wall style with a vertical structure
- Create and place a nested component family with family types based on size, utilizing parametric formulas

REVIT BUILDING COMMANDS AND SKILLS

Design Options

Option sets

Options

Stairs

Railings

Wall structure

Family creation

Family types

Parametric formulas

Conditional operators in formulas

Grid lines

DESIGN OPTIONS

In the previous exercises you developed various parts of building models. As most design projects progress, you will want, at some point, to explore multiple design schemes. Revit Building's Design Options allow you to develop alternate schemes, either simple concepts or detailed engineering solutions. Design Options coexist in the project file with the main model (all building elements that have not been assigned to a named option). You can study and develop each option independent of others, and easily create views to display option combinations. At any point you can resolve an option or option set into the main model to remove the alternates.

EXERCISE 1: DESIGN OPTION SETUP

1.1 Launch Revit Building. Open the file *Chapter 10 start.rvt*. The file will open to show the 3D View: Exterior Iso at the default southeast orientation.

This model is of the arts center for our hypothetical campus project. It contains performance halls, an art display area and cafe on the main floor, shop/storage space on the lower level, and classrooms/offices/rehearsal rooms on the upper levels. No seating has been created in the performance spaces, whose floors slope down to stages on the lower level.

The building owner wants two designs for a staircase from the main level down to the lower level: a simple version with brick sides to echo the walls inside the atrium, and a decorated version if fundraising exceeds expectations.

Save the file as **Chapter 10 options.rvt** in a location determined by your instructor.

1.2 Open the view Floor Plan: Lower Level. Zoom in to the upper half of the central hallway at the lower middle of the plan. Open the View Properties dialogue for the view. Change the Underlay value to Main Level. Choose OK. See Figure 10–1.

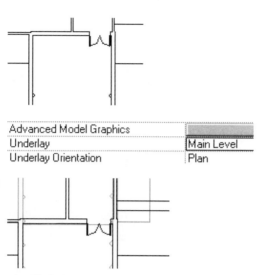

Figure 10–1 *Adjust the Underlay to see the main floor above*

1.3 From the Tools menu, choose Tools>Design Options>Design Options (see Figure 10–2).

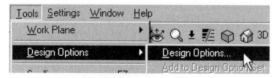

Figure 10–2 *Open the Design Options dialogue*

1.4 There are no Design Option Sets (the containers for Design Options) defined in this file. Select New under Option Set, as shown in Figure 10–3.

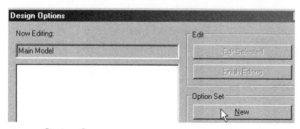

Figure 10–3 *Create an Option Set*

Revit Building will open the Design Options dialogue and create a new Option Set named Option Set 1, with Option 1 (primary) as its default subset. You can create as many Option Sets as you desire, each with a tree of Options within it.

1.5 Select Option Set 1 in the tree pane. Select Rename under Option Set. In the Rename dialogue, type **Stairs**. Choose OK. Select Option 1 (primary). Select

Rename under Option. In the Rename dialogue, enter **Enclosed**. Choose OK (see Figure 10–4).

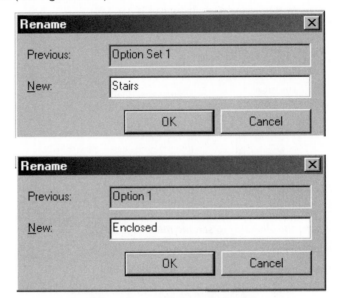

Figure 10–4 *Name the new Option Set and Option as shown*

1.6 Choose New under Option, then Rename. Name the second option **Open**. Select OK. See Figure 10–5.

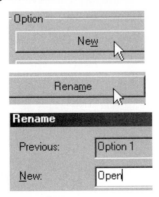

Figure 10–5 *Make and name a second option*

1.7 In the Design Options Dialogue, select Enclosed (primary), then choose Edit Selected. The field text under Now Editing will change from Main Model to Option Set 1: Enclosed (primary) Stairs. Choose Close to leave the dialogue and work on the Design Option (see Figure 10–6).

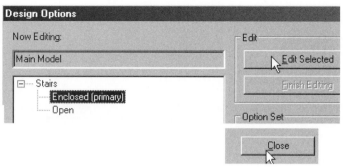

Figure 10–6 *Begin the Option Editing process and close the dialogue*

CREATE AN ENCLOSED STAIR AND LANDING

1.8 In the next steps you will create a stairway "out in space" away from walls, and then align it to the building. In preparation for that, you will first create a couple of Reference Planes as drawing aids. From the Basics Tab of the Design Bar, select Ref Plane. Draw a horizontal reference plane across the hallway space between the facing doors. Select Modify on the Design Bar. Copy the reference plane 7′ up (at 90°—see Figure 10–7).

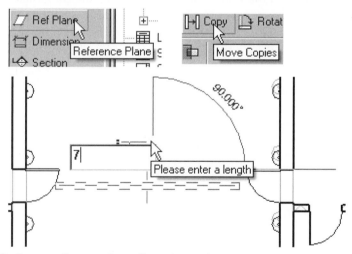

Figure 10–7 *Create reference planes for stair creation*

1.9 From the Modeling tab of the Design Bar, select Stairs. Select Stairs Properties from the Design Bar. In the Instance Parameters Section of the Element Properties dialogue, change the Width to **6′**, and change the Top Level to Main Level, as shown in Figure 10–8. Choose OK.

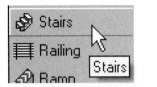

Constraints	
Base Level	Lower Level
Base Offset	0' 0"
Top Level	Main Level
Top Offset	0' 0"

Dimensions	
Width	6' 0"
Desired Number of Risers	23

Figure 10–8 *Adjust Width and Top Level for the new stairs*

1.10 The cursor will show a pencil for Sketch mode. Click on a starting point near the right end of the lower reference plane. Pull the cursor left, until the temporary dimension reads 10' 1" and the light gray riser counter reads 12 RISERS CREATED, 11 REMAINING, and click to establish the first run of stairs.

1.11 Pull the cursor directly up to the upper reference plane. An alignment line will appear. Left-click and pull the mouse to the right (9' 2" minimum) until the sketch outline does not grow any more and the riser counter indicates that the layout is complete, then click again to establish the sketch (see Figure 10–9).

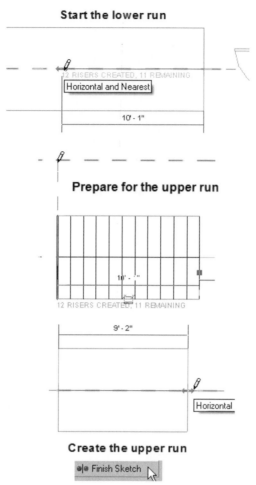

Figure 10–9 *Sketch the stairs, with space for a landing—read from top to bottom*

1.12   Choose Finish Sketch. Select the reference planes and delete them. Select the stair's outer railing, not the stair itself. (Select both and use the filter, if necessary.) Select the Properties icon.

1.13   In the Element Properties dialogue, choose Edit/New. Choose Duplicate. In the Name box, type **Single Wall Brick Cap**. Choose OK. Choose Edit in the Rail Structure value field. See Figure 10–10.

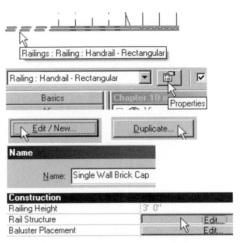

Figure 10–10 *Select the rails and start to define a new type*

1.14 In the Edit Rails dialogue, there is a single rail defined for this style. For the Profile value, assign Brick Cap: Brick Cap, a custom profile that has been loaded into this project. For the Material value, click in the cell, then click the dialogue arrow that appears in the cell.

1.15 In the Materials dialogue, find Concrete-Precast Concrete in the Name pane, and select it. Click the arrow to the right of the AccuRender Texture field. In the Material Library dialogue, navigate to folder *_ACCURENDER\Concrete* and select *Exposed Aggregate, Tan*, as shown in Figure 10–11. Choose OK three times to exit the Rail editor.

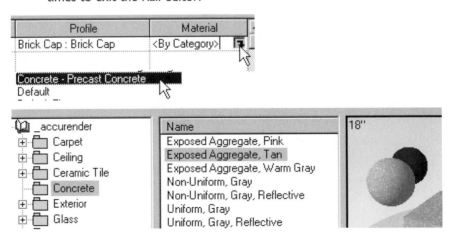

Figure 10–11 *Apply a material and rendering texture to the rail element in the railing*

1.16 Choose Edit in the Baluster Placement field of the Element Properties dialogue. In the Edit Baluster Placement dialogue, change the value for Baluster Family to None in row 2 of the Main Pattern and all rows of the

Posts, as shown in Figure 10–12. Choose OK three times to terminate the rail and stair editing.

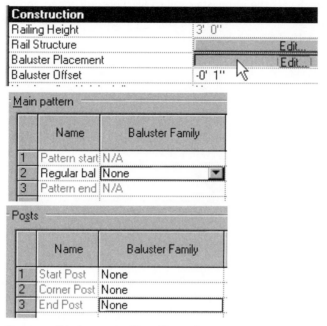

Figure 10–12 *Remove all balusters in this railing style*

1.17 Select the Railing and Stair together. Choose the Move tool from the Toolbar. Select the upper-left corner of the railing as the starting point for the Move command, and select the intersection of the left side wall and the edge of the Main Level floor in the halftone underlay, as shown in Figure 10–13. Zoom out until you can see the whole stair to check the alignment.

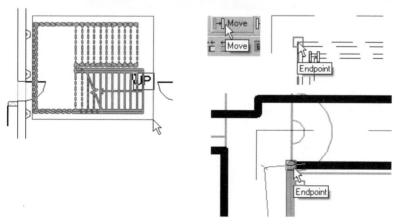

Figure 10–13 *Move the new railing and stair together*

1.18 Open the Main Level view. Zoom in around the new stairs. Select the Match Type icon from the Toolbar. The cursor will change to an empty eyedropper shape. Select the outer stair rail to fill the eyedropper, and then select the inner rail to change its type from Rectangular to Single Wall Brick Cap (see Figure 10–14).

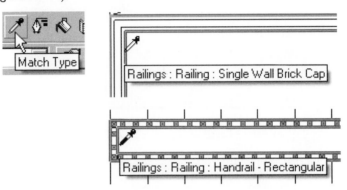

Figure 10–14 *Match properties from one railing to the next*

1.19 Select the Floor tool from the Modeling tab of the Design Bar. Choose Floor Properties from the Design Bar. In the Element Properties dialogue, change the Type to Generic – 12″ – Filled. In the Instance Parameters section, make sure the Level value is Main Level, and set the Height Offset From Level value to 0. (Revit Building will supply the units, as shown in Figure 10–15.) Choose OK.

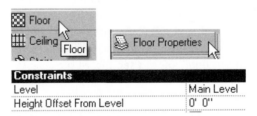

Figure 10–15 *Set the Floor Properties for the upper stair landing*

1.20 On the Sketch Bar choose Lines, and check the Rectangle tool on the Toolbar. Select two points as shown in Figure 10–16: the lower end of the inner rail, and the intersection of the right-hand wall and the floor edge. Choose Finish Sketch to draw the upper landing. Choose No in the question box about attaching walls.

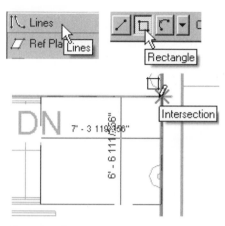

Figure 10–16 *Draw the floor sketch*

CREATE A BALCONY WALL

1.21 Select Wall from the Modeling tab of the Design Bar. Select the Properties icon from the Options Bar. In the Element Properties dialogue, make the current type Generic – 4″ brick. Select Edit/New.

1.22 Select Duplicate. In the Name dialogue, type **Double Wall Brick with Cap**. Choose OK. Select Edit in the Structure Value field. In the Edit Assembly dialogue, select Preview in the lower-left corner to open that pane. Change the view type to Section, as shown in Figure 10–17.

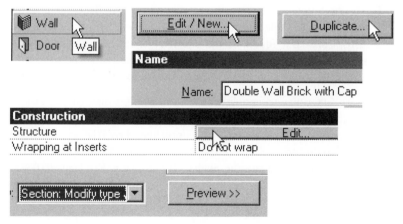

Figure 10–17 *Select the Section preview of this wall type to facilitate editing the new wall type*

1.23 Choose Row 2 in the Layers area. Choose Insert twice to create two new layers in the wall. Define the layers as shown in Figure 10–18: Set the Material for Row 2 Masonry – Brick with Thickness **3 5/8″**; assign Row 3

Function Thermal/Air Layer, Material Air Barrier – Air Infiltration Barrier with Thickness **1 3/4"**. Set the Thickness for Row 4 to **3 5/8"**. Revit Building will show the Total thickness as 9".

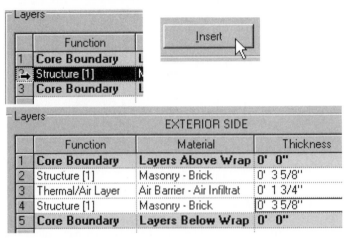

Figure 10–18 *Layers defined in the new wall type*

Revit Building allows you to create walls of many layers. The Core Boundary layer defines the point at which finish meets structure, and determines how far floors penetrate into walls, as shown in drafting sections. Since this wall type is for decoration only, you will not alter the placement of the Core Boundary layers.

1.24 Select the Sweeps button at the lower right of the dialogue. In the Wall Sweeps dialogue, select Add. Assign the Profile Stone Cap: Stone Cap. This is a custom profile loaded into the project file. Set its Material to Concrete: Precast Concrete. Set the AccuRender material value for Precast Concrete to _ACCURENDER\Concrete\Exposed Aggregate, Tan, as you did before.

1.25 Choose OK twice to return to the Wall Sweeps dialogue. Set the From Value to Top and the Offset to **–0 4.5** (Revit Building will supply the units and fraction). The offset value was determined by trial and error, to center the profile on the wall. Move the dialogue box off the preview if necessary and choose Apply to see the new sweep in place. See Figure 10–19. Choose OK.

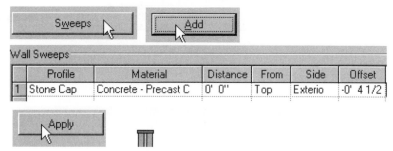

Figure 10–19 *Creating a Sweep at the top of the wall*

1.26   Select the Reveals button at the lower right of the Edit Assembly dialogue. In the Reveals dialogue, choose Add twice.

1.27   For both rows, make the Profile value Reveal-Brick Course: 1 Brick. Make the distance value **–0′ 8″** for both rows. Make the From value read Top. Make the Side value Interior for Row 1 and Exterior for Row 2. Move the dialogue box off the preview if necessary and choose Apply to see the new reveals in place, as shown in Figure 10–20.

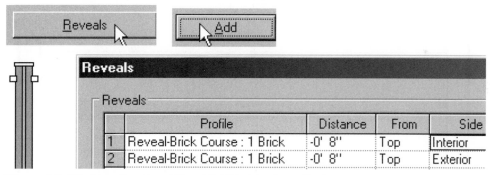

Figure 10–20 *New Sweep and Reveals added to the wall type*

1.28   Choose OK twice to return to the Type Properties dialogue. Change the Wrapping at Ends value to Exterior. Choose OK. In the Element Properties dialogue, set the Top Constraint to Unconnected, the Unconnected Height to **3′ 0″**, and the Location Line to Finish Face: Exterior, as shown in Figure 10–21. Choose OK.

Construction	
Structure	
Wrapping at Inserts	Do not wrap
Wrapping at Ends	Exterior

Constraints	
Location Line	Finish Face: Exterior
Base Constraint	Main Level
Base Offset	0' 0"
Base is Attached	☐
Base Extension Distance	0' 0"
Top Constraint	Unconnected
Unconnected Height	3' 0"

Figure 10–21 *Type and Instance Properties of the wall you are about to place*

1.29 Place two instances of the new wall type at the edge of the landing and the floor, as shown in Figure 10–22.

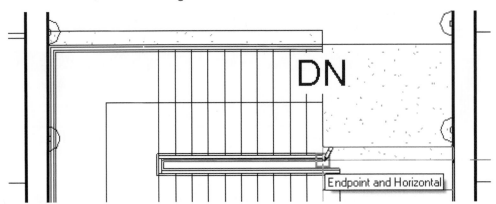

Figure 10–22 *Landing walls in place for this Design Option*

CREATE AN ENLOSED STAIR

1.30 Open the 3D View Atrium Staircase. Select the stair. Select the Properties icon on the Options Bar. Select Edit/New; and then select Duplicate. In the Name dialogue, type **Brick Sides**, as shown in Figure 10–23. Choose OK.

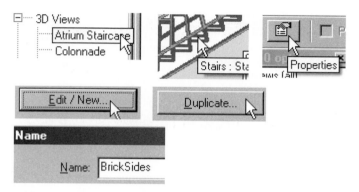

Figure 10–23 *Define a new style for the stairs*

1.31 In the Type Properties dialogue, adjust the values as shown in Figure 10–24. Make the material value for the Tread and Stringer Concrete – Cast In Place Concrete, and make the Stringer Material Masonry – Brick. Make the Stringer thickness **3 5/8″**; the Stringer Height **3′ 5″**; the Stringer Carriage Height **1″**; and the Landing Carriage Height **2″**. Choose OK twice to finish the stair edits.

Materials and Finishes	
Tread Material	Concrete - Cast-in-Place Concrete
Riser Material	Concrete - Cast-in-Place Concrete
Stringer Material	Masonry - Brick

Stringers	
Trim Stringers at Top	Do not trim
Right Stringer	Closed
Left Stringer	Closed
Middle Stringers	0
Stringer Thickness	0′ 3 5/8″
Stringer Height	3′ 5″
Open Stringer Offset	0′ 0″
Stringer Carriage Height	0′ 1″
Landing Carriage Height	0′ 2″

Figure 10–24 *Properties for the stairs*

1.32 Select the Rail. Use the TAB key to cycle through choices if necessary. Select the Properties icon from the Options Bar. Choose Edit/New. In the Type Properties dialogue, set the values for both Angled Joins and Tangent Joins to No Connector. Set Rail Connections to Trim (see Figure 10–25). Choose OK twice to finish the rail edits.

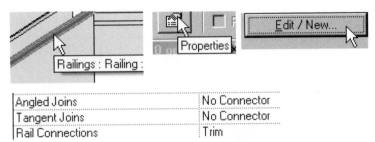

Angled Joins	No Connector
Tangent Joins	No Connector
Rail Connections	Trim

Figure 10–25 *Edit rail properties*

1.33 The first option for the stairs is now complete (see Figure 10–26).

Figure 10–26 *The first stair option is ready*

1.34 Place the cursor anywhere over the Tool Bar and right-click. Choose Design Options from the list of available toolbars. In the Design Options toolbar, select the Design Options dialogue icon, shown in Figure 10–27.

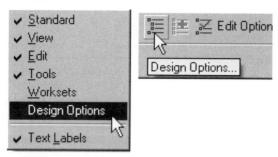

Figure 10–27 *Make the Design Options toolbar visible and open the Design Options dialogue*

1.35   Choose Enclosed (primary) in the Stairs tree. Choose Finish Editing, as shown in Figure 10–28.

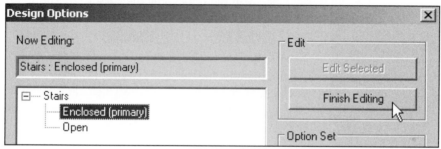

Figure 10–28 *Finish editing the Design Option—you can open it for more edits later if necessary*

1.36   The 3D view will change appearance, as components of the model become available for editing. Select Open in the Stairs tree. Choose Edit Selected, as shown in Figure 10–29. The enclosed stairs disappear, and the model becomes gray again. Choose Close. Save the file.

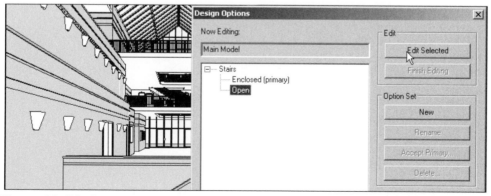

Figure 10–29 *Start editing another named option*

EXERCISE 2. SECOND DESIGN OPTION

CREATE AN OPEN STAIR

2.1 Open or continue working with the file from the previous exercise. Make view Lower Level current. Zoom in to the area you worked in before. Select Stairs from the Modeling tab of the Design Bar. Revit Building will go into Sketch mode.

2.2 Select Stairs Properties from the Sketch tab. Set the Stair properties as before: Width is **6'**, Base Level is Lower Level, Top Level is Main Level. Choose OK.

2.3 Start the run of stairs between the doors. Pull the cursor straight up (90°) until the temporary dimension reads 10' 1" and the riser counter shows 12 RISERS CREATED, 11 REMAINING. Left-click to establish the end of the run.

2.4 Pull the cursor straight up until a horizontal Reference Plane appears, along with the tooltip Midpoint, as shown in Figure 10–30. Left-click to start the second run, and continue pulling the cursor straight up until the stair outline is complete, as before. Click again to set the stair runs.

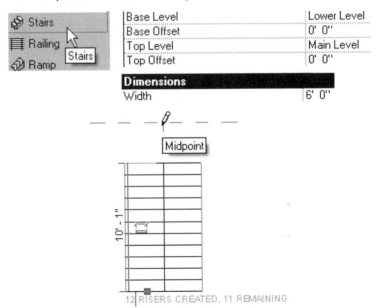

Figure 10–30 *Starting the upper stair run*

2.5 The Stair sketch consists of three types of Model lines: green boundaries, black risers, and blue runs. Select the left-hand boundary of the upper run. Copy it 2′ to the left.

2.6 Choose Boundary from the Sketch tab. Choose the 3-point arc tool from the Options Bar. Select the lower end of the new line for the first point of the arc; select the upper end of the original (copied) boundary for the end point of the arc. Pull the cursor to the right of the arc line until the arc bend snaps in place and click to establish the arc sketch, as shown in Figure 10–31.

2.7 The arc will display a radius dimension. Select the dimension and set it to 26. Revit Building will convert this to 26′ 0″.

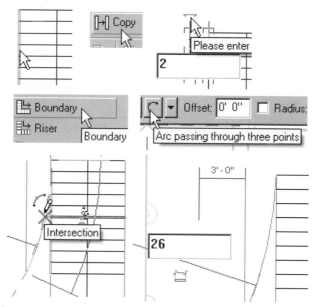

Figure 10–31 *Set the arc radius*

2.8 Erase the two straight boundary lines for the upper run of the stairs and the copied sketch line. Select the new arc. Choose the Mirror tool form the Options Bar and mirror the arc to the right, using the center line of the stairs as the mirror line (see Figure 10–32).

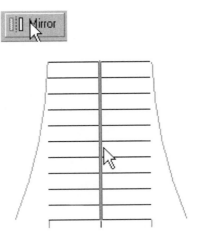

Figure 10–32 *Mirror the arc*

    2.9 Erase the boundaries of the landing. Move the boundaries of the lower stair run left and right 2′ to line up with the wide ends of the upper runs. See Figure 10–33.

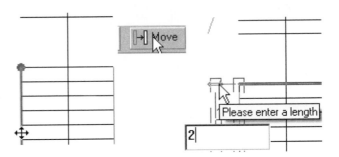

Figure 10–33 *Move the two lower boundary lines*

    2.10 Choose Boundary from the Sketch tab. Select the three-point arc tool. Select the lower end of the upper-left boundary and the upper end of the lower-left boundary as the start and end points, and pull the cursor to the left to define a 180° arc. Click to establish the arc. See Figure 10–34.

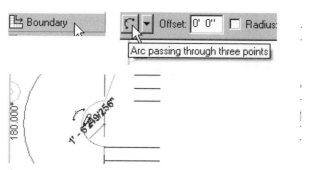

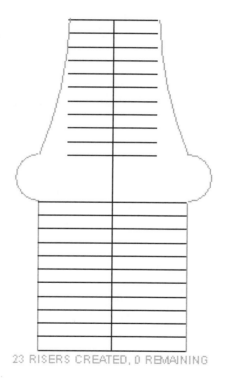

Figure 10–34 *Create a new arc boundary*

> 2.11 Mirror the arc to the right to create a landing with rounded sides, as shown in Figure 10–35.

Figure 10–35 *The completed stair sketch*

> 2.12 Select Stairs Properties on the Design Bar. Select Edit/New. Select Duplicate. In the Name box, type **Open Riser Steel**.

> 2.13 In the Type Properties dialogue, edit the following values: Tread Thickness 1″; Riser Type None; Stringer Thickness **2**″; Tread Material Metal – Steel;

and Stringer Material Metal – Steel, as shown in Figure 10–36. Choose OK twice.

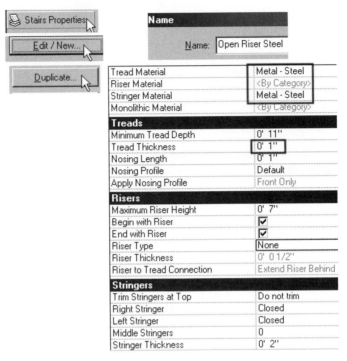

Figure 10–36 *Stair Properties*

2.14 Choose Finish Sketch. Select the stair and both railings. Choose Move from the Toolbar. Select the middle of the top of the stairs for the first point, and select the middle of the Main Level floor as the second point (see Figure 10–37).

 TIP If you are having difficulty getting Revit Building to snap to the middle of the floor edge, type SM (for Snap Middle) to force the Midpoint snap override.

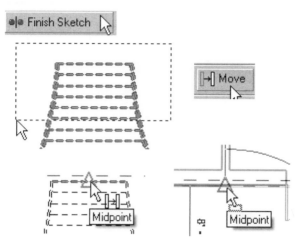

Figure 10–37 *Move the middle of the stair to the midpoint of the floor edge*

ADD A CUSTOM RAIL STYLE

2.15   Open the 3D view Atrium Staircase. Select one of the stair rails. Change the type in the Type Selector to Railing: Handrail – Pipe. Choose the Properties icon from the Options Bar. Select Edit/New. Select Duplicate. In the Name box, type **Handrail – Pipe – Copper and Cherry**. See Figure 10–38. Choose OK.

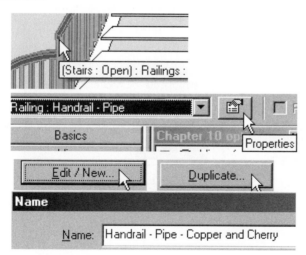

Figure 10–38 *Railing properties*

2.16   In the Type Properties dialogue, select Edit in the Rail Structure value.

2.17   In the Edit Rails dialogue, change Rail 1 Material to Wood – Cherry. Change the Material for Rail 2 to Metal – Trim. While in the Materials dialogue,

change the AccuRender Texture for Metal – Trim to _ACCURENDER\
Metals\Copper\Polished, Plain, as shown in Figure 10–39.

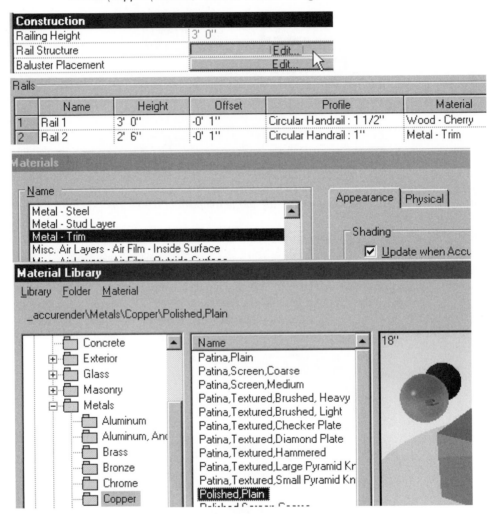

Figure 10–39 *Assign a material and texture to the rails*

2.18 Choose OK twice to exit the Materials Library and Materials. Change the
Material value for Rails 4 and 6 to match Rail 2.

2.19 Change the Profiles for Rails 3 and 5 to Fascia–Flat: 1" x 4". Make their
Materials Wood – Cherry, as shown in Figure 10–40. Choose OK.

Rails

	Name	Height	Offset	Profile	Material
1	Rail 1	3' 0"	-0' 1"	Circular Handrail : 1 1/2"	Wood - Cherry
2	Rail 2	2' 6"	-0' 1"	Circular Handrail : 1"	Metal - Trim
3	Rail 3	2' 0"	-0' 1"	Fascia-Flat : 1" x 4"	Wood - Cherry
4	Rail 4	1' 6"	-0' 1"	Circular Handrail : 1"	Metal - Trim
5	Rail 5	1' 0"	-0' 1"	Fascia-Flat : 1" x 4"	Wood - Cherry
6	Rail 6	0' 6"	-0' 1"	Circular Handrail : 1"	Metal - Trim

Figure 10–40 *Rails defined in the new style*

2.20 Choose Edit in the Baluster Placement field. Change the Dist. from previous value in Row 2 of the Main Pattern to **3'**, as shown in Figure 10–41. Choose OK three times.

Rail Structure	Edit...
Baluster Placement	Edit...
Baluster Offset	-0' 1"

Main pattern

	Name	Baluster Family	Base	Base offset	Top	Top offset	Dist. from previous
1	Pattern start	N/A	N/A	N/A	N/A	N/A	N/A
2	Regular bal	Baluster - Round : 1"	Host	0' 0"	Rail 1	0' 0"	3' 0"
3	Pattern end	N/A	N/A	N/A	N/A	N/A	2' 0"

Figure 10–41 *Edit the baluster spacing*

2.21 Select the rail you have not edited and change its type to Handrail – Pipe – Copper and Cherry. Right-click and select Zoom Out (2x).

2.22 Clear Active Option Only from the Options Bar. Select the bottom edge of the 3D View and drag it down to make more of the stair visible, as shown in Figure 10–42. Right-click and select Zoom to Fit.

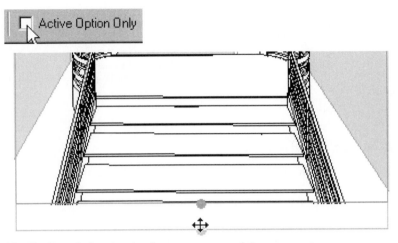

Figure 10–42 *Stretch the view border to see more of the open stairs*

2.23 Select the Design Option icon from the Toolbar. Choose Finish Editing. The 3D view will show the primary option, the enclosed stairs. Choose Close. See Figure 10–43. Save the file.

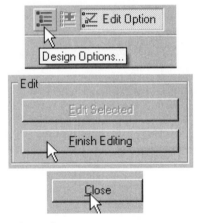

Figure 10–43 *Finish editing the option*

EXERCISE 3. CREATE BALCONY OPTIONS

3.1 Open or continue working with the file from the previous exercise. Open the Design Options dialogue. Select Open in the Stairs Tree. Choose Make Primary under Option, as shown in Figure 10–44. The 3D view will change to show the new primary option. A warning box will appear that you can ignore.

Figure 10–44 *Make the Open stairs the primary option*

3.2   Choose New under Option Set in the Design Options dialogue. Revit Building creates a new Option Set 1 and Option 1 (primary) under it. Choose New under Option to create a second Option in the new Option Set.

3.3   Select Option Set 1 and select Rename. In the Name box, type **Balconies**. Choose OK.

3.4   Select Option 1 and select Rename. In the Name box, type **Railings Only**. Choose OK.

3.5   Select Option 2 and select Rename. In the Name box, type **Twin Balconies**. Choose OK. Select the Railings Only Option and select Edit Selected. See Figure 10–45. Choose Close.

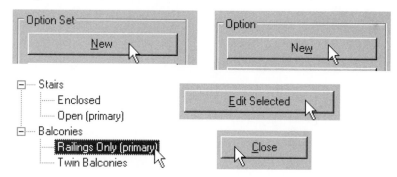

Figure 10–45 *New options for balconies*

3.6   Open the Floor Plan view Main Level. Select the Railing tool from the Modeling tab of the Design Bar. Select Railing Properties from the Sketch tab. Change the Type to Handrail – Pipe – Copper and Cherry. Choose OK.

3.7   On the Options Bar, change the Offset value to **0′ 2″**. Draw a line along the edge of the floor left to right from the left wall to the end of the left stair rail, as shown in Figure 10–46. Choose Finish Sketch.

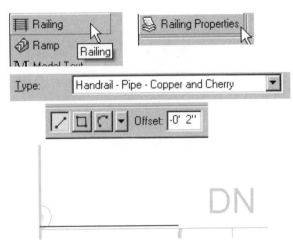

Figure 10–46 *Sketch a rail at the left side*

3.8 Select the new rail. Use the Mirror tool to copy it to the right side.

3.9 Open the Design Options dialogue. Choose Finish Editing. Select the Twin Balconies Option, and choose Edit Selected (see Figure 10–47). Select Close.

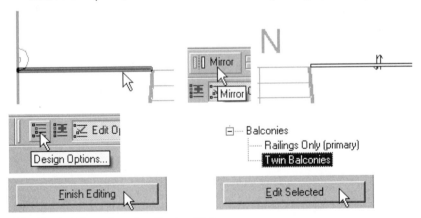

Figure 10–47 *Switch to the other Design Option*

3.10 From the Modeling tab of the Design Bar, select Floor. Select Lines from the Sketch tab. Select the Rectangle tool from the Options Bar.

3.11 Draw a rectangle 7′ 0″ wide by 6′ 0″ deep starting at the left end of the main floor, as shown in Figure 10–48.

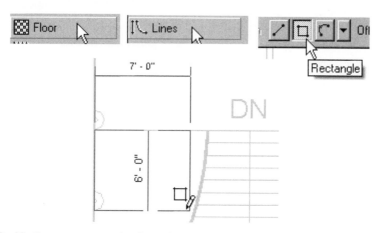

Figure 10–48 *Draw a rectangular floor sketch*

> 3.12 Select Modify on the Sketch tab. Erase the bottom line of the floor sketch. Select Lines on the Sketch tab. Draw a 180° three-point arc.

> 3.13 Select the entire floor sketch. Mirror it to the right, as shown in Figure 10–49. Select Finish Sketch. Answer No in the question box.

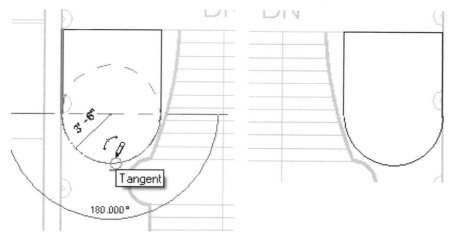

Figure 10–49 *Mirror the balcony floor sketch*

> 3.14 Select Railing from the Modeling tab of the Design Bar. Select the Pick arrow from the Options bar. Set the Offset to **0′ 2″** as before. Select the round edge of the left floor and the right side of that same floor, as shown in Figure 10–50.

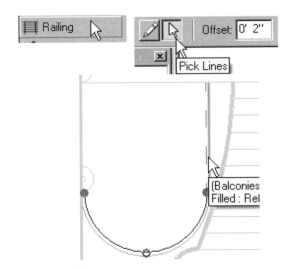

Figure 10–50 *Use Pick and Offset to create these lines*

3.15 Select the Draw (pencil) icon from the Options Bar. Keep the 2″ Offset. Draw a line from the end of the last line you created 1′ 6″ to the right. Revit Building will offset it from the floor edge. Use the Trim tool with the corner option to join the last two lines.

3.16 Select Finish Sketch. Mirror the new railing to the right (see Figure 10–51).

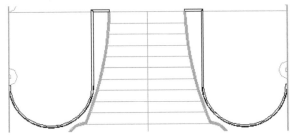

Figure 10–51 *Two new rails*

3.17 From the Toolbar select the Design Options icon. Select Finish Editing. The balconies will disappear and the primary option will appear in the view.

3.18 Under Options, select New. Rename the new Option None. Choose Edit Selected. Select Finish Editing. Choose Close. See Figure 10–52. This creates a balcony option with no railings or floor extensions, to be combined with the enclosed stair option.

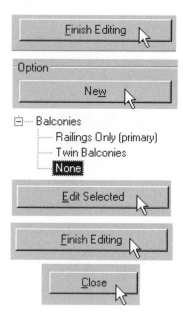

Figure 10–52 *Add a no-railing option*

3.19 Select the 3D view Atrium Staircase in the Project Browser. Right-click and select Duplicate. Repeat to make two copies of the Atrium Staircase view.

3.20 Select View Atrium Staircase, right-click, and choose Rename. In the Rename View box, type **Atrium Staircase, Enclosed Brick**. Choose OK.

3.21 The view name will still be highlighted. Right-click and select Properties. Choose Edit in the Visibility field. The Visibility/Graphics Override dialogue will now have a tab for Design Options. Select that tab. In the Design Option field for Option Set Stairs, choose Enclosed. In the Design Option Field for Option Set Balconies, choose None. See Figure 10–53. Choose OK twice to end the edits.

3D Views
├─ Atrium Staircase, Enclosed Brick

Model Categories	Annotation Categories	DWG/DXF/DGN Categories	Design Options

Design Option Set	Design Option
Stairs	Enclosed
Balconies	None

Figure 10–53 *Set the Design Option visibility in this view*

3.22 Select 3D view Copy of Atrium Staircase in the Project Browser. Right-click and select Rename. In the Rename View box, type **Atrium Staircase Open, No Balconies**. Choose OK.

3.23 Right-click and select Properties. Select Edit in the Visibility field. Select the Design Options tab. In the Design Option field for Option Set Stairs, choose Open (primary). In the Design Option Field for Option Set Balconies, choose Railings Only (primary). Choose OK twice to end the edits. See Figure 10–54.

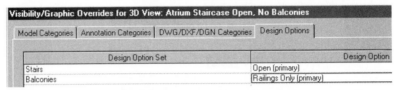

Figure 10–54 *Design Option visibility settings to show open staircase and railings*

3.24 Select view Copy (2) of Atrium Staircase in the Project Browser. Right-click and select Rename. In the Rename View box, type **Atrium Staircase Open, With Balconies**. Choose OK.

3.25 Right-click and select Properties. Select Edit in the Visibility field. Select the Design Options tab. In the Design Option field for Option Set Stairs, choose Open (primary). In the Design Option Field for Option Set Balconies, choose Twin Balconies. Choose OK twice to end the edits. See Figure 10–55.

Visibility/Graphic Overrides for 3D View: Atrium Staircase Open, with Balconies	
Model Categories Annotation Categories DWG/DXF/DGN Categories Design Options	
Design Option Set	**Design Option**
Stairs	Open (primary)
Balconies	Twin Balconies

Figure 10–55 *Settings to see open stairs with balconies*

3.26 Open each of the Atrium Staircase views in turn to check your work.

If you wish to print your work, place the Atrium Staircase views on a sheet side by side, as shown in Figure 10–56. If you wish to keep the separate design options to experiment with, save a copy of the project file under a different name.

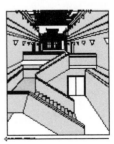

Figure 10–56 *Three options side by side*

3.27 Open the Design Options dialogue. Select Stairs in the left pane. Select Accept Primary under Option Set on the right side. Answer Yes in the question box (see Figure 10–57). Choose Delete in the option dialogue to delete the view dedicated to the eliminated option.

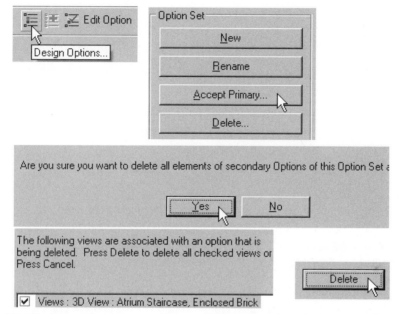

Figure 10–57 *Accept the primary option to eliminate the others in the Option Set*

3.28 Select Twin Balconies in the left pane. Select Make Primary under Option. Select Accept Primary under Option Set. Select Yes in the question box. Delete the out-of-date view. See Figure 10–58. Choose Close. Save the file.

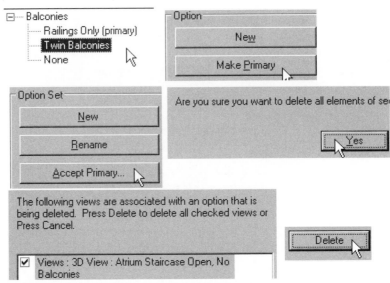

Figure 10–58 *Accept the twin balcony option and delete inaccurate views*

EXERCISE 4. CREATE AND PLACE A NESTED FAMILY WITH CONDITIONAL FORMULAS

Revit Building's Families hold and organize definitions for nearly everything in Revit Building. Expand the Families section in the Project Browser, as shown in Figure 10–59, to see a list of content by Family. In this exercise you will create a nested column family using very simple drafting, and then apply Parameters, which you have already studied in previous exercises, to create design rules for types and instances within this family. The Parameters will use Formulas—mathematical, yes/no, and conditional (i.e. logical operations such as "if-then-else")—to create smart switches that control the appearance of the column as it is inserted.

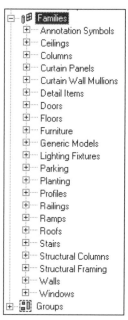

Figure 10–59 *Even a blank Revit Building file contains many Families*

CREATE SOME SIMPLE SOLIDS WITH MATERIALS

4.1 Open or continue working in the file from the previous exercise. Open view Floor Plans Main Level. Zoom to Fit. From the Window menu, select Window>Close Hidden Windows.

4.2 From the File menu, select File>New>Family, as shown in Figure 10–60. In the New dialogue box, select *Imperial Templates\Generic Model.rft*. Choose Open.

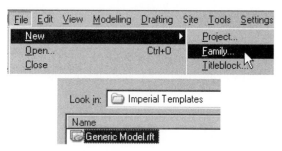

Figure 10–60 *Start a new Family*

4.3 In the new file, open view Elevations: Left. Zoom to Fit. Select Solid Form>Solid Revolve from the Family tab (the only one available) on the Design Bar as shown in Figure 10–61.

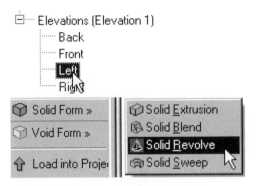

Figure 10–61 *Start a Revolve in the left elevation view*

> 4.4 Sketch the profile shown in Figure 10–62. Do not create the dimensions.
>
> The profile is basically a 2"-wide by 3"-tall rectangle set 4" to the right of the origin point, with 1" half-round stepped setbacks at 1/4". Use the Line and 3-point arc tool.

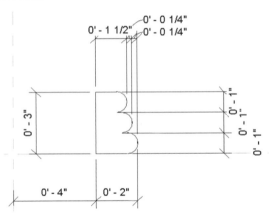

Figure 10–62 *The profile for the trim ring family*

> 4.5 When the profile is complete, select Axis from the Design Bar. Draw a short line vertically from the origin, as shown in Figure 10–63. This profile will create a stepped ring when spun around the axis.

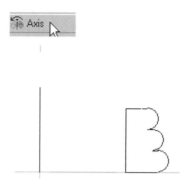

Figure 10–63 *The axis of revolution*

4.6 Select Revolution Properties from the Design Bar. Verify that the End Angle is 360° and the Start angle 0°. Select the Material value field to open the Materials dialogue.

4.7 In the Materials dialogue, select Default in the Name panel, then select Duplicate. In the New Material name box, type **Chrome**. Select the arrow next to the AccuRender Texture field. Navigate to _ACCURENDER\ Metals\Chrome\Satin, Plain, as shown in Figure 10–64. Choose OK three times to set the Revolution Properties.

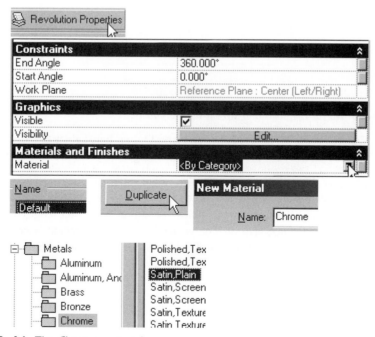

Figure 10–64 *The Chrome material*

4.8 Select Finish Sketch. Select the Default 3D view from the Toolbar to see the finished trim ring (see Figure 10–65).

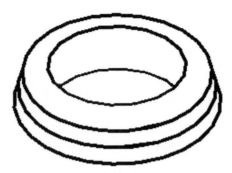

Figure 10–65 *The revolved solid*

4.9 Save the file as ***Trim Ring.rfa*** in a place where you can find it later.

4.10 Return to the Left Elevation view. Choose the solid. Make a Mirror copy below the horizontal reference plane, and erase the original, as shown in Figure 10–66. Save the file as ***Upper Trim Ring.rfa***.

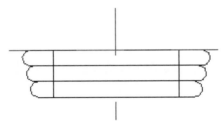

Figure 10–66 *Mirror the solid and save it as a separate file*

4.11 Delete the solid. Select Solid from the Design Bar. Select Extrusion and choose OK. Draw the profile shown in Figure 10–67. Do not create the dimensions.

The profile is a bracket of 1"-thick material, 36" x 13", with a 2" x 1" flange 1" from the right end and a 1/2"-thick arc brace. The exact length and placement of the arc are not critical. The vertical arm of the bracket is centered on the horizontal reference plane, and its left end is aligned with the vertical plane.

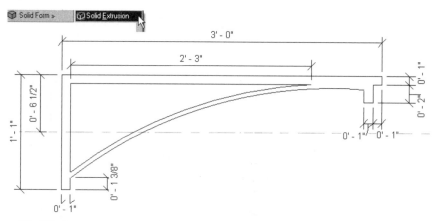

Figure 10–67 *Draw another profile for extruding*

4.12 When the profile is complete, select Extrusion Properties from the Design Bar. Set the Extrusion End to **–0 1/2″** and the Extrusion Start to **0 1/2″**. Select the Material value field to open the Materials dialogue.

4.13 Select the Default material and choose Duplicate. In the New Material name box, type **Black Iron**. Click the arrow next to the AccuRender Texture field. Navigate to *AR2_ACCUREND\matte, black*, as shown in Figure 10–68. Choose OK three times to set the Extrusion Properties.

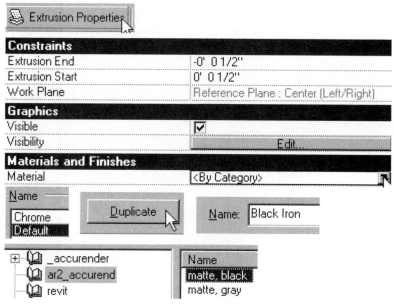

Figure 10–68 *Define the Black Iron material*

4.14 Choose Finish Sketch. Open the default 3D view to see the results, as shown in Figure 10–69.

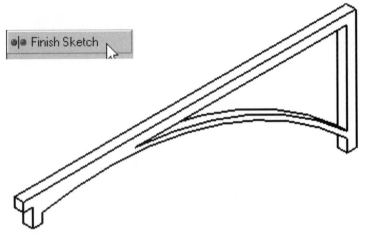

Figure 10–69 *The bracket profile, extruded*

4.15 Return to the Left Elevation view. Select Solid>Solid Revolve from the Design Bar.

4.16 Draw the profile shown in Figure 10–70. Do not create the dimensions. The profile is a 90° arc from the midpoint of the flange bottom face approximately 1' 9" to the left by 12" down, then offset 1/4". Connect the endpoints of the arc to close the profile. Actual dimensions are not critical. The revolved solid will be a translucent light cover.

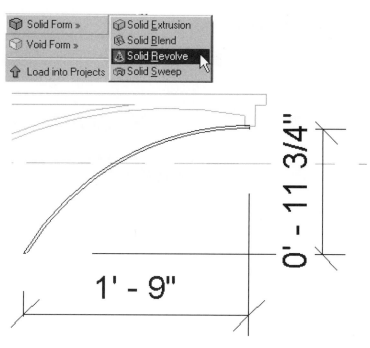

Figure 10–70 *A profile for revolution—this will be a light cover*

4.17 Select Revolution Properties from the Design Bar. Verify that the End Angle is 360° and the Start angle 0°. Select the Material value field to open the Materials dialogue.

4.18 Select Glass from the Name list. Choose the arrow next to the AccuRender Texture field. Navigate to _ACCURENDER\Glass\Tinted\White, Translucent, Frosted, as shown in Figure 10–71. Choose OK three times to set the Revolve Properties.

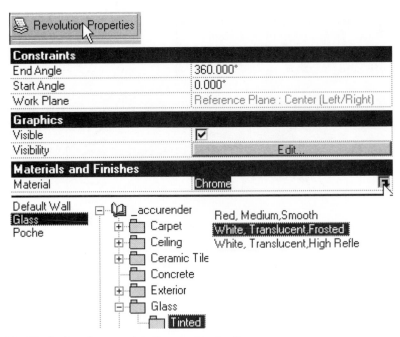

Figure 10–71 *Define the material Glass in this family*

4.19 Select Axis. Draw a line down from the right end of the profile, as shown in Figure 10–72.

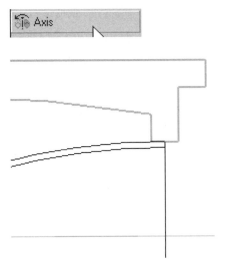

Figure 10–72 *Draw an axis*

4.20 Select Finish Sketch. Open the default 3D view (see Figure 10–73). Save this file as ***Bracket Light.rfa***. Close the file.

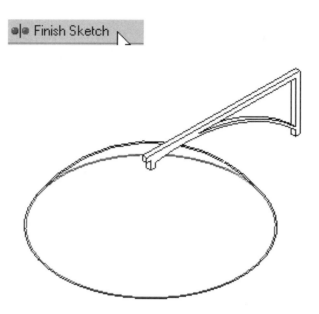

Figure 10–73 *The bracket and light shade*

EXERCISE 5. CREATE AND APPLY CONDITIONAL PARAMETERS IN A FAMILY

5.1 Select File>New>Family from the File menu. Select *Imperial Templates\Column.rft*, as shown in Figure 10–74, and choose OK.

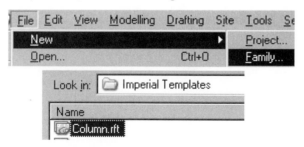

Figure 10–74 *Open the Column template*

5.2 The new file will open to a Plan view. Select Solid Form>Solid Extrusion from the Design Bar. Sketch a 2′ x 2′ square as shown in Figure 10–75, using the reference planes around the origin point.

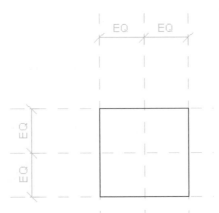

Figure 10–75 *Sketch a square for the first extrusion*

5.3 When the profile is complete, select Extrusion Properties from the Design Bar. Accept the default Extrusion Start and End values for now. Select the Material value field to open the Materials dialogue.

5.4 In the Materials dialogue, choose Default in the Name panel, then choose Duplicate. In the New Material name box, type **Brick**. Select the arrow next to the AccuRender Texture field. Navigate to *_ACCURENDER\ Masonry\Brown,_8",Running*, as shown in Figure 10–76. Choose OK three times to set the Revolution Properties. Choose Finish Sketch.

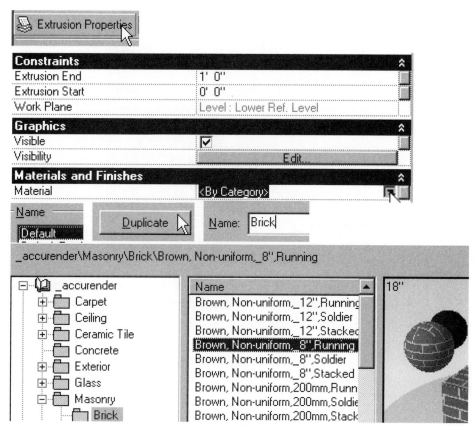

Figure 10–76 *Define the material for this extrusion*

    5.5 Choose Family Types from the Design Bar. Select Add in the Parameters section. In the Parameter Properties dialogue, verify that Family parameter is checked. Name the new parameter **Top Height,** select Length as the Type, Dimensions as the Group, Select Type, as shown in Figure 10–77. Choose OK.

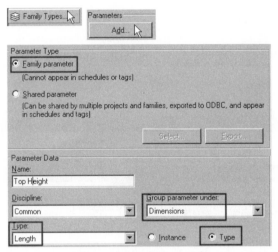

Figure 10–77 *Define the Top height as a Length parameter*

5.6 Select Add in the Parameters section. Name the new parameter **Base Height**, and select Length as the Type. Leave the other options set as for the previous parameter. Choose OK.

5.7 In the Formula field for the new Base height parameter, type **Top Height/6**. Choose Apply. See Figure 10–78. Revit Building will space the text according to its formula syntax. Revit Building will notify you if a formula is not correct. Formula text is case-sensitive.

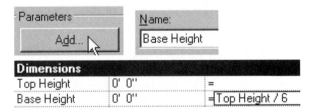

Figure 10–78 *Apply a simple formula to the Base Height parameter*

 TIP This dialogue box will resize. Stretch its borders left and right to give the Formula field plenty of space for typing. Adjust the width of the other columns as necessary.

5.8 Select New under Family Types. In the Name Box, type **12′** and choose OK. Repeat this step twice more, creating Family Types named **17′** and **23′**.

There are many ways to organize family types. For this column style you will differentiate the types by height. In this project, a 12′ column will fit under the wing roofs at the front (North side) of the building file, and a 23′ column will fit under the 3rd story overhang at the main entry. You will also create a 17′

column, perhaps to use as a parking lot or walkway light. Columns of different heights will have different characteristics, which will not be drafted, but driven by formulas that use the column height as their basis.

5.9   Put the 23′ column name in the Name list. In the Value field for Top height, type **23**. Choose Apply. Revit Building will set the units to 23′ 0″ and display a calculated value for the Base height, as shown in Figure 10–79.

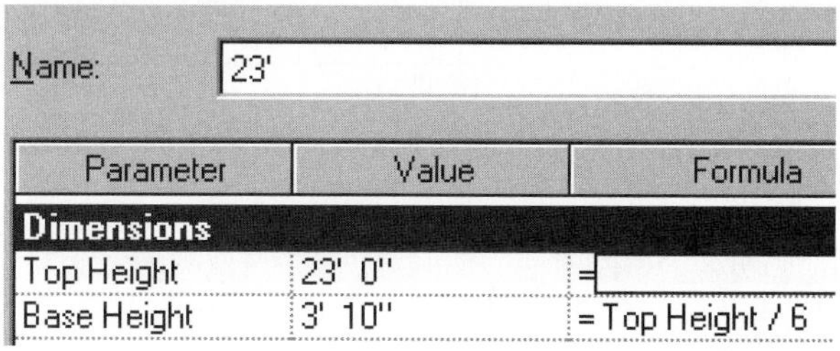

Name:	23′	
Parameter	Value	Formula
Dimensions		
Top Height	23′ 0″	=
Base Height	3′ 10″	= Top Height / 6

Figure 10–79 *Create a Family Type and set its Top height—Revit Building will calculate the Base height*

5.10   Repeat this step for the 17′ and 12′ Family Types. Make the 23′ Family Type the current Type. Choose OK.

5.11   Select the extrusion. Select Edit from the Options Bar. Select Extrusion Properties from the Design Bar. Click the small button to the right of the value field for Extrusion End. In the Associate Family Parameter dialogue, select Base height and choose OK.

5.12   In the Element Properties dialogue, the Extrusion End value will display a different number, which will be grayed, indicating it is not editable by using this field. The small button to the right of the field will hold an equals (=) sign, indicating that a parameter is driving this value (see Figure 10–80). Choose OK. Choose Finish Sketch.

576

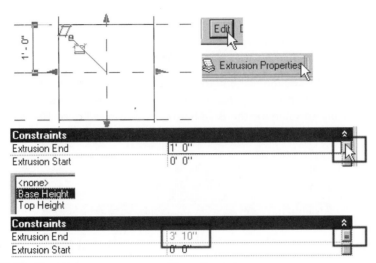

Figure 10–80 *Associating a parameter with an Extrusion*

> 5.13 Choose Solid From>Solid Extrusion in the Design Bar. Select the Circle tool from the Options Bar, and draw a circle of 4″ radius at the origin, as shown in Figure 10–81.

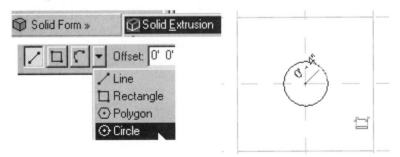

Figure 10–81 *Draw the next extrusion using the Circle tool*

> 5.14 Select Extrusion Properties from the Design Bar. Select the Associate Parameter button as before. Choose Top height and choose OK.
>
> 5.15 Repeat this step for the Extrusion Start value, using Base height as the Associated Parameter. Choose OK. See Figure 10–82.

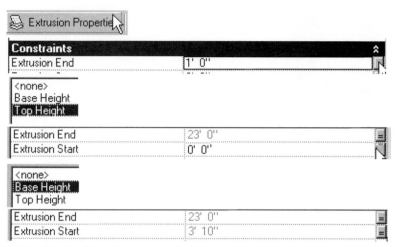

Figure 10–82 *Set the Extrusion start and end to parametric values*

5.16 Select the Material value field. Select Brick and choose Duplicate. Name the new material **Aluminum**. Use the AccuRender Texture field to set the material to _ACCURENDER\Metals\Aluminum\Satin, Plain, as shown in Figure 10–83. Choose OK three times to exit the dialogues. Select Finish Sketch. Except for placing components, drafting for this family has now finished, and the rest of the work will be with parameters and formulas.

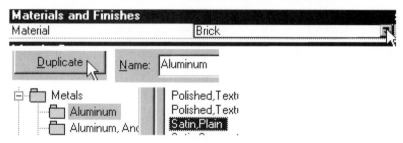

Figure 10–83 *Define the column's Aluminum material*

5.17 Open the default 3D view. Choose Family Types from the Design Bar. Cycle the Family Types as you did when you created them, using the Apply button for each Type even though you did not make changes. Move the Family Types dialogue box so you can see the column shaft and base change height. See Figure 10–84.

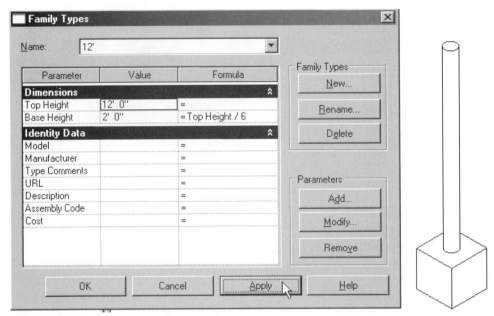

Figure 10–84 *Flex the model by checking all the family types for unexpected behavior*

NOTE Checking family types is an important part of creating them effectively. Check your work often when adding parameters, formulas, or reference planes with dimensions, to make sure that your families work as expected.

5.18 Move the Family Types dialogue to the center of the screen. Stretch its sides left and right to make typing easier.

5.19 Select Add in the Parameter section of the dialogue. Create a Family parameter named **Light Height** as a Length value stored by Type. Choose OK.

5.20 Add a Family Parameter named **Trim Height** as a Length value stored by Type. See Figure 10–85. Choose OK.

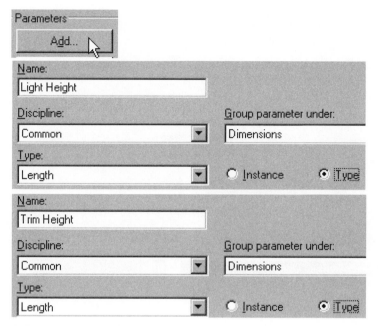

Figure 10–85 *Add height parameters*

5.21 Repeat the Add step to create a Family parameter named **Lights 1** as a Yes/No value stored by Type. Choose OK.

5.22 Add a Family parameter named **Lights 2** as a Yes/No value stored by Type. Choose OK.

5.23 Add a Family parameter named **Middle Trim** as a Yes/No value stored by Type. Choose OK. The Family Types dialogue will display the five new parameters you have created. They will apply to all the Family Types, not just the type that was active when you started adding parameters (see Figure 10–86).

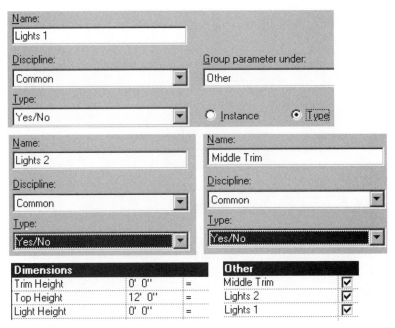

Figure 10–86 *The new parameters defined*

5.24 Enter formulas into the appropriate fields as shown in Figure 10–87. Choose Apply after entering each one to check your formula.

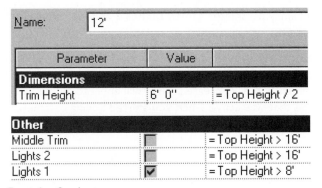

Figure 10–87 *Formulas for the parameters*

The formulas mean the following: The Trim height will always be 1/2 the height of the column. The Middle Trim value will be Yes when the column height is greater than (>) 16'. This condition is satisfied in the family types 17' and 23' tall, but not in the 12' type, so any element keyed to the Middle Trim parameter will not appear in that type. The same condition applies to Lights 2. Lights 1 will appear in all column types taller than 8'. The three column types you have defined all meet this condition—a shorter column style could be defined, maybe for interior use, that would not have any elements keyed to Lights appear.

5.25   Carefully enter the following text into the formula field for Light height:

If (Top Height > 20′, Top Height –4′, If (Top Height < 15′, Top Height –2′, Top Height –3′))

Choose Apply to check your formula.

NOTE Revit Building will supply spaces in formulas, as you have seen before, but be very careful with spelling, case, and punctuation. Open and closing parentheses must match.

This formula means: If the column is taller than 20′, the Light height will be 4′ down from the top. If the column height is less than 15′, the Light height will be 2′ down from the top. For all heights between 20′ and 15′, the Light height will be 3′ down from the top. A complete study of conditional syntax is beyond the scope of this book.

5.26   Cycle through the Family Types to see changes, as shown in Figure 10–88. In the 12′ type, the Middle Trim and Lights 2 value check boxes are empty, since this height is below the threshold value defined by the formula (see Figure 10–87). Choose OK to finish the parameter and formula setup.

Name:	17′	
Parameter	**Value**	
Dimensions		
Trim Height	8′ 6″	= Top Height / 2
Top Height	17′ 0″	=
Light Height	14′ 0″	= if(Top Height >
Base Height	2′ 10″	= Top Height / 6

Figure 10–88 *Check the Family Types*

5.27   Save the file as ***decor column.rfa***. From the File menu, select File>Load From Library>Load Family. Navigate to the folder(s) where you saved the *bracket light* and *trim ring* files, select them, and choose Open.

TIP If the files are in the same folder, you can hold down the **CTRL** key and select multiple file names.

5.28 Open the Lower Reference Level plan view. Select Component from the Design Bar. Put Trim Ring in the type selector. Locate an instance of the trim ring at the origin point, as shown in Figure 10–89.

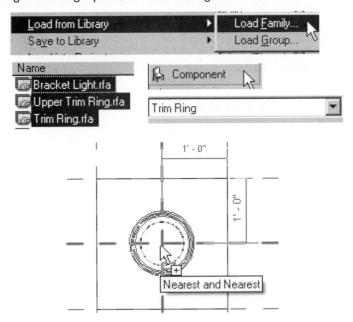

Figure 10–89 *Locate the trim ring at the origin*

5.29 Select Modify to terminate the insert. Open the Left Elevation view. Select the new trim ring. Select the Properties icon from the Options Bar.

5.30 In the Element Properties dialogue, select the Associate Parameter button next to the Offset value field. Select Base height and choose OK (see Figure 10–90).

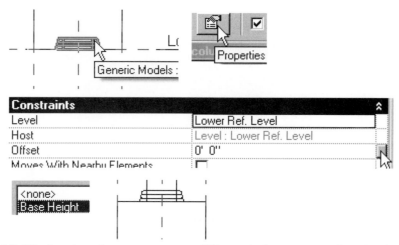

Figure 10–90 Associate the new component Offset with the parameter Base height

5.31 The trim ring component will still be highlighted. Copy it some distance straight up (the actual distance does not matter.) Choose the Properties icon from the Options Bar.

5.32 In the Element Properties dialogue for the new instance of the trim ring, associate its Offset value with the parameter Trim height. Choose OK. Associate the Visible value for this instance with the parameter Middle Trim. Choose OK twice. See Figure 10–91. The component will change height on the column.

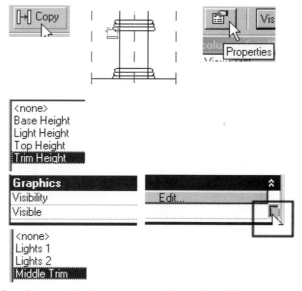

Figure 10–91 Give the new trim parametric height and visibility

5.33 Open the plan view again. Select Component from the Design Bar. Put the upper trim ring in the type selector. Locate an instance as the origin, as before. See Figure 10–92. A warning will appear, which you can ignore. Choose Modify.

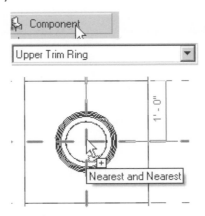

Figure 10–92 *Add a different trim ring*

5.34 Open the Left Elevation view. Select the new upper trim ring instance. Select the Properties icon from the Options Bar. In the Element Properties for this component, select the Associate Parameter button for its Offset and associate its value with Top height. Choose OK twice.

5.35 Copy the upper trim ring some distance down, as before. Repeat step 5.32 to associate the new copy's elevation with the Trim height parameter and associate its Visible value with the Middle Trim parameter (see Figure 10–93).

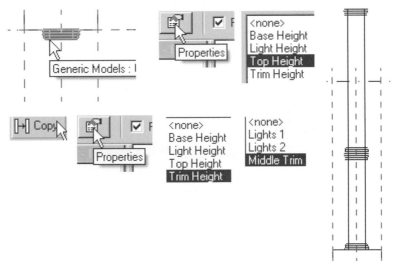

Figure 10–93 *Trim rings in place*

5.36 Open the Lower Ref. Level plan view. Select Component from the Design Bar and choose bracket light in the Type Selector. It will appear with the bracket facing up (plan north). Place the instance at the face of the column extrusion, as shown in Figure 10–94. Select Modify.

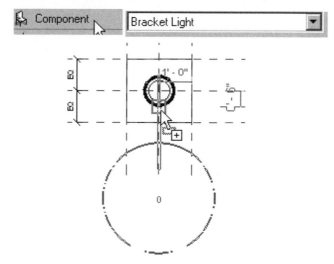

Figure 10–94 *The first light component placed in the plan*

5.37 Select the light. Use the Mirror tool and the horizontal reference plane to make a copy at the other side of the column.

5.38 Copy the new instance to the left, rotate it 90°, and place the new instance at the column face, as shown in Figure 10–95.

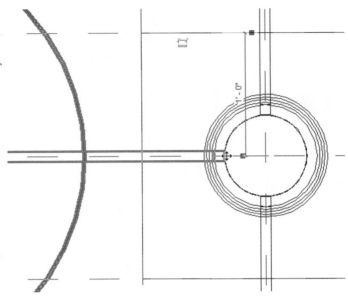

Figure 10–95 *Place the bracket endpoint at the intersection of the column and reference plane*

    5.39 Mirror the new light as before, for a total of four lights in place around the column (see Figure 10–96).

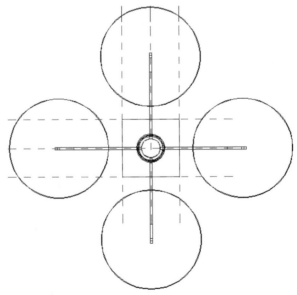

Figure 10–96 *Four lights placed in the plan*

    5.40 Open the Left Elevation view. Select the four lights and choose the Properties icon from the Design Bar. In the Element Properties dialogue, associate the Elevation value of the lights with the Parameter Light height. Choose OK twice. The lights will move up.

5.41 Open the default 3D view. Select the front and rear lights, as shown in Figure 10–97. Choose the Properties icon. Associate the Visible parameter of those components with Lights 1. Choose OK twice.

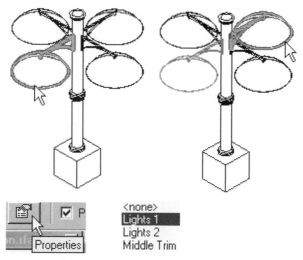

Figure 10–97 *Associate these lights with parameter Lights 1*

5.42 Select the other two lights. Associate their Visible parameter with Lights 2. Choose OK twice.

5.43 Cycle through the Family Types to make sure all components obey the elevation formulas. The Visible condition does not show in the family model, so lights and trim will not turn on and off. Save the file. Close the file.

EXERCISE 6. PLACE FAMILIES IN A PROJECT

6.1 In the Main Level view of the Chapter 10 project file, make the Basics tab of the Design Bar active. Select Grid.

We will not provide a complete examination of grids and their uses in this text. Columns snap to grids, so you will quickly place grid lines to locate columns, and then delete the grid lines. Therefore the location or numbering of grid-line bubbles does not matter.

6.2 Select the 3-point tool on the Options Bar. Set the Offset value to **2′ 0″**. Select the endpoints right to left and then the middle of the curved front plaza, as shown in Figure 10–98.

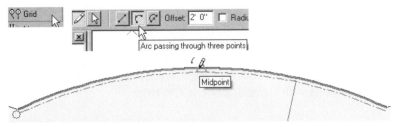

Figure 10–98 *The first curved grid line follows the front walkway*

    6.3 Select the Line tool on the Options Bar. Type **SC** at the keyboard to force a Center snap. Select the curved curtain wall at the main foyer.

    6.4 Pull the cursor straight up at 90° until it crosses the first grid line, as shown in Figure 10–99. Left click to create the grid line. Choose Modify.

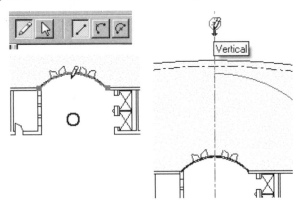

Figure 10–99 *Pull the new grid line vertically*

    6.5 Array the new vertical grid line four times left and right at a distance of 24′, as shown in Figure 10–100.

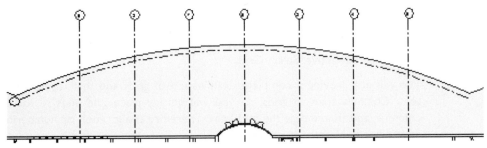

Figure 10–100 *Array grid lines*

    6.6 Right-click to the View Properties dialogue. Set the Underlay value to Balcony Level, as shown in Figure 10–101. Choose OK. Make the Modeling tab active on the Design Bar.

Figure 10–101 *Show the Balcony level as Underlay*

6.7 Choose the Column tool on the Design Bar. Select Load on the Options Bar. Navigate to the folder where you saved the *decor column.rfa* file, and choose Open. In the Type selector, make decor column: 23′ the active type. See Figure 10–102.

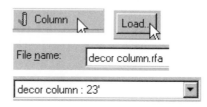

Figure 10–102 *Load the décor column file*

6.8 Place instances at the intersections of the grid lines along the rim of the front plaza, as shown in Figure 10–103.

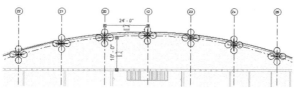

Figure 10–103 *Columns placed along the front*

6.9 Change the column type to decor column: 12′ and check Rotate after placement on the Options Bar. Place four instances, rotated 110° or −70°, at the corners and middle of the roof on the front right of the building. Precise placement is not critical for this exercise (see Figure 10–104). Note that these columns only have two lights.

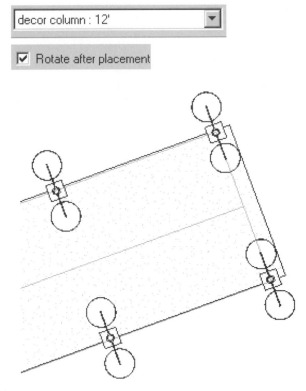

Figure 10–104 *Columns under the right roof*

6.10 Change the column type to decor column: 17′ and place two instances about 24′ apart near the front entrance, as shown in Figure 10–105. Precise placement is not critical.

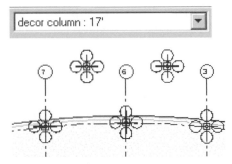

Figure 10–105 *17′ columns near the front*

6.11 Mirror the four 12′ (rotated, two-light) columns to the left of the building using the middle grid line. Delete the grid lines. Open the View Properties dialogue and return the Underlay value to None (see Figure 10–106).

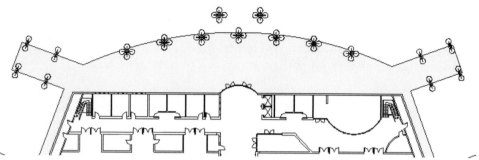

Figure 10–106 *Columns in place*

6.12 Open the 3D view Exterior Iso. From the View menu, select View>Orient>Northeast. Zoom in a little to study the three types of columns. The 12′ columns under the wing roof have two lights and no middle trim, as you defined in the family type formulas (see Figure 10–107).

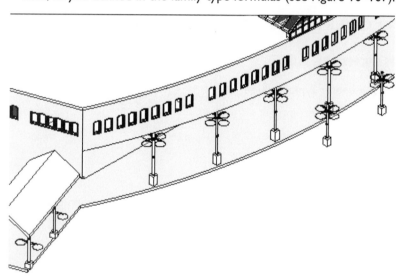

Figure 10–107 *Three types of columns from one family*

6.13 There are other 3D views set up in this file to examine the columns and other parts of the model. Save the file. Close the file.

> ## SUMMARY
>
> This final chapter has showed you some of the most powerful and useful features in Revit Building, of particular concern to busy designers.
>
> Design Options allow you to create many alternatives inside any project on the fly, and then view them, change them, and ultimately resolve them into your preferred options.
>
> Family creation in Revit Building is an extremely powerful toolset—practice creating your own component families to suit your purposes. Experiment with writing simple formulas that will take advantage of the design rules inherent in any component so that you can make many versions of the same basic family structure without extensive drafting.

REVIEW QUESTIONS – CHAPTER 10

MULTIPLE CHOICE

1. Family parameter types include

    a) Text, Number, Integer

    b) Length, Area, Volume

    c) Angle, URL, Material

    d) all of the above, plus Yes/No

2. Which of the following is not available when editing a Wall Structure?

    a) Sweeps

    b) Reveals

    c) Balusters

    d) Layers

3. A Design Option Set holds

    a) Family Parameters

    b) as many Design Options as you define, one of which will always be primary

    c) only the Design Option that is currently being edited

    d) the key to a happy life

4. Stair Parameters that you cannot edit include

    a) Nosing, Thickness, and Profile

     b) Riser Type, Thickness, Material, and Tread Connection

     c) Stringer Thickness, Height, Material, and Trim at Top value

     d) all of the above can be edited

5. Revit Building does not provide family template files (*.rft) for

     a) Doors, windows, furniture

     b) Roofs, floors, walls

     c) Lights, columns, electrical fixtures

     d) Profiles, plumbing, mechanical equipment

TRUE/FALSE

6. Revit Building keeps Design Options as separate files from the main project model.

7. Extrusion Start and End values can be set by parameters.

8. Length parameters cannot use formulas.

9. You can edit a Railing baluster placement value only once if the railing is a Design Option.

10. Stair boundaries can be edited to change the shape of a stair.

 Answers will be found on the CD.

Congratulations! You have studied and experimented with the major features of a powerful, efficient building-design application, and you have joined the growing ranks of those able to use the latest technology that the market has to offer. Practice and experience with Revit Building will ensure speed in your work and the confidence to explore its extensive capabilities.

For those of you in business, it's now up to you and your organization to determine how best to implement what Revit Building has to offer. Take a look at the Appendix for a discussion of factors that affect the implementation process in design firms. You can blend Revit Building into your workflow by combining it with standard CAD applications, or start out with a project planned and executed entirely using BIM (Building Information Modeling). Either way, you can refer back to the concepts and exercises in this text to help you understand and make effective use of Revit Building's techniques for massing, modeling, drafting, data management and illustration in your own work process. It's important to match your expectations with your resources. It will be hard to ask too much of the program, but it can be easy to underestimate the time and effort necessary to make it do exactly what you think you are asking for.

If you worked through this text on Revit Building as part of a school curriculum, you have now explored the only architectural design application in wide use that combines drafting, massing, element modeling, illustration and data extraction in one interface, with one common set of tools throughout. Exposure in the classroom is not proficiency in the workplace and simple understanding is not technical competence, but your expectations heading into the job market should be high.

Those of you who develop expertise with Revit Building can and should lead the way in providing guidance, support and training for others in your class, team or company. Patience and sympathy can be key—not everyone learns at the same pace, hits the same stumbling blocks, or finds the same solutions to procedural problems. Revit Building is well-designed software in that there is more than one way to do just about anything. This can be confusing for both instructors and students at first, but practice will help everyone develop flexible working methods and the knack for finding answers to questions unaided. Don't forget online resources such as user groups and their ongoing discussion sites. You can engage enthusiastic, experienced, articulate Revit Building users worldwide with very little effort.

The authors wish you goodspeed moving on from this point. It is our sincere hope that you will soon find that, just like the fabled Mr. Natural, you can Keep on Trucking through your design work with your hands firmly on The Right Tool For The Job.

What it Really Takes to Implement Revit Building

 Author's Note This appendix is taken from a highly-rated presentation that co-author Jim Balding gave at Autodesk University 2003. His audience consisted of executives, managers, and engineers from firms seeking to understand what they would be facing while migrating to Revit Building. The term Building Information Modeling (BIM) refers to database systems such as Revit Building, rather than vector-based CAD applications.

Readers from educational institutions or other organizations with in-house design departments may readily identify similar concerns that their programs or departments may have when facing upgrade, deployment, and training issues.

WHY CHANGE?

There are many reasons to change to the Building Information Model (BIM). In the beginning, the most compelling is the three-dimensional aspect of designing and documenting a project. In addition to that, many find the constant, complete coordination of the single model to be a compelling reason. If that doesn't convince you, the fact that firms using Revit Building are consistently providing better service and deliverables in less time with less staff—and occasionally both—may sway you. The bottom line is that a totally efficient design process has always been the "Holy Grail" of design and documentation, and only recently have the hardware and software caught up and become able to deliver on this elusive promise.

INTRODUCTION—POINTERS, NOT PRESCRIPTION

We have been asked countless times, "How do I implement Revit Building at my company?" The answer to this question is very simple—we have absolutely no idea. That may seem a little odd coming from two people writing a book about the subject, but it is very true. All firms and institutions are different in so many ways that there is no way to create a one-size-fits-all, step-by-step instruction book on how to implement a software such as Revit Building in any given firm. With that in mind we have identified seven factors that will influence the implementation process in just about any office, and have outlined the issues to look for while implementing Revit Building.

THE SEVEN FACTORS OF IMPLEMENTATION

While there are many different factors that will affect the success of implementation efforts, most can be grouped into one of seven categories. The following list is in order of importance.

- Firm size
- Project size
- Architectural style
- Project type
- Scope of services
- Firm locations
- Firm culture and age

FIRM SIZE

The size of the firm is the number-one issue when it comes to implementing the Building Information Model. While there are many factors that firm size will affect during implementation, the key factors are project size, technology, training/support/R&D, personnel, decision making, and standardization. Below are *general* observations regarding key factors related to company size. Each category has been rated as holding an advantage towards either larger or smaller firms. This is by no means an absolute system, and the advantage could be great or slight—use your own experience when considering these issues and evaluating your own situation.

PROJECT SIZE—ADVANTAGE: SMALLER FIRMS

Large firms generally have larger projects, and when it comes to Revit Building, project size can be viewed as an obstacle with regards to implementation. If you consider that you are adding all of the building data to one file, that file can certainly grow, and becomes exponentially larger than the single referenced files you may be presently using. BIM products including Revit Building have strategies to alleviate the issue of large projects; however, many firms feel that they need more time to get to a comfort level using full Revit Building implementation on such large-scale projects. It is important to note that this does not count out larger projects, they will simply require a little more planning and more experience.

This is where the smaller firms tend to shine. They typically have smaller projects that are ideal for Revit Building. There is not as much information to be communicated in a smaller project, which translates to smaller files and project teams.

One thing to keep in mind when selecting projects based on project size: It is advisable not to base the decision entirely on the floor plan area but the "spatial size," including both the amount of mass and detail to be built. For example, designing and documenting a 750,000—square foot (sq. ft.) warehouse using Revit Building can be far easier than creating a 20,000—sq. ft. gothic cathedral.

TECHNOLOGY—ADVANTAGE: LARGER FIRMS

Due to the fact that Revit Building models are larger and more complex than ordinary 2D drafting files, the software and models require fast machines with plenty of RAM and good video cards. When it comes to technology, the larger firms usually have the upper hand. They typically have the latest hardware, fast Local Area and Wide Area Networks, and perhaps even a dedicated Information Systems team to support the systems.

The small firm generally finds themselves behind the eight-ball in this arena. They may have made a recent, significant investment in connectivity, hardware, or software, and can't afford to budget large-ticket items such as these every year. A small firm that has made a substantial investment in technology could find themselves in a very advantageous position when implementing Revit Building.

Firms of any size should pay close attention to the hardware requirements of BIM software, as well as performance on typical firm project types. Maintaining up-to-date hardware and software often pays dividends not only in the speed of deliverables, but also in perceivable pride of ownership, which can provide incentives to learn and maintain knowledge, thus reinforcing a company's commitment to technology.

TRAINING/SUPPORT/R & D—ADVANTAGE: LARGER FIRMS

The larger the firm, the larger the operating expenses. With this in mind, the ability to create and maintain in-house expertise to provide training and support, as well as perform research and development roles for firm-specific tasks, becomes significant. This phenomenon often runs parallel with a firm's CAD manager role. If the firm has a full time CAD manager, typically it will have a full time Revit Building manager; who is often one and the same person.

The smaller firm will tend to have a part-time CAD manager, often an architect who knows the ins and outs of computers better than the rest of the firm. Firms can expect additional overhead and responsibilities when it comes to implementing and steadily using BIM. The return on investment over time is proving to be much larger proportionately when comparing results from Revit Building to 2D CAD.

Training and support are two of the keys to success, discussed in greater detail later, and they should be given serious thought when implementing Revit Building.

PERSONNEL—ADVANTAGE: LARGER FIRMS

When planning the implementation of BIM, one of the first things you need is a "champion," the person responsible for heading up the change. However, one person cannot do everything required to make the move to Revit Building. You will need talented people to carry out the plan and use the software effectively enough to deliver on the promise.

Because larger firms have larger talent pools to select from, they generally have the luxury of hand-selecting the users that best suit the chosen tool. One thing to note, however—the proportion of talent and roles within a firm will tend to be about the same for both larger and smaller firms.

DECISION MAKING—ADVANTAGE: SMALLER FIRMS

When it comes to making the decision to change, it seems to be better to have a single entity make that decision rather than a committee. Nowhere does the saying "too many chefs spoil the broth" apply more appropriately than in implementing change. Larger firms with multiple locations will have more challenges in the way of purchasing, planning, and organizing an implementation plan.

Smaller firms, generally based in one location, have just one "chef" who makes the decision to implement BIM, and then it is time to move on. When changes in the plan occur, there isn't another round of negotiations, planning, etc., to slow progress down.

STANDARDIZATION—ADVANTAGE: SMALLER FIRMS

Here again, the larger firm will tend to have many "chefs," from the top to the bottom. Somewhere in between might be a local CAD manager or two with individual CAD standards and techniques. If your firm is international you have, potentially, different languages, building techniques, documentation techniques, cultures, and infinitely more issues of diversity to face.

Smaller firms get the nod in this arena. With one location, standards can be established, and even voted on, in one location, perhaps at one time. When changes to the standard occur, communicating that change is simple, and most users will understand the reasoning behind it as it undoubtedly has arisen from issues within the local firm.

FIRM SIZE—CONCLUSION

While there are many factors listed here, there are many more that will be firm-specific. As stated above, firm size is the number-one consideration when deciding how to implement Revit Building at a firm. And while each subject has been labeled with a score giving an advantage to one size firm or the other, the intention of this section is to make firms aware of the issues regarding firm size and to assist firms in their implementation process.

Firm Size Final Score: Larger firms   3

                                    Smaller firms   3

PROJECT SIZE

While project size is a major factor in Revit Building implementation, it also is influenced more generally by firm size. Refer to the Project Size section under the Firm Size discussion above.

ARCHITECTURAL STYLE

When referring to architectural style and the implementation of Revit Building, the foundation of the distinction is the size, shape, and amount of detail involved. Revit Building does not recognize the literal difference, for instance, between Romanesque, Renaissance, and Post-Modern architecture.

Common sense, while not always common, does, in fact, make sense here. When it comes to modeling a building, straight, square forms are easier to model/build than that of the organic styles of Mr. Frank Gehry. This is not to say that it is impossible to build a sweeping tilted wave of a facade, just that more time and technique are required. If you can imagine the effort required to physically build the building, there is not a great difference to building it virtually using Revit Building. Simply put, more detail, shapes, and mass equals more time to develop and build.

While modeling all of the components in a project has desirable effects, keep in mind that it is not always necessary. During the planning stages of a project it should be determined what will be modeled and what can be "represented" with lines, arcs, and circles. A simple example: if you have raised wood paneling on a door, there is no harm in drawing lines on the surface of the door to represent the paneling and moving on with the project. Some might argue that it would not render properly or that there is a different material in the panel. The point is that there needs to be an understanding, on a per project basis, of what is important enough to model and what isn't. Three-dimensional modeling in Revit Building is not an all or nothing proposition.

PROJECT TYPE

Project types often affect different firms in different ways. Generally speaking, however, some project types are better suited for Revit Building than others. Again, this is not to say that a certain project or project type cannot be done effectively. Some of the aspects of project types that dictate the ease of implementation include elements that are repetitive, component-driven, area-driven, or those that make use of—and find great value in—the different visualization and/or scheduling advantages of Revit Building.

Below is a *general* outline of the advantages and challenges typically presented by various project types.

ENTERTAINMENT/RETAIL

Some of the advantages of Revit Building in the entertainment and retail industries might be the presentation value gained in color fills for area or room types, or the coordination of area calculations. Perhaps there is a need to schedule and color the tenant spaces by lease expiration, retail type, or rent-per-foot. Bi-directional, live scheduling may also have advantages, whether that includes display racks, flooring, or parking stalls.

Some of the challenges facing the entertainment/retail architect might include custom furniture and fixtures, or project size and scale. It has also been noted that due to the nature of a retail mall, individual retailers often have architects of their own, so there can be many different designs and ideas flying about at any given time.

HEALTHCARE/HOSPITALITY

Healthcare and hospitality designs, of course, are going to have significant gains using the repetitive, modular, or component-driven aspects of Revit Building. It could be argued that this is also the case with CAD drawings, where xrefs, blocks, or cells are used, but Revit Building goes beyond that. When duplicating the data along with graphic representation of that data, you have the benefit of scheduling, coordination, and visualization. When a single change to one guest room sink and vanity can affect the plans, interior elevations, sections, scheduling, and specification information for 500 rooms, there is great efficiency in place.

Generally, healthcare and hospitality projects are large, very complex designs with large project teams. Incorporating and coordinating the necessary data set can become a significant burden. Planning a project of this nature is possible, but requires advanced planning and experienced Revit Building-capable staff.

RESIDENTIAL

Residential projects gain tremendously from using Revit Building since they are generally smaller in size, as discussed above. Beyond that there is also the fact that residential clientele may not read or understand floor plans, sections, and elevations, but certainly would be able to understand a perspective or a rendering.

Challenges facing a residential architect might include the smaller scale of the project as well, surprisingly enough. Residential work tends to be shown at a larger scale, and therefore includes more detailing in each drawing. This tendency can give rise to the notion that all of the information must therefore be modeled. The residential architect would be wise to evaluate and plan the model simply.

PUBLIC/RELIGIOUS

This is a very broad spectrum, and includes projects large and small, simple and complex. What they have in common is that there is a large body of critics to please. Whether it is the congregation of a church or the citizens of a city, communication is the key ingredient here. This brings the visualization aspect of Revit Building to the forefront. City councils can be shown the proposed building model within the proposed setting. The public can readily understand the aesthetics of a rendering and the color-fill diagrams showing the location of the services to be provided.

Challenges facing an architect specializing in these areas might include the fact that many of these building types can be rather large and ornate in nature, requiring additional time spent modeling. The need to provide many different presentation-level images requires additional time working out materials, and possibly significant computer rendering time.

SCOPE OF SERVICES

Due to the nature of Revit Building—building a virtual model of the project rather that representing one with lines—it has been noted that more of the effort is front-loaded in the schematic design and design development phases. In other words, you are working out constructability issues earlier in the design process. This generally provides great gains in efficiency in the construction-documentation phase. Architects need to be cognizant of this issue, and perhaps make adjustments in the area. When the scope of services ends at what could be described as the traditional design-development phase, there is a greater level of information within the model than before. Architects are beginning to understand this, and respond in many ways, such as marketing additional services, adjusting fees (front-loading) to maintain projects through their entirety, and using a project information base as a sales tool for current and future work.

FIRM LOCATIONS

It may go without saying, but the more office locations there are in a firm, the more complex the implementation process gets. There are many things to consider, as noted above. Training, support, and standardization also become more important (and slightly more difficult) when there are multiple office locations. It is not impossible to accomplish widespread implementation efficiently; it will, however, require a little more planning. The planning will need to take into account the different office cultures, personnel, and so on.

FIRM CULTURE AND AGE

Firm culture and age may not seem important in the planning stages of implementation; however, it will become readily apparent during the implementation itself. How well does the firm accept and anticipate change? Is it a corporate culture? How do the lines of communication operate and what is the management organization? Who will need to back and support the initiative?

When age is considered, generally speaking, younger firms are more open to change, while older firms might tend to be more set in their ways and resistant to a change of this nature.

FOUR KEYS TO SUCCESS

In addition to the seven factors of implementation outlined above, we have developed four keys to success in implementation, listed below. We have found that when focusing on these steps, the potential for success increases greatly.

Plan

Communicate

Train

Support

PLAN

The first stage of implementation is planning. Of course, each firm is as individual as the employees within the firm. Careful planning is the foundation to a successful implementation. Carefully considering the seven factors of implementation outlined above can supply a firm with guidelines, but each firm should consider its own unique characteristics at every step. Planning the implementation of any strategy or technology should take into account all of the individuals and groups that the change will affect. It is advisable to discuss these issues with key individuals in these groups.

COMMUNICATE

Once the plan is in place, it will become imperative that everyone knows what the plan is. It is very important to share the plan with the entire office. This does not mean inviting the office as a whole to a meeting to show them the latest tool. There are many different roles within any firm; any new tool will affect users in different ways, and therefore users will be interested in different aspects of the tool.

The principals or senior leadership will be interested in how the new technology affects the bottom line—what is the net gain? They will also be interested in what kind of investment in capital and staff that it will require. This group will also be interested in why the firm would make the change from the process currently in place. Generally speaking, their question will fall along the lines of "what does this mean to the business of architecture"?

The project managers will want to know how it affects their work flow. Will it really allow them to get work out sooner or with less staff? Revit Building's coordination of drawings, details, and grid bubbles is greatly appreciated by those who experience it. Managers are also interested in the coordination between consultants and how a team might share the data. What does this mean to the process of architecture?

The architects and designers that will be using the tool will generally be concerned with the user interface, toolsets, and learning curve. They will also be asking questions about how to create specific detailed models. What does this mean to the design and documentation of architecture?

TRAIN

When implementing a technology like Revit Building, it is advisable to have dedicated training. While on-the-job training can be a great learning experience, it can also be detrimental. When there are deadlines and revisions flying about, trainees tend to want to jump back to their comfort zone, 2D CAD.

The training should be organized and tailored to cover a firm's specific topics and issues. Having general training to cover the basics of the tool can be helpful initially; however, having an expert in the firm covering firm-specific issues will pay dividends later.

SUPPORT

Once the plan is complete and disseminated to your staff, and when training has been finished, the implementation is not yet complete. Firms should plan on maintaining and updating information through ongoing support. It is advisable to have regular meetings to discuss issues as they arise, from topics including new techniques to changes in structure and beyond.

WHAT HAS WORKED

- **Planning**—There is no substitute for good planning. Take the time to consider the issues within your firm and plan on addressing each and every one. It is a good idea to plan on re-planning too, as firm-specific issues arise and need to be addressed.

- **Formal training**—While on-the-job training can be the best-quality training for the real world, real deadlines, revisions, and owners can often frustrate users and prompt them to want to give up and go back to good old CAD.

- **On-the-job training**—Occasionally, with the right personnel, project, schedule, team, and support, a user can get up to speed while working on an active project. Great care needs to be taken if this is the path you choose to take.

- **Baby steps**—Take a few of the overall benefits of Revit Building and focus on those issues on a particular project, mastering those, and then take on a few more on the next project. Some firms use pilot projects to take these steps.

- **Partial projects**—Consider the "horizontal" approach; use Revit Building to design and document the plan view and scheduling while using CAD to document the vertical drawings, like the sections, elevations, and perhaps the details.

- **Small projects with experienced users**—It may seem obvious, but overloading the first few small projects with experienced users works well and serves purpose down the road. The theory here is that these users will gain real-world experience and confidence. These users can then go on to work with others and spread the knowledge.

- **Management buy-in**—It will be important for the entire company to have the management and upper management understand the issues and lend support to the projects and users.

- **Rewards and recognition**—The early adopters of these technologies often have to endure the nay-sayers, critics, and the frustration of learning a new software and design process. With rewards and recognition they will, at the very least, feel that it is worth the effort. Some firms reward new users by upgrading their computers first, or doling out preferred seating in meetings or recognition in newsletters.

- **On-going training**—Many firms, recognizing the fact that trainees cannot learn a large application overnight, organize and support weekly mini-training sessions. These sessions generally last one hour and cover a single topic.

- **Exit Strategy**—On the rare occasion that the BIM is not working out on a particular project, it may be necessary to exit and return to AutoCAD. It will be helpful if that exit is prepared ahead of time.

WHAT TO AVOID

- **Indifferent communication**—As noted above, users need to know not only how the tools work, but the concepts of Revit Building and why the firm is going in this direction. Understanding what BIM is, how it is to be implemented, and why, can be just as important as learning how to use it.

- **On-the-job training without full support**—If it is impossible to get formalized, non-project-related training, it will become essential that the new users have a support mechanism available as close to full time as possible. When it comes down to crunch time with new functionality and there's no one in-house to assist, a deadline can become a breakdown point in implementation. The support team could be an in-house expert or a reseller/training center.

- **Attempting to hit a home run**—Some parts of Revit Building can be easier to learn than 2D CAD. It is, however, incredibly comprehensive. Take baby steps into the software. After all, most people did not learn CAD on the first project. Generally speaking, users become very comfortable and highly productive on their 2nd or 3rd project.

- **Isolating users**—This goes back to training and support. Users that are set free on a project without consistent support will tend to lose interest and become increasingly frustrated while they attempt to figure everything out themselves.

- **The oversell**—Avoid selling all of the benefits of Revit Building to clients prior to actually having a few projects under your belt. Be certain that the firm can deliver on promises. Avoid repeating vendor marketing promises without in-house verification.

WRAP UP

This appendix is, just that, a little bit of extra information at the end of this book. There any number of combinations and additional factors that could, and very well will, affect the implementation of Revit Building at any given firm. The intent of this appendix is to give general guidance to those on the implementation path. Good luck, and good hunting.

INDEX

LICENSE AGREEMENT FOR AUTODESK PRESS
A Thomson Learning Company

Educational Software/Data

You the customer, and Autodesk Press incur certain benefits, rights, and obligations to each other when you open this package and use the software/data it contains. BE SURE YOU READ THE LICENSE AGREEMENT CAREFULLY, SINCE BY USING THE SOFTWARE/DATA YOU INDICATE YOU HAVE READ, UNDERSTOOD, AND ACCEPTED THE TERMS OF THIS AGREEMENT.

Your rights:

1. You enjoy a non-exclusive license to use the enclosed software/data on a single microcomputer that is not part of a network or multi-machine system in consideration for payment of the required license fee, (which may be included in the purchase price of an accompanying print component), or receipt of this software/data, and your acceptance of the terms and conditions of this agreement.

2. You own the media on which the software/data is recorded, but you acknowledge that you do not own the software/data recorded on them. You also acknowledge that the software/data is furnished "as is," and contains copyrighted and/or proprietary and confidential information of Autodesk Press or its licensors.

3. If you do not accept the terms of this license agreement you may return the media within 30 days. However, you may not use the software during this period.

There are limitations on your rights:

1. You may not copy or print the software/data for any reason whatsoever, except to install it on a hard drive on a single microcomputer and to make one archival copy, unless copying or printing is expressly permitted in writing or statements recorded on the diskette(s).

2. You may not revise, translate, convert, disassemble or otherwise reverse engineer the software/data except that you may add to or rearrange any data recorded on the media as part of the normal use of the software/data.

3. You may not sell, license, lease, rent, loan, or otherwise distribute or network the software/data except that you may give the software/data to a student or and instructor for use at school or, temporarily at home.

Should you fail to abide by the Copyright Law of the United States as it applies to this software/data your license to use it will become invalid. You agree to erase or otherwise destroy the software/data immediately after receiving note of Autodesk Press' termination of this agreement for violation of its provisions.

Autodesk Press gives you a LIMITED WARRANTY covering the enclosed software/data. The LIMITED WARRANTY can be found in this product and/or the instructor's manual that accompanies it.

This license is the entire agreement between you and Autodesk Press interpreted and enforced under New York law.

Limited Warranty

Autodesk Press warrants to the original licensee/ purchaser of this copy of microcomputer software/ data and the media on which it is recorded that the media will be free from defects in material and workmanship for ninety (90) days from the date of original purchase. All implied warranties are limited in duration to this ninety (90) day period. THEREAFTER, ANY IMPLIED WARRANTIES, INCLUDING IMPLIED WARRANTIES OF MERCHANTABILITY AND FITNESS FOR A PARTICULAR PURPOSE ARE EXCLUDED. THIS WARRANTY IS IN LIEU OF ALL OTHER WARRANTIES, WHETHER ORAL OR WRITTEN, EXPRESSED OR IMPLIED.

If you believe the media is defective, please return it during the ninety day period to the address shown below. A defective diskette will be replaced without charge provided that it has not been subjected to misuse or damage.

This warranty does not extend to the software or information recorded on the media. The software and information are provided "AS IS." Any statements made about the utility of the software or information are not to be considered as express or implied warranties. Delmar will not be liable for incidental or consequential damages of any kind incurred by you, the consumer, or any other user.

Some states do not allow the exclusion or limitation of incidental or consequential damages, or limitations on the duration of implied warranties, so the above limitation or exclusion may not apply to you. This warranty gives you specific legal rights, and you may also have other rights which vary from state to state. Address all correspondence to:

AutodeskPress
Executive Woods
5 Maxwell Drive
Clifton Park, New York 12065-2919